$ 114.00

Synthetic Peptides

Volume 3

Synthetic Peptides

Volume 3

George R. Pettit

Department of Chemistry
Arizona State University
Tempe, Arizona

Academic Press New York San Francisco London 1975

A Subsidiary of Harcourt Brace Jovanovich, Publishers

ACADEMIC PRESS, INC.
111 Fifth Avenue, New York, New York 10003

United Kingdom Edition published by
ACADEMIC PRESS, INC. (LONDON) LTD.
24/28 Oval Road, London NW1

Library of Congress Cataloging in Publication Data

Pettit, George R
Synthetic peptides.

Vols. 3- published by Academic Press
Includes bibliography.
1. Peptides–Tables, etc. 2. Peptide synthesis.
I. Title. [DNLM: 1. Peptides–Chemical synthesis.
QU60 P511s]
QD431.7.P46 547'.756 74-132182
ISBN 0-12-552403-X (v. 3)

PRINTED IN THE UNITED STATES OF AMERICA

Contents

Preface

When the survey of synthetic peptide literature was prepared in the fall of 1970 for Volume 2, the annual appearance of such papers was estimated to be about 250. By fall 1972, the annual rate had increased to approximately 400 and gave every evidence of continued acceleration. The magnitude of this medically and scientifically most welcome expansion of effort may be evaluated by considering that the literature of synthetic peptides summarized for Volume 1 covered the 10-year period from 1960 to 1970 and has almost been matched by the nearly 800 papers reviewed for this volume for just the 2-year period January 1971 to January 1973.

Because of the obvious difficulties and complexity associated with keeping abreast of this rapidly expanding field of bio-organic chemistry, an attempt was made to expand the detailed coverage of this volume by increasing the tabular data to include notations for various chromatographic procedures (including an extensive listing of TLC solvent systems), countercurrent distribution techniques, electrophoresis, and physical methods of characterization such as infrared, proton magnetic resonance, and mass spectral interpretations. A special feature of this volume is the structural presentation of representative or otherwise interesting peptide syntheses, particularly relating to the chemistry of cysteine and cystine.

I am especially grateful to Mrs. Christine H. Duplissa for dedicated and expert assistance with preparation of the tabular survey and to Mrs. Marie D. Baughman and Misses Sally Keehl and E. Gail Dubie for their willing and helpful assistance in the final stages of manuscript preparation.

George R. Pettit

Introduction and Glossary

The extensive investigations of recent years directed at total syntheses and structure–activity relationships among the peptide and protein-type hormones, such as, oxytocin, vasopressin, ACTH,[515] and insulin, have been extended recently, for example, to the important luteinizing hormone-releasing factor (LH-RF),[541] calcitonin M,[574] and secretin.[695] Analogous synthetic programs concerned with other animal and plant polypeptide and protein constituents (which have also been on the increase) vary, for example, from melittin,[416] the hemolytic peptide of honey bee (*Apis mellifica*) venom, to yeast cytochrome c,[421] to *E. coli* acyl carrier protein,[207] to tobacco mosaic virus protein.[486a,486b] Comparable increases are apparent in programs concerned with uncovering useful amino acid derivatives and relatively small peptides for uses ranging from cancer chemotherapy to sweetening agents. All of these synthetic peptide problems are being reinforced and enhanced by still expanding efforts to further refine the protecting group techniques and coupling methods of peptide chemistry. In the latter area, the past two years have witnessed major contributions to further refinements in the solid phase technique and its automation.

An interesting and authoritative review of peptide syntheses was prepared by Geiger.[170] In his discussion mention is made of the fact that a living organism is able to synthesize peptides and proteins in a matter of seconds while chemists' achievements to date have required anywhere from weeks to years. The bio-organic chemist should be of good cheer, however, as Geiger also mentions that the time required to complete a gene mutation involving only the replacement of a single amino acid unit in a protein such as hemoglobin is estimated to require seven to ten million years. The review of peptide syntheses and leading references to recent reviews contained therein by Jones[281] provides a summary of the literature for 1969 and 1970.

Appendix C of the Jones[281] review in Volume 3 of the Chemical Society's *Specialists Periodical Reports* contains a partial list of solvents recently used with Sephadex LH-20 and as noted there, the use of solvent mixtures implies for the most part partition chromatography where the most polar solvent is more strongly adsorbed and not gel permeation chromatography in the usual sense. Here it should be mentioned that Sephadex QAE-A 25 has recently been recommended for the separation of amino acid and peptide mixtures in the form of copper(II) complexes.[499] Also in this general analytical area, methods for the characterization and sequential analysis of peptides by mass spectrometry continue to improve.[424] Our initial study[76] of peptide analysis by field ionization mass spectrometry has recently been joined by the chemical ionization technique.[321] Both of these more gentle ionization techniques when applied to peptide sequential analysis nicely complement the usual electron impact-induced fragmentation data. Field ionization mass spectrometry is especially useful for obtaining molecular ions when the electron impact mode has proved useless for this purpose. An interesting application of mass spectral sequential analysis of histidine-containing peptides was illustrated using the octapeptide Pro-Phe-His-Leu-Leu-Val-Tyr-Ser.[497] After acetylation and controlled permethylation (using a modified sodium hydride-dimethyl sulfoxide-methyl iodide procedure to avoid N-methylimidazolium iodide formation) the complete sequence was deduced by mass spectrometry. Similarly α-MSH, the histidine-containing tridecapeptide, yielded to structural elucidation on the basis of mass spectral analysis of protected cyanogen bromide and chymotryptic cleavage products.[497] The electron impact fragmentation of permethylated Tyr- and Trp-containing peptides[118] and Cys peptides[446] have also been studied.

A summary review of the current status in each major field of synthetic peptide endeavors appears in the introductions to the tabular surveys of subsequent chapters.

Because of importance to the whole field of peptide syntheses, recent advances pertaining to the subject of racemization have been treated as the subject of Chapter 1. Recent advances in protecting group techniques and peptide bond-forming methodology have been discussed, respectively, in Chapter 2 concerned with amino acids and in Chapter 3 concerned with dipeptide syntheses. Examples of recent progress in syntheses of peptides ranging in size from 3 units to polymers appear in Chapters 4 to 15. Because of the present and future potential of solid phase synthesis methods, a more extensive treatment of this subject appears in Chapter 14. The remaining coverage is devoted to the more specialized topics from cyclopeptides to nucleopeptides and includes a new chapter devoted to chromopeptides of the actinomycin D type. These latter areas hold promise of great importance to chemotherapy but unfortunately still remain relatively neglected. Hopefully the medical potential, for example, of the depsipeptides and nucleopeptides will become more generally recognized.

The introduction to each table includes not only a survey of recent advances, but also includes a brief outline of pertinent experimental techniques. The list of amino acids in the glossary (see p. 4) corresponds to those counted as one unit when determining location of a peptide in the tables. One exception to this organization involves derivatives of glycolic acid, hydroxyproline, and hydroxyisovaleric acid which have been included in the depsipeptide tables.

The tables have been organized so that when proceeding from right of the method column, the next entry is reaction solvent followed, in order of appearance, by the solvent employed for precipitation, crystallization and/or recrystallization, melting point of the presumed pure peptide, optical rotation at the sodium D line, solvent used in obtaining the optical data and, lastly, reference to the original literature. Referral to the paper cited will, in many cases, lead to earlier references to that particular peptide which may contain data not included in the tables. Only synthetic products are included in the tables. From the dipeptide chapter onward, the method column of each table summarizes, where known, the peptide bond-forming method and whether, for example, an N-terminal protected amino acid was condensed with a hexapeptide (1 + 6) or whether the same heptapeptide was prepared by condensing an N-protected tetrapeptide with a dipeptide (4 + 2) corresponding to the C-terminal end.

The tabular surveys include headings for chromatography, countercurrent distribution, electrophoresis, and spectral characterization methods. Chromatography has been subdivided into column (Col) which covers methods ranging from the usual column absorption techniques with silica gel or alumina to the newer techniques employing Sephadex LH-20 and gel permeation methods. TL refers to thin-layer chromatography primarily used to assess purity and the number entered here corresponds to a solvent mixture represented by number in an extensive listing of tlc solvent systems which follows the glossary of this chapter. Referral to this tlc solvent list should expedite finding the proper solvent system for analogous new compounds. Unless noted otherwise, all thin-layer chromatographic procedures were conducted with silica gel on glass plates. The next entry under chromatography reflects an ion-exchange method of the column type ranging from the commonly employed commercial resins to the newer Sephadex types. The last entry designated paper corresponds to one of the usual paper chromatography systems. The number recorded for the countercurrent distribution (CCD) entry corresponds to a list of solvents presented after the thin-layer solvent systems. The Elect. heading has simply been checked wherever an electrophoresis procedure was employed. The heading marked Other indicates whether or not the amino acid or peptide was characterized by amino acid analysis (AA) or a physical method of the infrared, proton magnetic resonance, or mass spectral type. Thus, by consulting the tabular summary it should generally be possible to substantially reduce the overall time required to locate an efficient procedure for a known compound or devise one for a new substance.

The introduction to each table includes selected examples in that particular area for the period January 1971 to July 1972. For the period July 1972 through literature available January 3, 1973, each paper containing synthetic peptides related to that area was cited. The tabular survey of synthetic peptides has been continued from October 15, 1970 to July 1, 1972. In another volume I plan to extend the tabular summary to July, 1973 and the literature of synthetic peptides to January, 1974.

Nearly all of the nomenclature recorded in the literature cited has been changed to conform, where considered appropriate, with current IUPAC-IUB Commission recommendations,[258] or correspond to abbreviations in current use or suggested by this reviewer. Attempts to employ a unified and abbreviated nomenclature involved considerable effort which should in the long term be well worthwhile for the reader. Each amino acid unit is oriented such that the carboxyl end is on the right and substituents are therefore named reading from the left terminus of the molecule to the right terminus. To simplify reordering nomenclature in future volumes, I wish to make a plea to my colleagues to more closely follow IUPAC-IUB Biochemical Nomenclature recommendations and other suggestions in the following glossary for abbreviations. Also it would be most helpful to order the sequence of substituents in a left to right fashion as indicated for each of the following amino acid examples where X and Y represent the various protecting groups.

N^G-X-N^α-Y-Arg
N^β-X-N^α-Y-Asn
β-OX-α-OY-Asp
S-X-N-Y-Cys
N^γ-X-N^α-Y-Gln
γ-OX-α-OY-Glu
N^{im}-X-N^α-Y-His
N^ϵ-X-N^α-Y-Lys
N^ϵ-X-N^β-Y-Lys
N^δ-X-N^α-Y-Orn
O-X-N-Y-Ser
O-X-N-Y-Thr
O-X-N-Y-Tyr

The reader should become acquainted with the nomenclature and abbreviations of the glossary which follows to derive maximum benefit from quick scans of the tabular material.

Amino Acids

Ala	Alanine
β-Ala	β-Alanine
Abu	α-Aminobutyric acid
ϵ-Acp	ϵ-Aminocaproic acid
Aib	α-Aminoisobutyric acid
Arg	Arginine
Asn	Asparagine
Asp	Aspartic acid
δ-Ave	δ-Aminovaleric acid
Cit	Citrullene

Cysta	Cystathionine
Cys	Cysteine
Cyst	Cystine
Dbu	α,γ-Diaminobutyric acid
α,β-Dbu	*Threo*-α,β-diaminobutyric acid
Ete	Ethionine
Glu	Glutamic acid
Gln	Glutamine
Gly	Glycine
His	Histidine
Hyv	D-α-Hydroxyisovaleric acid
Ile	Isoleucine
Leu	Leucine
Lys	Lysine
β-Lys	β-Lysine
Met	Methionine
MeIle	N-Methylisoleucine
MeVal	N-Methylvaline
Nle	Norleucine
Nva	Norvaline
Orn	Ornithine
Phe	Phenylalanine
Phenyl-Gly	Phenylglycine
Pro	Proline
3-Hyp	3-Hydroxyproline
4-Hyp	4-Hydroxyproline
Sar	Sarcosine
Ser	Serine
Thr	Threonine
Trp	Tryptophan
Tyr	Tyrosine
Val	Valine

Urethane Type-Protecting Groups

Adoc-	Adamantyloxycarbonyl
Aoc-	*t*-Amyloxycarbonyl
Bzhoc-	Benzhydryloxycarbonyl
Z-	Benzyloxycarbonyl
Z(Br)-	*p*-Bromobenzyloxycarbonyl
Boc-	*t*-Butyloxycarbonyl
Dmpoc-	1,1-Dimethyl-2-propynyloxycarbonyl
DNbzloc-	2,2′-Dinitrodiphenylmethoxycarbonyl
Bpoc-	2-(*p*-Diphenyl)isopropyloxycarbonyl
Iboc-	Isobornyloxycarbonyl
Z(OMe)-	*p*-Methoxybenzyloxycarbonyl
Mz-	*p*-Methoxyphenylazobenzyloxycarbonyl
Z(NO_2)-	*p*-Nitrobenzyloxycarbonyl
Nbzloc-	2-Nitrobenzyloxycarbonyl
Nvoc-	6-Nitroveratryloxycarbonyl
Pz-	*p*-Phenylazobenzyloxycarbonyl
Pipoc-	Piperidinooxycarbonyl
TrOC-	β,β,β-Trichloroethyloxycarbonyl

Z-TF-	2,2,2-Trifluoro-1-benzyloxycarbonylaminoethyl
Boc-TF-	2,2,2-Trifluoro-1-*t*-butyloxycarbonylaminoethyl

Other N- and O-Protecting Groups

Ac-	Acetyl
Bz-	Benzoyl
Bzl-	Benzyl
Btm-	Benzylthiomethyl
CHla-	Chlorambucil
DNS	Dansyl
Dca-	Dichloroacetyl
MBh-	4,4′-Dimethoxybenzhydryl
Dmb-	2,4-Dimethoxybenzyl
DNT-	N-(3,5-Dinitro-2-thienyl)
Dpm-	Diphenylmethyl
For-	Formyl
Imaa-	(7-Isopropyl-1-methylazulen-4-yl)-acetyl
DAZ-	α-Methyl-α-(4,5-dimethyl-2-phenylazophenoxy)-propionyl
MAZ-	α-Methyl-α-(4-methyl-2-phenylazophenoxy)propionyl
Nps-	*o*-Nitrophenylsulfenyl
Pfp-	Pentafluorophenyl
Pmb-	Pentamethylbenzyl
Pcp-	Perchlorophenyl
Pht-	Phthalyl
Pic-	α-Picolinyl
Thp-	Tetrahydropyranyl
Tos-	Tosyl
Tca-	Trichloroacetyl
Tfa-	Trifluoracetyl
Boc-TF-	2,2,2-Trifluoro-1-*t*-butyloxycarbonylaminoethyl
Trt-	Trityl
Trs-	Tritylsulfenyl

S-Protecting Groups

-Ame	Acetamidomethyl
-Bzh	Diphenylmethyl
-Eac	Ethylaminocarbonyl
-iBM	Isobutyloxymethyl
-Pac	Phenylaminocarbonyl
-Trt	Triphenylmethyl

Carboxyl-Protecting Groups and Active Esters

-OBhae	Benzhydroxamic acid ester
-OBzl	Benzyloxy (benzyl ester)
-OPh(OBzl)	2-Benzyloxyphenyl ester
-OChlBzl	2-(4-Chlorobenzyloxy)phenyl ester

-OPH(Ph)CN	*o*-Cyanobenzyl ester
$-OCH_2CN$	Cyanomethoxy ester
-ODnp	Dinitrophenyl ester
-OBzh	Diphenylmethoxy (benzhydryl ester)
-OEt	Ethoxy (ethyl ester)
-OHPh	*o*-Hydroxyphenyl ester
-OHq	8-Hydroxyquinoline ester
-O-*i*-Bu	Isobutyl ester
-OMe	Methoxy (methyl ester)
-OBzl(OMe)	*p*-Methoxybenzyl ester
-OPh(OMe)	*o*-Methoxyphenyl ester
-OP(SMe)	4-Methylthiophenyl ester
-ONBzl	*p*-Nitrobenzyl ester
-SNBzl	*p*-Nitrobenzylthio ester
-ONp	*p*-Nitrophenoxy (*p*-nitrophenyl ester)
-SNp	*p*-Nitrophenylthio ester
-OPfp	Pentafluorophenyl ester
-OPmb	Pentamethylbenzyl ester
-OPcp	Pentachlorophenyl ester
-SPh	Phenylthiol ester
-OPht	N-Phthalimide ester
-OPic	4-Picolyl ester
-OPn	Piperonyl ester
-SPy	2-Pyridinethiol ester
-OSu	N-Succinimide ester
$-OBu^t$	Tertiary butoxy (*t*-butyl ester)
-OTcp	2,4,5-Trichlorophenyl ester
-OTMB	2,4,6-Trimethylbenzyl ester

Miscellaneous Derivatives

T	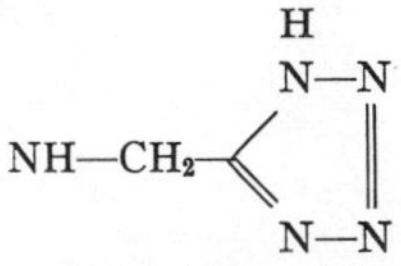
Cha	Cyclohexylamine salt
Dcha	Dicyclohexylamine salt

Solvents

Ac	Acetic acid
Acn	Acetonitrile
Alc	Ethyl alcohol
Am	Amyl alcohol
t-Am	*t*-Amyl alcohol
i-Am	Isoamyl alcohol
An	Acetone
Aq	Water
Be	Benzene
Bu	Butyl alcohol
t-Bu	*t*-Butanol
2-Bu	2-Butanol

Chf	Chloroform
ClAc	Chloroacetic acid
Cte	Carbon tetrachloride
Cy	Cyclohexane
DCA	Dichloroacetic acid
Di	Dioxane
Dipr-Eth	Diisopropyl ether
DMA	Dimethylacetamide
DMF	Dimethylformamide
DMSO	Dimethyl sulfoxide
Glyme	1,2-Dimethoxyethane
EtAc	Ethyl acetate
EtCl	Ethylene chloride
Eth	Ethyl ether
Fm	Formic acid
Fipr	Hexafluoroisopropanol
He	Hexane
Hep	Heptane
HMPA	Hexamethylphosphotriamide
i-Bu	Isobutyl alcohol
i-Pr	Isopropyl alcohol
i-PrAc	Isopropyl acetate
i-PrEth	Isopropyl ether
Me	Methyl alcohol
MeCl	Methylene chloride
MEK	Methyl ethyl ketone
MeOAlc	2-Methoxyethanol
NMe	Nitromethane
NMM	N-Methylmorpholine
Pe	Petroleum ether or ligroin
Pen	Pentane
Pr	Propyl alcohol
Pr-Eth	Di-*n*-propyl ether
Py	Pyridine
TEA	Triethylamine
TFA	Trifluoroacetic acid
TFE	Trifluoroethanol
THF	Tetrahydrofuran
THP	Tetrahydropyran
Tol	Toluene

Peptide Bond-Forming Methods

A	Azide
AC	Acid chloride
BBDQ	2-Isobutoxy-1-isobutoxycarbonyl-1,2-dihydroquinoline
CDI	N,N′-Carbonyldiimidazole
DCCI	Dicyclohexylcarbodiimide
DCCI-HBT	Dicyclohexylcarbodiimide-1-hydroxybenzotriazole
DCCI-NHS	Dicyclohexylcarbodiimide and N-hydroxysuccinimide
2,4-DNPO	2,4-Dinitrophenyl ester
EA	Ethoxyacetylene

EDCI	1-Ethyl-3-(3′-dimethylaminopropyl)-carbodiimide hydrochloride
EEDQ	N-Ethoxycarbonyl-2-ethoxy-1,2-dihydroquinoline
MA	Mixed anhydride (e.g. with pivaloyl chloride diethyl chlorophosphite or diphenylphosphoryl chloride)
MCA	Mixed carbonic anhydride
NCA	N-Carboxyanhydride
NTA	N-Thiocarboxyamino acid anhydride
OBHAE	Benzhydroxamic acid ester
OCH_2CN	Cyanomethyl ester
OHQ	8-Hydroxyquinoline ester
ONP	*p*-Nitrophenyl ester
OPAP	Phenylazophenyl ester
OPCP	Pentachlorophenyl ester
OPFP	Pentafluorophenyl ester
OPh	Phenyl ester
OPhBzl	2-Benzyloxyphenyl ester
OPic	4-Picolyl ester
$OP(SO_2Me)$	4-(Methylsulfonyl)phenyl ester
OPy	2-Pyridyl ester
OQ(Cl)	5-Chloroquinolin-8-yl ester
OSU	N-Hydroxysuccinimide ester
OTCP	2,4,5-Trichlorophenyl ester
PPP	Bis-*o*-phenylene pyrophosphate
$SiCl_4$	Silicon tetrachloride
SNP	2-Nitrothiophenyl ester
SP	Solid phase
SPh	Phenythio ester
SPy	2-Pyridylthiol ester (2-Pyridine thiol ester)
TMP	Trimetaphosphate (sodium)
TP	Triphenylphosphite (with imidazole) or triethylphosphite
TPP-CTe	Triphenylphosphine-carbon tetrachloride
TPP-DDD	Triphenylphosphine with O,O′-dinitrodiphenyldisulfide or with dipyridyl disulfide
TPP-DNBS	Triphenylphosphine-2,4-dinitrobenzenesulfonate
TPP-DPDS	Triphenylphosphine-dipyridyl-2,2′-disulfide
WRK	Woodward's Reagent K (N-ethyl-5-phenylisoxazolium 3′-sulfonate)
Δ	Heat

Other

AA	Amino acid analysis
ESR	Electron spin resonance
IR	Infrared
Mol. wt.	Molecular weight
PMR	Proton magnetic resonance
UV	Ultraviolet

Tlc Solvent Systems[a]

No.	*Solvents*	*Ratios*
1	Be	
2	Eth	
3	EtAc	
4	Me	
5	Thf	
6	Acn-Aq	3:1
7	Alc-Aq	7:3
8	Be-Ac	7:1
9	Be-An	1:1
10	Be-Chf	1:1
11	Be-Eth	1:1
12		5:2
13	Be-Me	9:1
14	Bu-3% Ammonia	70:30
15	2-Bu-Ammonia	3:1
17	2-Bu-Ammonia	25:11
18	2-Bu-Aq	3:1
19	Chf-An	1:1
20	Chf-An	4:1
21	Chf-Bu	1:1
22	Chf-EtAc	1:1
23	Chf-Me	1:1
23a	Chf-Me	2:1
24	Chf-Me	4:1
25	Chf-Me	5:1
26	Chf-Me	8:2
27	Chf-Me	9:1
28	Chf-Me	9:4
29	Chf-Me	19:1
30	Chf-Me	97.5:2.5
31	Cy-EtAc	1:1
32	EtAc-Ac	5:1
33	EtAc-Ac	25:0.3
34	EtAc-Ac	99:1
35	EtAc-Pe	1:1
36	He-Be	1:3
36a	MeCl-Me	9:1
36b	PhOH-Aq	3:1
37	*i*-Pr-25% Ammonia	9:2
39	*i*-Pr-Aq	7:3
40	Pr-Aq	2:1
41	Pr-Aq	7:3
42	Py-Aq	4:1
43	Tol-An	1:1
44	Tol-An	7:3
45	*t*-Am-*i*-Pr-Aq	51:21:28
46	*t*-Am-*i*-Pr-Aq	67:26:7
47	Am-Py-Aq	7:6:6

Tlc Solvent Systems (continued)

No.	*Solvents*	*Ratios*
48	Am-Py-Aq	7:7:6
49	*i*-Am-Py-Aq	7:7:6
51	Be-EtAc-Ac	12:4:1
51a	Be-Me-Ac	10:2:1
52	Be-Py-Ac	16:4:1
53	Bu-Ac-Aq	2:6:5
54	Bu-Ac-Aq	3:1:1
55	Bu-Ac-Aq	4:1:1
56	Bu-Ac-Aq	4:1:2
57	Bu-Ac-Aq	4:1:5
58	Bu-Ac-Aq	7:1:2
59	Bu-Ac-Aq	8:1:1
60	Bu-Ac-Aq	10:1:3
61	Bu-Alc-Aq	4:1:5
62	Bu-Alc-Aq	7:2:9
63	Bu-Alc-Aq	8:3:10
64	Bu-Ammonia-Aq	85:7.5:7.5
65	Bu-EtAc-0.2 *N* Ammonia	1:1:2
66	Bu-EtAc-Aq	1:4:1
67	Bu-EtAc-Aq	3:1:1
68	Bu-EtAc-Aq	10:2.3:6
69	Bu-EtAc-Aq	14:12:5
70	Bu-Fm-Aq	15:13.5:11.5
71	Bu-Fm-Aq	75:13.5:11.5
72	Bu-Py-Aq	13:7:13
73	2-Bu-Fm-Aq	15:3:2
75	2-Bu-90% Fm-Aq	14:3:3
76	2-Bu-90% Fm-Aq	75:13.5:13.5
77	2-Bu-Py-Aq	2:2:1
78	*t*-Bu-Ac-Aq	1:1:1
79	*t*-Bu-Ac-Aq	4:1:1
80	*t*-Bu-Ac-Aq	4:1:2
81	*t*-Bu-Ac-Aq	4:1:5
82	*t*-Bu-Ac-Aq	10:1:3
83	*t*-Bu-Ac-Aq	10:3:9
84	*t*-Bu-Ac-Aq	12:3:5
85	Chf-Alc-Aq	5:4:1
86	Chf-Cy-Ac	9:9:2
87	Chf-Me-Ac	5:4:1
88	Chf-Me-Ac	7:1:2
89	Chf-Me-Ac	17:2:1
91	Chf-Me-Ac	45:4:1
92	Chf-Me-Ac	82:15:3
93	Chf-Me-Ac	95:3:2
94	Chf-Me-Ac	95:5:1
95	Chf-Me-Ac	95:5:3
96	Chf-Me-An	10:1:1
97	Chf-Me-An	13:1:1

Tlc Solvent Systems (continued)

No.	*Solvents*	*Ratios*
98	Chf-Me-An	13:2:2
99	Chf-Me-Ammonia	12:9:4
100	Chf-Me-Ammonia	50:47:3
101	Chf-Me-Py	95:5:3
102	EtAc-Chf-Me	5:3:2
103	EtAc-Chf-Me	6:3:1
104	Fm-Ac-Aq	1:1:3
105	Fm-Ac-Aq	1:1:8
106	Fm-Ac-Aq	3:3:4
107	Hep-*t*-Bu-Ac	3:1:1
108	Hep-*t*-Bu-Ac	5:1:1
109	He-Bu-Ac	3:2:1
110	He-Bu-Py	3:2:1
110a	MEK-Py-Aq	10:5:4
111	*i*-Pr-Ac-Aq	7:1:2
112	*i*-Pr-Ac-Aq	10:1:1
113	*i*-Pr-Ammonia-Aq	7:1:2
114	*i*-Pr-Py-Aq	2:1:1
115	Pr-Ac-Aq	4:1:1
116	Py-Ac-Aq	10:6:3
117	Tol-Ac-Aq	5:5:1
118	Alc-Aq-Be-Ac	4:4:2:1
119	Be-An-Me-Ac	14:1:4:1
120	Bu-Ac-EtAc-Aq	1:1:1:1
121	Bu-Be-Py-0.1% Ac	6:2:1:9
122	Bu-*i*-Pr-ClAc-Aq	65:15:3:20
123	Bu-Py-Ac-Aq	3:2:0.6:1.2
124	Bu-Py-Ac-Aq	4:1:1:2
125	Bu-Py-Ac-Aq	15:10:3:6
126	Bu-Py-Ac-Aq	15:10:3:12
128	Bu-Py-Ac-Aq	15:12:3:10
130	Bu-Py-Ac-Aq	30:10:3:12
131	2-Bu-*i*-Pr-ClAc-Aq	65:15:3:20
132	*t*-Bu-MEK-Aq-Ammonia	4:3:2:1
133	Cy-EtAc-Me-Ac	10:10:5:1
134	EtAc-MEK-Ac-Aq	5:3:1:1
135	EtAc-MEK-Fm-Aq	5:3:1:1
136	EtAc-Py-Ac-Aq	5:5:1:3
137	EtAc-Py-Ac-Aq	60:20:6:11
138	Bu-Ac-Me-Py-MeCl-Aq	4:1:1:1:2:1
	CCD Solvents	
1	Me-Buffer-Chf	3:2:4
2	Me-Buffer[b]-Chf-Cte	10:3:7:4
3	Bu-Pr-Aq	3:0.5:4
4	Pe-EtAc-Me-Aq	8:16:9:6
5	Chf-Cte-Me-0.1 *M* NH_4Ac	5:5:8:2
6	2-Bu-0.05% Ac	

Tlc Solvent Systems (continued)

7	Ac-Chf-0.01 *M* NH_4Ac	7:7:4
8	*t*-Bu-Aq-Ac	3:4:0.025
9	Bu-Be-Aq (1.5% PyAc)	1:1:2
10	Me-Buffer-Chf-Cte	11:3:6:7
11	Bu-Aq	

[a] Nos. 16, 38, 50, 74, 90, 127, and 129 corresponded to data which was found inappropriate for inclusion in the final draft and were deleted.

[b] Buffer prepared from 29 ml of acetic acid, 19 g ammonium acetate and 11 ml of water.

Chapter 1

Racemization

The onset of racemization can be a capricious and unwanted event in any reaction sequence involving optically active amino acids and their derivatives. In peptide bond-forming reactions the extent of racemization will vary depending on the type of coupling reaction, the type of N-protecting group, type of base, solvent, salts present, and the temperature. Generally, as already discussed in the introduction to Volume 2, the formation of an oxazolinone and subsequent enolization at the asymmetric carbon is responsible for racemization in peptide-coupling reactions. The rate of enolization is in part base dependent and in one example studied in detail racemization with triethylamine was found to be twelve times faster than with diisopropylethylamine.[191] Exposure of amino acid or peptide azide derivatives to a base can markedly enhance enolization and racemization can become extensive as nicely illustrated by Rittel and colleagues.[572] The generally racemization-safe azide and mixed carbonic anhydride-coupling techniques can lead to a substantial loss of optical purity when excess base is employed. This matter has been discussed in some detail on pages 2 to 5 of Volume 2.

With new coupling methods and/or reaction techniques, the extent of racemization has usually been determined using one or more standardized evaluation methods. One of the earliest was due to Anderson[14] where the condensation of Z-Gly-Phe with Gly-OEt was employed and specific rotation of the product measured. Also, racemate may be isolated by fractional crystallization. By this means down to 1–2% racemization may be detected. Next in 1963 appeared the Weygand[672] (Z-Leu-Phe + Val-OBu^t, partial hydrolysis and glc determination of D-Phe-Val, sensitive to 0.1–1% racemization) and Young[683] (Bz-Leu + Gly-OEt, specific rotation of crude material and isolation of racemate, can detect 1–2% racemization) procedures. In 1969 Izumiya's amino acid analyzer method was reported and details were available last year.[263] With this technique, preparation of the tripeptide Gly-Ala-Leu is preferred as elution volumes for the LL and LD are distinctly different. As an illustration, Z-Gly-Ala azide was condensed with Leu-OBzl and the product was directly hydrogenated and submitted to amino acid analysis. Only the optically pure tripeptide was detected and this procedure is assumed useful down to 0.1–1% racemate. On the other hand, the use of DCCI in THF with triethylamine gave 22% racemate. When a mixed carbonic anhydride procedure was used with isobutylchloroformate and N-hydroxysuccinimide in tetrahydrofuran with N-methylmorpholine, only 0.2% racemate was found. With the DCCI–NHS procedure no racemate was detected.

The isotopic dilution technique of Kemp[312] is sensitive to approximately 0.001–0.01% racemate. In this procedure a ^{14}C label at the carboxylic acid carbon is employed and the radioactive racemate is isolated following dilution with unlabeled racemate. The newest racemization test depends on the proton magnetic resonance properties of N-Ac-Phe-Ala-OMe and has been recommended by Weinstein.[666] Here, it is assumed that magnetic nonequivalence of the LL isomer and the DL isomer is due to the aliphatic side chain of the former being more extended and deshielded as compared to the more compact DL conformation. The shift in the Ala-methyl signal provides a method for determining the extent of racemization to $\pm 3\%$ with CAT or ^{13}C side-band measurements. The sensitivity can be increased to $\pm 0.3\%$. The N-AC blocking group is well known to favor racemization and therefore the dipeptide used here leads to a wide range in degree of racemization. For example, MCA → 50%, DCCI → 30%, EDCI-NHS less than 3%, and azide less than 3% diastereoisomer formation. The simplicity of the amino acid analyzer and proton magnetic resonance methods are certainly attractive for most purposes and the isotopic dilution procedure seems the best choice for careful comparisons involving less than 1% racemate.

The base strengths of six commonly employed amines in nine commonly employed solvents ranging from dioxane to nitromethane have been measured.[682] The effect of solvent on the rate of racemization using the ONP procedure was evaluated in detail. The rapid increase in racemization rate by base in highly polar solvents was illustrated, e.g., the rate from dioxane to nitromethane was 95-fold. In the reaction studied, THF, DMF, and DMSO were found safe but Chf, MeCl, Acn, and NMe gave clearly unfavorable results. A combination of N-methylmorpholine in THF seems a reasonable choice for coupling reactions where oxazolinone formation is possible. The basicity of chloride ion in organic solvents combined with the increase in ionic strength of the solution is sufficient to cause racemization of active esters. For example, 0.05 *M* triethylamine hydrochloride in acetonitrile caused 9% racemization of Bz-Leu-ONp over a 22-hr period. However, racemization was not observed with a perchlorate salt.

In other new developments N-Tfa-Abu-Abu cyclohexyl ester has been recommended as an optically active stationary Glc column substrate for determining the amount of racemate in N-Tfa isopropyl ester derivatives prepared from 16 of the common naturally occurring amino acids.[475] The limiting detection range was 0.1%. For suppressing racemization with N^{im}-His the use of 1-hydroxybenzotriazole has been recommended.[299]

Chapter 2

Amino Acids and Derivatives

Methods for selective protection and deprotection of amino acid units comprise an expanding series of reactions of fundamental importance in peptide synthesis. In most approaches to peptide syntheses, the proper choice of protecting groups can mean the difference between success and failure. Fortunately, further refinements in the methanology, of well-established protecting groups, as well as the introduction of new blocking procedures, continues to be of interest to organic chemists. Recent illustrations of this chemistry appear in Charts 1 and 2 (reference numbers in these and subsequent structure pages have been entered in the left margin).

The Boc N-protecting group is widely employed at present and most of the common Boc-protected amino acids are commercially available. However, frequently it is desirable from a cost standpoint, or because of unavailability of certain derivatives, to rely heavily still on the use of *t*-butyl azidoformate as starting material. The preparation of this useful reagent has been further simplified by application of tetramethylguanidinium azide. The phenyl and *t*-amyl azidoformates were also obtained in high yields.[533] Reaction of the azide with representative amino acids has been simplified by use of 1,1,3,3-tetramethylguanidine as base.[8,638] Cleavage of the Boc group using boron trifluoride etherate in either acetic acid or acetic acid–chloroform mixtures has recently been introduced. The reaction is generally complete in 5 min at room temperature or in about 1 hr at ice-bath temperature.[234] A mixture of ethanethiol sulfonic acid in acetic acid has been suggested for the same purpose.[366]

Preparation of Bpoc-protected amino acids has been shown feasible in aqueous solvents using the corresponding fluorocarbonate illustrated in Chart 1.[545] A series of even newer urethane type-protecting groups are presented in Chart 2. In histidine chemistry, protection of the N-imidazole position with the adamantyloxycarbonyl group has been shown to offer some advantages.[617] A further study of the 2,2,2-trichloroethyloxycarbonyl (TrOC) group has also been made.[716] This protecting group is resistant to cleavage by trifluoroacetic acid or hydrogen chloride in dioxane. Of particular interest is its ready removal by treatment with zinc in acetic acid or in hot methanol. An electrolytic reduction procedure has also been explored for removing such protecting groups.[295] The same approach has been applied in a new cleavage method for N-tosyl groups.[457] Here, sodium or potassium ion is used as an electron transfer agent with a lead cathode in methanol, aqueous methanol, or water. In practice, the N-tosyl amino acid or peptide in solution was placed in a cathode compartment containing two equivalents of concentrated sodium hydroxide solution and a lead electrode. The anode compartment with a carbon electrode contained 20% hydrochloric acid. Electrolysis was conducted below 30° and a 20–30% current efficiency was realized.

The two newest urethane type-protecting groups comprise the 2-cyano-*t*-butyloxycarbonyl group (Cyoc) which can be removed by β-elimination reactions[706] and the isobornyloxycarbonyl group (Iboc).[162] The chloroformate precursor of the Iboc group is reasonably stable upon storage. The 1,1-dimethyl-2-propynyloxycarbonyl (DMPOC) group has been employed as a useful urethane type-blocking unit which can be removed from sulfur-containing peptides employing palladium(5% on carbon)-catalyzed hydrogenolysis in methanol.[590]

Among the nonurethane-N-acyl type derivatives 2-methoxy-[2-ethoxy-(2-ethoxy)]acetic acid has been condensed at the amino-group position, and in conjunction with a morpholine amide at the carboxyl end has been used to increase solubility of alanine peptides to the nonapeptide stage.[186] Two other new N-acyl amino acid derivatives seem of particular interest for certain specialized applications. As indicated in Chart 2 an azulene acetic acid has been used to prepare a series of blue marker groups.[704] Such chromophoric groups are being used to study the specificity of collagenase[703] and the specificity of carboxypeptidase B.[702] The positions of cleavage which result from enzymatic hydrolysis can be located by such chromophoric markers, and structures of the cleavage products have been precisely identified using mass spectrometry.[700] The ready hydrolysis of N-picolinyl amides promoted by copper(II) chleate formation appears to be of further interest but will require more detailed study, particularly in respect to racemization.[335]

A new series of Schiff bases have been prepared from twelve of the common amino acids and these particular ketimines were found to be stable, optically pure yellow solids, usable in peptide bond-forming reactions.[241] Of particular utility is the observation that the protecting group can be cleaved using 80% acetic acid at 80° in 20 min or at room temperature in 10 hr. Under the same conditions, a Boc group is essentially resistant to cleavage. The reaction of aldehyde and ketones with dipeptide sodium salts in methanol has been shown to easily afford imidazolidinyl peptide derivatives.[467] Water-soluble amine derivatives have been obtained using the N-acetylpyridinium group.[166]

An important contribution to cysteine protection has been provided by introduction of the S-acetamidomethyl group.[638] As outlined in Chart 2, S-Acm-Cys can be prepared and subsequently cleaved in nearly quantitative yields. Furthermore, the group is stable to the usual mild acid Boc-cleavage reactions and to treatment with hydrogen chloride at about 0° which will remove a benzyloxycarbonyl group. This thiol acetal has proved stable to hydrazide and azide formation as well as the weak base reaction conditions generally employed in peptide-bond formation. However, it can be easily removed with mercury(II) ion under mild conditions. Nine equivalents of mercuric acetate in 50% acetic acid for 70 min appears to be the maximum requirement. With peptides of low molecular weight, hydrogen sulfide is sufficient to remove excess mercury(II) ion but thorough removal of the sulfide reagent is necessary to eliminate oxidation of hydrogen sulfide to sulfur dioxide with subsequent S-sulfonate formation. The use of hydrogen sulfide was found unsatisfactory with larger peptides due to their absorption on the mercury(II) sulfide but this short-coming was eliminated by adding mercaptoethanol and removing the corresponding mercury sulfide by gel filtration.

At the carboxyl end efforts have also continued to find specific protecting groups. In the general area of esterification, triethyloxonium fluoroborate in neutral or slightly basic aqueous solution has been used at room temperature to esterify amino acids and peptides.[205] Yields of the derived ethyl esters ranged from 82 to 96%. However, the imidazole ring of His and the sulfide of Met are also readily alkylated with formation of the corresponding ammonium and sulfonium salts. The Schiff bases prepared from amino acids and ethyl acetoacetate are readily alkylated at the carboxyl group, for example, by benzyl halides, and the protecting group is then easily removed by mild acid hydrolysis to yield the amino acid ester.[379] The oxazolidin-5-one derivatives of such Schiff basis have now been prepared.[144]

Stewart [595] has extended his exploration of acid-labile substituted benzyl esters to those of the piperonyl type. This protecting group was removed from Z-Gly-OPn with trifluoroacetic acid in 5 min at room temperature to give Z-Gly in 91% yield. At present this type benzyl ester may present a workable alternative to the analogous 4-methoxybenzyl derivatives. For blocking the amide group of Gln and Asn in peptide syntheses, the N-2,4-dimethoxybenzyl group has been added to available methods.[491] Such dmb derivatives have been obtained using DCCI and removed by means of trifluoroacetic acid or hydrogen chloride.

In the more general area of amino acid chemistry, 14 new unnatural amino acids have been synthesized and explored for biological activity.[2] The crystal structure of L-Trp has been determined[428] and substances such as ninhydrin and isatin have been shown to oxidatively deaminate and decarboxylate α-amino acids to yield aldehydes and isatogen reduction products.[240] The decarboxylation (Strecker degradation) of α-amino acids has been studied with benzil and with benzoin.[11] Both the amino and carboxylic acid groups have been induced to enter into spirophosphorane formation.[430] Various methods continue to be proposed for the separation of protected amino acids by gas liquid chromatography.[128,192,252,427] A neutron diffraction study of His has shown this amino acid to be in the zwitterion form but not by protonation of the imidazole ring.[350] The molecular conformation of His was found to be stabilized by an intramolecular hydrogen bond between ammonium group hydrogen and the imidazole ring as well as by an electrostatic interaction between the ammonium group hydrogen and a carboxyl oxygen.

Other miscellaneous contributions to amino acid chemistry which appeared after the tables of this chapter were completed were dehydrogenation of 1,4-cyclohexadiene 1-Ala hydrate to Phe,[510] β-elimination reactions with β-hydroxy-Asp,[418] tritylation of amino acids using trityl chloride, and triethylamine in anhydrous pyridine,[204] a CD study of N-(3-nitro-2-pyridyl)amino acids,[624] a resolution of DL-Asp using a copper chleate of, for example, Ala,[211] preparation of β-(5-tetrazolyl)-Ala[193] a betaine derivative of Cys,[256] preparation of the four isomers of α,β-Dbu,[28] synthesis of His isoalloxazine,[169] synthesis of *trans*-4,5-didehydro-DL-Lys and 4-Oxo-Lys,[229] resolution of *cis*- and *trans*-5-Me-Pro,[462] acetylation of N^{δ}-benzyloxy-Orn,[254] synthesis of N^{δ}-hydroxy-Orn,[255] a synthesis of phenylglycine by carboxylation of α-lithiobenzylisocyanide,[633] synthesis of succinimides and glutarimides derived from amino acids for anticonvulsant study,[686] and a number of new carbohydrate derivatives of amino acids were prepared[135,167,556,629] including a new route to such derivatives based on addition of the amino group of an amino acid ester to an α-nitroolefin sugar.[507]

Chart 1

533

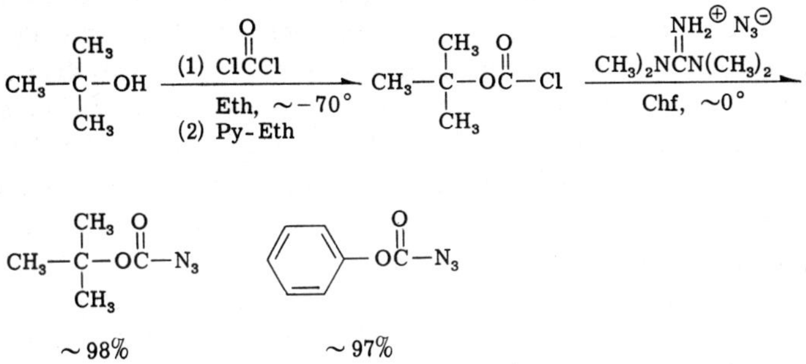

8

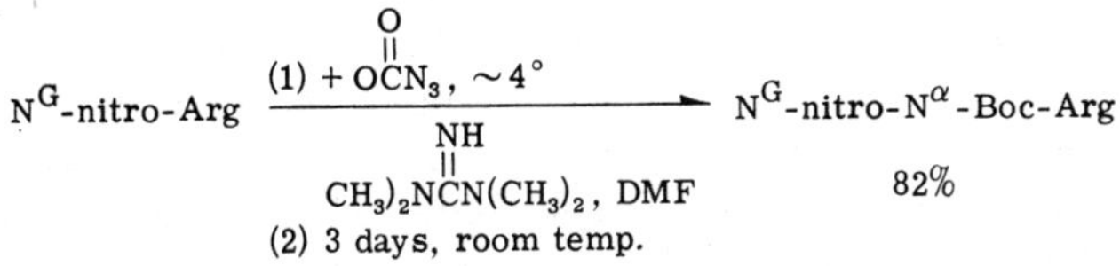

545

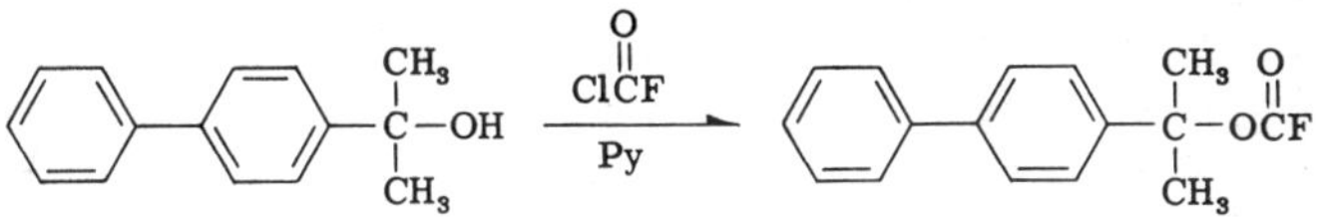

Chart 2

617

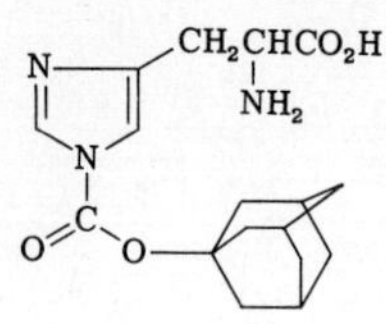

716

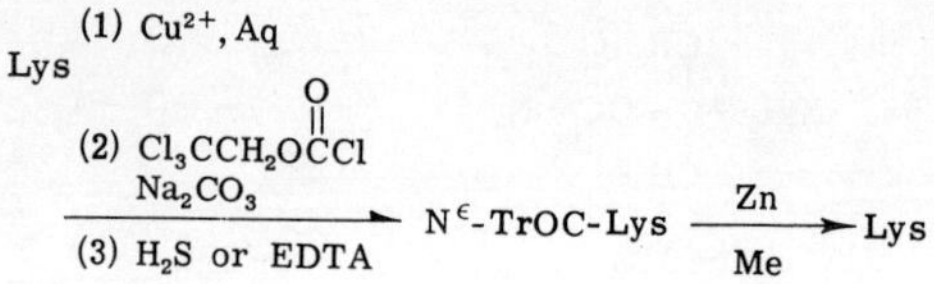

457

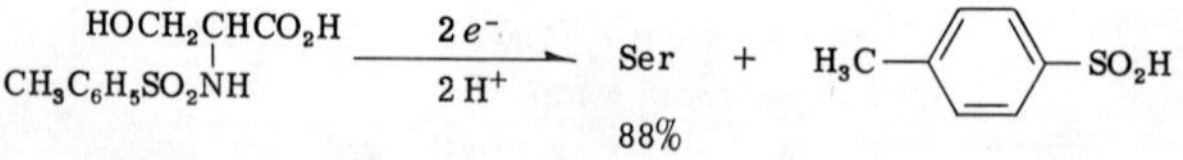

706

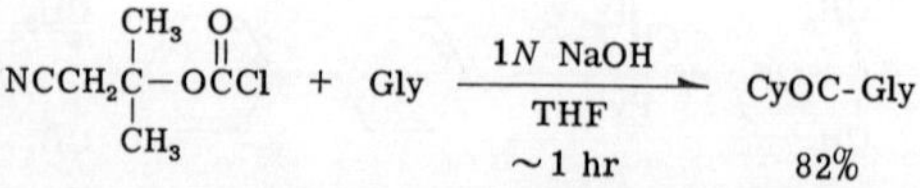

162

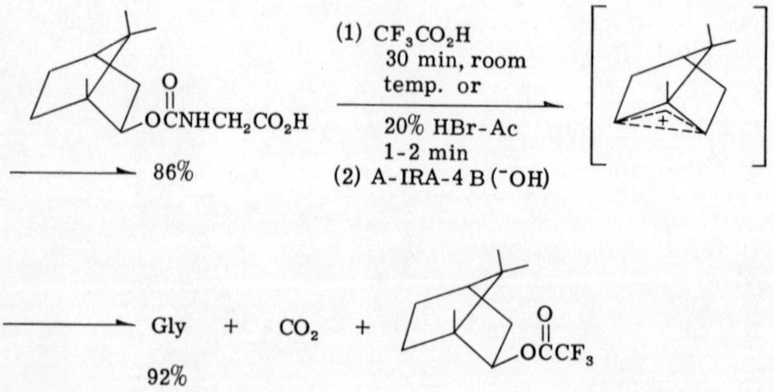

Chart 2 (cont.)

704

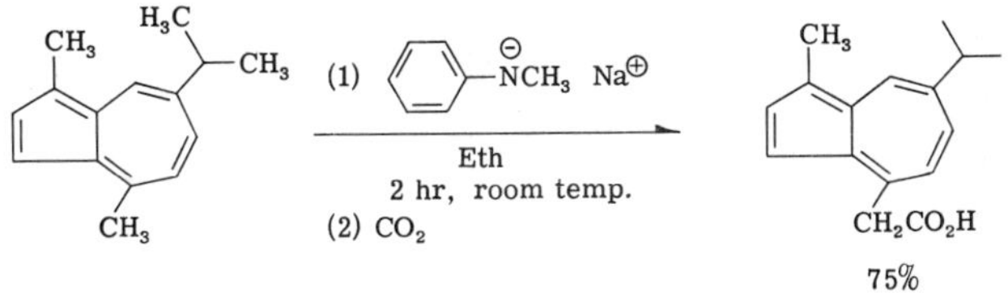

(1) Pro-OMe
MeCl
DCCI
1 day, 0°
(2) di-Aq
1*N* NaOH
→ N-Imaa-Pro-OMe
69%
blue powder
m.p. 103-105°

335

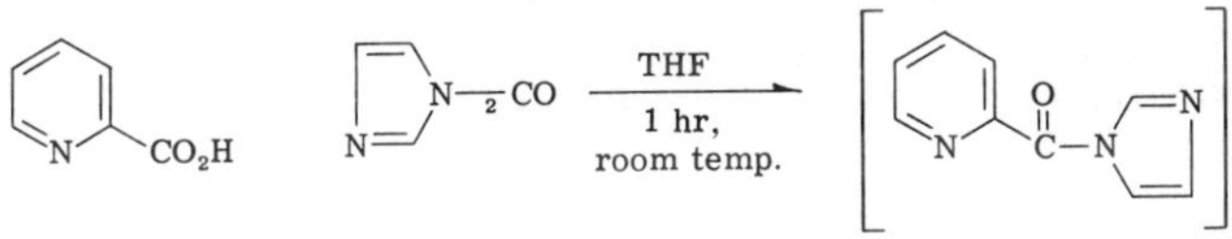

$\xrightarrow{\text{Gly-OEt}}$ Pic-Gly-OEt
95%

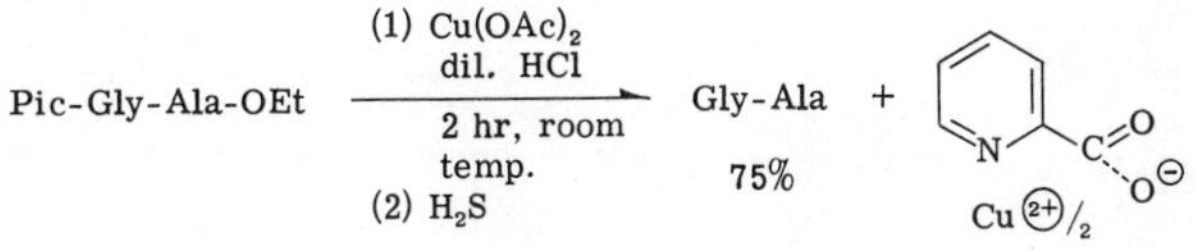

590

$HC{\equiv}C{-}C(CH_3)_2{-}OC(=O)NH{-}CH(CH_2CH_2SCH_3)CO_2^{\ominus}\ Dcha^{\oplus}$

——→ 79%

186

$CH_3OCH_2CH_2OCH_2CH_2OCH_2CO_2H$

Chart 2 (cont.)

241

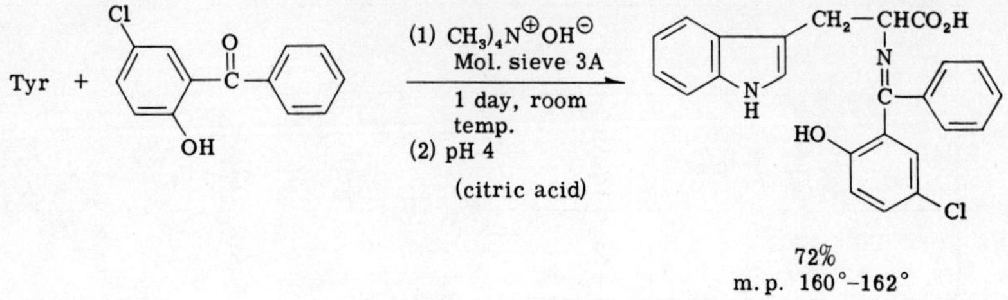

638

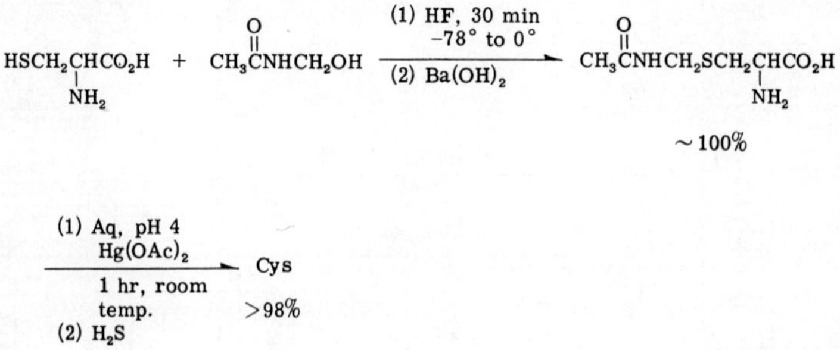

Table 1 Amino Acid Derivatives

Compound	Method	Solvent	% Yield	Chromatography Col	Tl	Ion	Paper	CCD	Recrys. Sol.	M.P. °C	$[\alpha]_D$	Solvent Rotation	Elect.	Other	Ref.
Alanine															
N-Ac-Ala															553
															647
N-Ac-Ala-OEt										34–35	−63	Aq-Alc		IR	439
N-Ac-β-(1,3-dioxy-4,4,5,5-tetra-methyldihydroimidazol-2-yl)-DL-Ala															664 665
N-Ac-β-(1,3-dioxy-4,4,5,5-tetra-methyldihydroimidazol-2-yl)-DL-Ala-OEt														Mol. wt.	665
N-Ac-3-[3,4-bis-Bz-oxy-phenyl]-Ala-OMe					117				Eth	156–158	+74	Chf			646
N-Ac-3-(3,4-dihydroxyphenyl)-Ala											+43	Me			291
N-Ac-N-ethyl-Ala															647
N-Ac-N-methyl-Ala										110–112	−50	DMF			101
Aceto-Ac-Ala										92–93	−4.6				42
(*R*)(−)-Ala					13	DO-1	×			295 (dec)	−6.1	6 *N* HCl			117 334
(*S*)(+)-Ala											+0.32	5 *N* HCl			65
DL-Ala-OBzl Tos salt					124					109					263
Ala-OPhBzl HCl									Alc-Eth	160–170	−15	Chf			110
Ala-2-(4-OChlBzl)-phenyl HCl									Chf-Eth	114.5–116.5	−12	Chf			110
Ala-OEt HCl										76					186

(*continued*)

Table 1 Amino Acid Derivatives (continued)

Compound	Method	Solvent	% Yield	Chromatography Col	Tl	Ion	Paper	CCD	Recrys. Sol.	M.P. °C	$[\alpha]_D$	Solvent Rotation	Elect.	Other	Ref
(*S*)-Ala-O-*i*-Bu Tos salt										120–121	−1.1	Alc			210
Ala-NCA										92.5					127
															185
Ala-O-2-(2,4-dichlorobenzyloxy)-phenyl HBr										135–136	−13	Chf			110
Ala methylthiohydantoin derivative														CD	622
D-Ala methylthiohydantoin derivative														CD	622
Ala-Sph										73–74					444
Ala-Thp-hydroxy-OEt										oil					585
Ala-thio-NCA										93					185
									EtAc-He	91–93				IR	127
N-Bz-Ala															647
															553
N-Bz-Ala anilide										178–179	−9	Py			534
N-Bz-3-[3,4-bis-Bz-oxy-phenyl]-Ala-OMe					117				Me	163–165	+101	Chf		NMR	646
N-Bz-N-methyl-Ala										133	−32	DMF			101
N-(2-Bz-1-methylvinyl)-Ala									MeCl-Eth	163–164	+159	Alc			589
N-Bz-3-pyridine-Ala									Aq-Me						3
Z-Ala									Chf-Pe	87	−14	Ac			186
Z-Ala amide									Alc-He	133	−1.8	Alc			400
Z-Ala-NHBzl									EtAc	137–138					371
Z-Ala cyclohexane									EtAc	156–157					134
Z-Ala-OBhae									EtAc	152–153					371

Z-Ala-O-2-(2,4-dichlorobenzyloxy)-phenyl									EtAc-Pe	94–96	−22	Chf			110
Z-Ala *m*-fluorophenylhydrazide										156–158	−27	Py			1
Z-Ala *o*-fluorophenylhydrazide										156–158	−23	Py			1
Z-Ala *p*-fluorophenylhydrazide										174–176	−24	Py			12
Z-Ala-O-hydroxyethyl										58					585
Z-Ala *o*-methoxyphenylhydrazide										159–160	−25	Py			1
Z-Ala *p*-methoxyphenylhydrazide										162–165	−29	Py			1
Z-Ala N^1-methyl-N^1-phenylhydrazide										152–154	−16	Py			1
Z-Ala *m*-nitrophenylhydrazide										154–156	−34	MeOAlc			1
Z-Ala *o*-nitrophenylhydrazide										190–191	−55	MeOAlc			1
Z-Ala phenylhydrazide										153–155	−32	Py			1
Z-Ala-SPy										55–59	−35	DMF			365
Z-Ala *m*-tolylhydrazide										143–144	−27	Py			1
Z-Ala *o*-tolylhydrazide										165–166	−34	Py			1
Z-Ala *p*-tolylhydrazide										160–162	−29	Py			1
Z-D-Ala 2-Boc-hydrazide									EtAc-Pe						259
Z-D-Ala-ONp										77–78	+54	Alc			658
Z-DL-Ala-OBhae									EtAc	139–140					371
Z-DL-Ala-$NHCH_2CN$										123–125					195
Z-DL-3-(3-Pyridyl)-Ala anilide										176–178					339
Z-DL-3-(3-Pyridyl)-Ala hydrazide									Aq	125–126					339
N-Z-3-[3,4-bis-Bz-oxy-phenyl]-Ala-OMe					117				Eth	109–111	+34	Chf			646
Z-dehydro-Ala-OBzh															652
N-(Z)-3-(3,4-dihydroxyphenyl)-Ala											+1.5	Me			291
N-(Z-β-(1,3-dioxy-4,4,5,5-tetra-methyldihydroimidazol-2-yl)-Ala															665
N-Z-β-(1,3-dioxy-4,4,5,5-tetra-methyldihydroimidazol-2-yl)-Ala														Mass spec.	664
N-Z-β-(1-oxy-4,4,5,5-tetramethyl-dihydroimidazol-2-yl)-Ala															665

(*continued*)

Table 1 Amino Acid Derivatives (continued)

Compound	Method	Solvent	% Yield	Chromatography Col	Chromatography Tl	Chromatography Ion	Chromatography Paper	CCD	Recrys. Sol.	M.P. °C	[α]D	Solvent Rotation	Elect.	Other	Ref.
N-Z-N-methyl-Ala										65–66	−29	DMF			101
N-(Z)-2-pyridine-Ala															3
Z-β-(8-theobrominyl)-α-Ala-OCH_2CN										185–186					637
Z-β-(8-theobrominyl)-α-Ala-ONp										182–183					637
β-(2′-Bromo-3′-pyridyl)-Ala									Aq-An	200–201 (dec)					603
β-(2′-Bromo-4′-pyridyl)-Ala									Aq-An	187–189 (dec)					603
β-(2′-Bromo-5′-pyridyl)-Ala									Aq	253–255 (dec)					603
β-(2′-Bromo-6′-pyridyl)-Ala									Aq	230–232 (dec)					603
Boc-Ala									Eth-Pe	80–82	−22	Ac			330 544
DL-Boc-Ala-OMe									Eth-Pe	54–55					544
Boc-Ala-O-2,2′-dinitrodiphenyl-methyl										129–130	−19				478
Boc-Ala-NHC_6H_{11}									EtAc	151–152					134
Boc-dehydro-Ala									EtAc-Pe	190–192 (dec)					546
N-(Boc)-3-(3,4-dihydroxy-phenyl)-Ala									EtAc-Cy		+16	Me			291
Boc-Ala-OSu										166–167					367

Boc-Ala-OMe										33–34					8
Boc-Ala-OPy									Eth-Pe	65–66	−50	DMF			134
Boc-Ala-SPy										98	−66	EtAc			365
Boc-cyclohexyl-Ala										45–50	−13	Ac			316
Boc-N-methyl-Ala										91–92	−32	EtAc			17
N(-Boc)-2-pyridine-Ala									Alc-Aq						3
β-(2′-Chloro-3′-pyridyl)-Ala									Aq-An	200–201 (dec)					603
β-(2′-Chloro-4′-pyridyl)-Ala									Aq-An	203–206 (dec)					603
β-(2′-Chloro-5′-pyridyl)-Ala									Aq	260–262 (dec)					603
β-(2′-Chloro-6′-pyridyl)-Ala									Aq-An	182–184					603
β-Cyano-Ala									Eth-Pe	93.5–94.5	−14	Me	×	IR	511
Cyclohexyl-Ala HCl					24									IR UV	316
α-Deuterio-DL-Ala							×							NMR	615
3-[3,4-Dihydroxy-phenyl]-Ala					55				Aq	278 (dec)	−10	1 *N* HCl			646
3-[3,4-Dihydroxy-phenyl]-Ala-OMe HCl										172–174	+9.8	Me			646
DNbzloc-Ala										172–179	−39				478
DNT-Ala										165–167					227
β-(1,3-Dioxy-4,4,5,5-tetramethyl-dihydroimidazol-2-yl)-Ala															665
β-(2′-Fluoro-3′-pyridyl)-Ala										198–200 (dec)					604
β-(2′-Fluoro-5′-pyridyl)-Ala										252–253 (dec)					604
β-(2′-Fluoro-6′-pyridyl)-Ala										220–221 (dec)					604

(*continued*)

Table 1 Amino Acid Derivatives (continued)

Compound	Method	Solvent	% Yield	Chromatography Col	Chromatography Tl	Chromatography Ion	Chromatography Paper	CCD	Recrys. Sol.	M.P. °C	$[\alpha]_D$	Solvent Rotation	Elect.	Other	Ref.
[αR]-(+)-N-[2-Hydroxy-1-hydroxymethyl-2-phenylethyl]-α-methyl-β-[3,4-di-OMe-phenyl]-Ala										225–227					662
β-(2′-Hydroxy-3′-pyridyl)-Ala										227–229 (dec)					604
β-(2′-Hydroxy-4′-pyridyl)-Ala										283–284 (dec)					604
β-(2′-Hydroxy-5′-pyridyl)-Ala										215–216 (dec)					604
β-(2′-Hydroxy-6′-pyridyl)-Ala										257–258 (dec)					604
3-{[*p*-(Hydroxymethyl)-α,α-diphenyl-Bzl]thio}-Ala									Me-Aq	145–147 (dec)					731
[αS]-(+)-N-[2-Hydroxy-1-hydroxymethyl-2-phenylethyl]-α-methyl-β-[3,4-di-OMe-phenyl]-Ala									Me	178					662
N-Lauroyl-Ala															647
Z(OMe) Ala OSu										78–81					715
Z(OMe) Ala ONp									Alc-Eth-Pe	92–93	−46	Me			325
Z(OMe)-Ala-Pcp carbonate										81.5–82.5					715
Z(OMe)-Ala-quinolyl carbonate										82–82.5					715
Z(OMe)-Ala-Tcp carbonate										73–75					715

Methyl-3-(N-Z-Ala)-2-oxothiazolidine-4-carboxylate									An-Pe	172–173	−58	Aq			250
DAZ-Ala-OEt															468
MAZ-Ala-OEt															468
Methoxythiocarbonyl-Ala									EtAc-He	114–115				IR	127
N-Methyl-Ala				Anion-ex-change resin AG1-X2					Alc		+4.8	Aq			17
N-Methyl-Ala HBr									Me-Eth	146–148	+5.8	Eth			17
N-Methyl-Ala-OMe HBr										109–109.5	−5.7	DMF			101
[S]-(−)-α-Methyl-β-[3,4-dihydroxyphenyl]-Ala									Aq	307	−14	Aq			662
[S]-(−)-α-Methyl-β-[3,4-di-OMe-phenyl]-Ala HCl									Me	174–175	−4.3	Aq		Mass spec.	662
Nbzloc-Ala									EtAc-Pe	132–134	−11				478
Nvoc-Ala									Be-Di	180	−12	DMF			478
β-(1-Oxy-4,4,5,5-tetramethyldihydroimidazol-2-yl)-Ala													×		665
Pht-Ala-OCH_2CN										114–116	−34	An			351
Pht-Ala-$OCOOCH_2CH(CH_3)_2$										67.5–68.5	−36	Di			16
2-Pyridine-Ala										203				IR	3
2-Pyridine-Ala-OEt Di-HCl															3
3-Pyridine-Ala						A-IR-4B			Alc-Aq	260–263					3
4-Pyridine-Ala						A-IR-4B				236					3

(*continued*)

Table 1 Amino Acid Derivatives (continued)

Compound	Method	Solvent	% Yield	Chromatography Col	Tl	Ion	Paper	CCD	Recrys. Sol.	M.P. °C	$[\alpha]_D$	Solvent Rotation	Elect.	Other	Ref.
N-Tos-dehydro-Ala-OEt									Eth-Pe	73–75.5				NMR	437
Tos-N-methyl-Ala									Chf-He	121–122	−8.8	Eth			17
3-(2,4,5-Trihydroxyphenyl)-Ala										252–253 (dec)	+12	HCl-Me			291
N-[Tris(hydroxymethyl)methyl]Ala															165
α-Tritio-DL-Ala															615
β-Alanine															
N-Ac-β-Ala-OEt										b.p. 116–117 (1.1–1.2 mm)				IR	439
β-Ala-α-methyltryptamine tartrate									Alc-Eth	98–99					442
Z-β-Ala-cyanomethyl															195
N-Z-β-cyano-β-Ala															511
N-Z-β-cyano-β-Ala-OMe															511
β-Cyano-β-Ala										122.5 (dec)	−12	Aq		IR Mass spec.	511
Z-β-Ala-α-methyltryptamine									Be	139–140					442
Z-β-Ala-SPy										68–70					365
N-Methyl-β-Ala									Acn·Alc	68–70					531
β-(1-Pyrazolyl)-Ala						A-IR-120			50% Alc	243–246					432
β-(Pyrazolyl-3)-Ala Boc-hydrazide									Me-Eth	156–157	+6.4	Me			239

α-Aminobutyric acid															
N-Ac-Abu															647
Abu											+20	6 *N* HCl			669
γ-Abu-α-methyltryptamine tartrate									Alc-Eth	110–114					442
N-Bz-Abu anilide										169–170	−16	Py			534 669
Z-Abu									EtAc-Pe	69–70					249
Z-γ-Abu-α-methyltryptamine									Aq-Alc	92–94					442
N-Z-γ-Cyano-γ-Abu															511
N-Z-γ-Cyano-γ-Abu-OMe															511
γ-Bromo-Abu-HBr					43 125				Alc-Eth	188–190	+5.3	DMF			286
Boc-Abu									Eth	119–120					249
Boc-γ-Abu									Pe	52–54					544
γ-Cyano-γ-Abu							×			134.5–136.5 (dec)	+25	Aq			511
β-Hydroxy-γ-Abu-α-methyltryptamine tartrate									Alc-Eth	143–146					442
Isopropyl-4-Cyano-D-1-Tos-Abu									*i*-Pr-Pe	69–70	+21	DMF			200
N-Z-(OMe)-γ-Cyano-γ-Abu															511
N-Z-(OMe)-γ-Cyano-γ-Abu-OMe															511
N-Pht-β-hydroxy-γ-Abu									Be-EtAc-Pe	128–129					442
Pht-β-hydroxy-γ-Abu-α-methyltryptamine										Amorph					442
ε-Aminocaproic acid															
Pht-ε-Acp										105–106					156
Pht-ε-Acp acid chloride										66–67					156
α-Aminoisobutyric acid															
Pht-Aib-SPy										82–84					365
Boc-Aib			**12**		×				Eth	117–118					282

(*continued*)

Table 1 Amino Acid Derivatives (continued)

Compound	Method	Solvent	% Yield	Chromatography Col	Tl	Ion	Paper	CCD	Recrys. Sol	M.P. °C	$[\alpha]_D$	Solvent Rotation	Elect.	Other	Ref.
Arginine															
Ac-Arg-OMe acetate salt															640
Ac-N^G-nitro-Arg-OMe									An-Pe	147–149	−8	Me			640
N^6-[(4-Amino-4-carboxybutyl)carbamoyl]-Arg acetate				Ce					Aq-Alc	178–183				NMR	18
Aoc-Arg Tos salt									EtAc	79.5–83	−13	Py			157
Arg amide															668
Arg amide Di-HCl monohydrate										72–73					668
Arg-N-Bzl amide Di-HCl											$[\alpha]^{20}_{436}$ −31	Aq			668
Arg-N-Bzl-N-methylamide Di-HCl									Alc	112	$[\alpha]^{20}_{436}$ −10	Aq			668
Arg-N-Bu amide Di-HCl											$[\alpha]^{20}_{436}$ +17	Aq			668
Arg-Obu^t Di-HCl											$[\alpha]^{20}_{436}$ +28	Alc			668
Arg-N-ethylamide Di-HCl											$[\alpha]^{20}_{436}$ +49	Aq			668
Arg-OEt Di-HCl									Alc-Eth	115	$[\alpha]^{20}_{436}$ +36	Alc			668
Arg-NCA HBr															232 661
Arg N-isopropylamide Di-HCl											$[\alpha]^{20}_{436}$ +47	Aq			668

Arg-OMe Di-HCl															668
Arg methylthiohydantoin derivative														CD	622
Arg-N-phenethylamide Di-HCl											$[\alpha]^{20}_{436}$ −56	Aq			668
Arg-ThioNCA HBr									*i*-Pr-Eth	115–117	−19	MeCl			127
D-Arg															387
D-Arg-OEt Di-HCl															668
N-Bz-Arg-amide															458
Bz-DL-Arg-*p*-nitroanilide															605
N-(2-Bz-1-methylvinyl)-Arg										265–266					589
N-(2-Bz-1-methylvinyl)-nitro-Arg									MeCl-Eth	103	+12	DMF			589
N-(2-Bz-1-methylvinyl)-nitro-Arg Dcha										101–107	+12	DMF			589
Z-Arg-OEt										Oil				IR NMR	205
Z-Boc-Arg									Me	151–152	−2.5	Me			544
N^G-Tos-N^α-Z-Arg															451
Boc-Arg									THF	181–182					544
N^G-Boc-N^α-Boc-Arg									Acn	182–184 (dec)					544
N^G-Nitro-N^α-Boc-Arg										122–124					8
N^G-Nitro-N^α-Boc-homo-Arg															642
N^G-Nitro-N^α-Boc-homo-Arg-OPcp										156–157	−25	DMF		UV	642
N^α-Boc-Arg									Alc-Eth	117–119 (dec)	−8.8	Aq			722
N^G-Tos-N^α-Boc-Arg									EtAc	96–97	−3.6	DMF			722
Arg chloride															151
Decanoyl-Arg									Me	168–169	−4	Ac			640
N^G-Nitro-Z(OMe)-Arg-OPcp									Acn-Eth	153–154	−11	DMF			325
N^G-Nitro-Boc-Arg									EtAc	123–125	−5.9	DMF			544
N^G-Nitro-homo-Arg										228–230 (dec)	+20	N-HCl			642

(*continued*)

Table 1 Amino Acid Derivatives (continued)

Compound	Method	Solvent	% Yield	Chromatography				CCD	Recrys. Sol.	M.P. °C	$[\alpha]_D$	Solvent Rotation	Elect.	Other	Ref.
				Col	Tl	Ion	Paper								
N^G-Nitro-N^α-Nps-Arg Dcha										150–151	−4.5	DMF			197
N^G-Nitro-N^α-Nps-Arg	2,4-DNPO									67–70					197
N^G-Nitro-N^α-Nps-Arg-OPcp									EtAc-Alc	166–168	−32	DMF			196
N^α-Tos-Arg-NCA															451
N^G-Tos-Arg-NCA HCl									An-Eth	75 (dec)	−9.0	An			451
N^G-Tos-Boc-Arg															544
Tris-(Z)-Arg amide									Alc	182					668
Tris(Z)-Arg-N-Bzl amide									Alc	167	$[\alpha]^{20}_{436}$ +35	Chf			668
Tris(Z)-Arg-N-Bzl-N-methylamide									Alc	78	$[\alpha]^{20}_{436}$ +45	Chf			668
Tris(Z)-Arg-N-Bu^t amide									EtAc	165	$[\alpha]^{20}_{436}$ +47	Chf			668
Tris(Z)-Arg-N-ethylamide									Alc	162	$[\alpha]^{22}_{436}$ +53	Chf			668
Tris(Z)-Arg-N-isopropylamide									Alc	175	$[\alpha]^{22}_{436}$ +32	Chf			668
Tris(Z)-Arg-N-phenethylamide									Alc	170	$[\alpha]^{20}_{436}$ +34	Chf			668
Asparagine															
Asn											−12	1 *N* NaOH			491
Asn-OBu^t									He	96–98	+2	Me			293

Asn methylthiohydantoin derivative														CD	622
N-(2-Bz-1-methylvinyl)-Asn									MeCl-Eth	166–167	+38	Alc			589
N-(2-Bz-1-methylvinyl)-N-xanthyl-Asn									DMF-Eth	181–183	−80	Alc			589
N^{β}-Bzl-Asn									Alc-Aq	253	+44	2 *N* HCl			225
N^{β}-Bzl-Asn-NCA									EtAc-Eth	130 (dec)					225
N^{β}-Bzl-N^{β}-Z-Asn-Obu^t									EtAc-He	97.5	+6.3	EtAc			225
(Z)-Asn-OPht									DMF-Aq	153–155					306
Z-Asn-OSu									DMF-*i*-Pr	129–130	−28	DMF			219
Boc-Asn										174–176	−7.8	DMF			544
									Me	176–177	−8.2	DMF			546
											+12	2 *N* NaOH			
Boc-Asn-ONp									Eth	145–147	−34	DMF			546
Boc-Asn-OPy									EtAc-Pe	72–73	−49	DMF			134
Boc-Asn-SPy										100–103	−89	DMF			365
Boc-β-cyano-Ala-ONp									EtAc	152–153	−32	DMF			546
N^{β}-(Dmb)-Asn									Me	230–231					491
N^{β}-(Dmb)-N^{α}-Z-Asn									EtAc	152–154	+3.2	Me			491
N^{β}-(Dmb)-N^{α}-Z-Asn-α-OMe									EtAc	162–163	+60	DMF			491
N^{β}-(Dmb)-N^{α}-Z-Asn-ONp									Alc	148–149					491
N^{β}-(Dmb)-N^{α}-Boc-Asn-α-OBzl									EtAc-Pe	102–103	−6.4	Me			491
N^{β}-(Dmb)-N^{α}-Boc-Asn Dcha									EtAc	124–125					491
N^{β}-(Dmb)-N^{α}-Boc-iso-Asn-β-OBzl									EtAc-Pe	99–100	−10	Me			491
N^{β}-(Dmb)-N^{α}-Boc-iso-Asn Dcha										166–167					491
N^{β}-(Dmb)-iso-Asn									Alc-Eth	160–162					491
Bpoc-Asn									Eth	176–178	−8				573
N^{α}-Bpoc-Asn									Acn	218–220	−14	DMF			545
N-Ethyl-Asn										258	−5.1	Aq			320
Z(OMe)-Asn															511
Z(OMe)-iso-Asn										156	−25	DMF			511

(*continued*)

Table 1 Amino Acid Derivatives (continued)

Compound	Method	Solvent	% Yield	Chromatography Col	Tl	Ion	Paper	CCD	Recrys. Sol.	M.P. °C	$[\alpha]_D$	Solvent Rotation	Elect.	Other	Ref.
Iso-Asn															491
N-β-Methyl-Asn										252	−3.3	Aq			320
N^{β}-Methyl-N^{α}-Z-Asn									Alc	142–144					224
N^{β}-Methyl-N^{α}-Z-Asn-ONp									Me	155–156	−29	DMF			224
Nps-Asn-OSu										150–151	−53	Di			413
N^{β}-(*d*)-(1)-Phenethyl-Asn										215	+153	2 *N* HCl			225
N^{β}-(*d*)-(1)-Phenethyl-Asn-NCA										75					225
N^{β}-(*l*)-(1)-Phenethyl-Asn									Me-*i*-Pr-Eth	215					225
N^{β}-(*l*)-(1)-Phenethyl-Asn HBr									Alc-Eth	184	−60	Aq			225
N^{β}-(*l*)-(1)-Phenethyl-Asn NCA															225
N^{β}-(*d*)-(1)-Phenethyl-N^{α}-Z-Asn-OBu^t										116	+59	EtAc			225
N^{β}-(*l*)-(1)-Phenethyl-N^{α}-Z-Asn-OBu^t									EtAc-He	130	−47	EtAc			225
N^{β}-Bzl-N^{α}-Pht-Asn									Eth-EtAc	186	−53	Me			225
N^{β}-(*l*)-(1)-Phenethyl-N^{α}-Pht-Asn									EtAc-He	160					225
Z-Asn-OTcp									THF-Eth	184–185	−28	DMF			39
Aspartic acid															
N-Ac-DL-Asp-β-semialdehyde-OEt															665
Asp-α-OBzl										168–169.5	−16	HCl			425
Asp-β-OBzl										222	+7.6	Ac			185

Asp-β-OBzl-NCA										125.5 (dec)	−67	EtAc			
Asp-OBzl(OMe) HCl									Alc	159	+6.6	DMF			380
Asp β-Boc-hydrazide									Eth-Pe	134–136					544
Asp-α-OEt										180.5	+26	Aq			425
										181 (dec)					
D-Asp-β-OMe ephedrine salt										141	−30	Alc			456
Asp methylthiohydantoin derivative														CD	622
DL-Asp β-*p*-nitroanilide															608
(*S*)-Asp-O-Di-*i*-Bu Tos										128–129	+12	Alc			210
Asp-$NHCH_2CH_2$-*c*-C_6H_{11}									Me-Aq	193–202	+7	Ac			404
D-Asp-$NHCH(CH_3)CH_2$-*c*-C_6H_{11}									Me-Aq	207–208	+16	Me			404
Asp-$NHCH(CH_3)CH_2$-*c*-C_6H_{11}									Me-Aq	184–185	−19	Me			404
Asp-$NHCH(CH_3)(CH_2)_3CH_3$									An-Aq	188–193	+11	Ac			404
Asp-$NHCH(CH_3)(CH_2)_5CH_3$										180–190	−8	Me			404
Asp-$NHCH(CH_3)CH_2C_6H_5$									An-Aq	164–166	+34	Me			404
Asp-$NHCH_2CH_2C_6H_5$									*i*-Pr-Aq	212–214	−15	Aq			404
Asp-$NHCH(CH_3)CH_2CH(CH_3)CH_3$									Aq	166–168	−17	Me			404
D-Asp-$NHCH(CH_3)CH_2CH(CH_3)$-CH_3									Aq	201–202	+9	Me			404
Asp-$NHCH(CH_3)CH_2CH(CH_3)$-CH_2CH_3									Aq	162–166	−2	Me			404
Asp-$NHCH(C_2H_5)CH_2C_6H_5$									Me-Eth	158–163	+8	Me			404
Asp-$NHCH(C_2H_5)CH_2CH_2CH_3$									An-Aq	196–200	+12	Ac			404
Asp-$NHCH(C_2H_5)(CH_2)_3CH_3$										190–195	+9	Ac			404
Asp-$NHCH_2CH_2C_6H_4OH$-4									Aq	209–210	−21	Aq			404
Asp-$NHCH(CH_3)CH_2C_6H_4NHSO_2$-$CH_3$-4									Aq	199–208	+14	1 *N* HCl			404
Asp-$NHCH(CH_2OH)CH_2C_6H_4OH$-4									Me	212–213	−10	1 *N* HCl			404
Asp-$NH(CH_2)_6CH_3$									Aq	200–201	+8	Ac			404
Asp-$NH(CH_2)_3CH(CH_3)CH_3$									Me-Aq	222–223					404

(*continued*)

Table 1 Amino Acid Derivatives (continued)

Compound	Method	Solvent	% Yield	Chromatography: Col	Tl	Ion	Paper	CCD	Recrys. Sol.	M.P. °C	$[\alpha]_D$	Solvent Rotation	Elect.	Other	Ref.
Asp-$NHCH(CH_2OH)CH_2C_6H_5$									Aq	237–238	−26	Ac			404
Asp-$NHCH(CH_3)CH_2C_6H_3$-3,4-(OCH_2O)										189–192	+6	Me			404
Asp-$NH(CH_2)_5CH_3$									Aq	200–201	+9	Ac			404
Asp-$NHCH(CH_3)CH_2CH_2CH_3$									An-Aq	190–194	+9	Ac			404
Asp-$NHCH(CH_3)CH_2CH_2CH(CH_3)$-$CH_3$									Aq	187–190	+23	1 *N* HCl			404
									An-Aq	181–188	+9	Ac			404
									Me-Aq	215–216	+2	Me			404
D-Asp-$NHCH(CH_3)CH_2CH_2$-									Me-Aq	210–213	−3	Me			404
$CH(CH_3)CH_3$									Aq	192–195	−26	1 *N* HCl			404
Asp-$NHCH(CH_3)CH_2CH_2OCH_2CH_3$									Me-Eth	166–170	+3	Me			404
Asp-$N(CH_3)CH(CH_3)CH_2$-*c*-C_6H_{11}									*i*-Pr-Eth	179–180	−14	Aq			404
D-Asp-$N(CH_3)CH(CH_3)CH_2$-*c*-C_6H_{11}										194–196	+1	Aq			404
Asp-NH-*c*-C_6H_{11}									Aq	224–225	+16	1 *N* HCl			404
Asp-$NHCH_2CH_2OC_6H_5$									Aq	184–185	−13	1 *N* HCl			404
Asp-$NHCH(CH_3)CH_2C_6H_4F$-4									Me-Eth	203–209	+9	1 *N* HCl			404
Asp-$NHCH_2CH_2C_6H_4F$-4									Aq	208–209	−6	Me			404
Asp-$NHCH(CH_3)CH(OH)C_6H_5$									Me	188–190	+10	Me			404
Asp-$NHCH(CH_3)CH_2C_6H_4OH$-4										160–185	+5	Aq			404
Asp-$NHCH(CH_3)CH_2OC_6H_5$									Me-Aq	180–184	+11	1 *N* HCl			404
D-Asp-$NHCH(CH_3)CH_2C_6H_5$									Alc-Aq	222–225	+14	Aq			404
Asp-$NHCH_2CH(CH_3)C_6H_5$									Aq	182–188	−20	Aq			404

Asp-N(CH_3)CH(CH_3)$CH_2C_6H_5$										164–166	+47	Aq			404
D-Asp-N(CH_3)CH(CH_3)$CH_2C_6H_5$										185–187	+12	Aq			404
Asp-NHCH(CH_3)$CH_2CH_2C_6H_5$									Me-Aq	190–196	+16	1 *N* HCl			404
Asp-NHCH(CH_3)$CH_2C_6H_5$									Aq	197–198	−12	Me			404
Asp-NHC$(CH_3)_2CH_2C_6H_5$									Aq	159–161	−16	Me			404
D-Asp-NHCH(CH_3)$(CH_2)_4CH_3$									Aq	189–192	+6	Me			404
Asp-NHCH(CH_3)$(CH_2)_4CH_3$										190–194	+7	Ac			404
									Aq	187–189	−5	Me			404
									Me-Aq	213–214	−5	Me			404
D-Asp-NHCH(CH_3)$(CH_2)_4CH_3$									Me-Aq	217–218	+5	Me			404
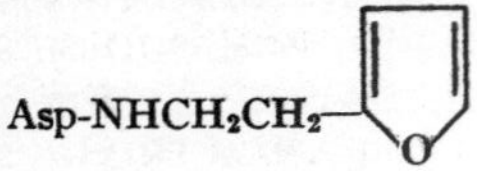									Me	195–196	−17	Me			404
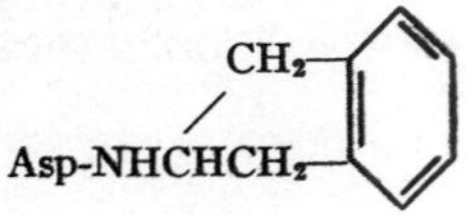									Me-Aq	223–224	−6	1 *N* HCl			404
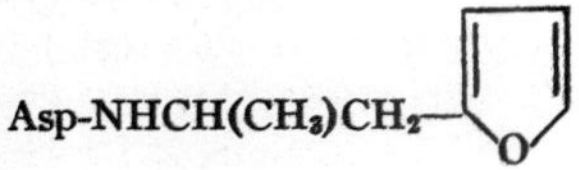									Me-Eth	168–180	+6	1 *N* HCl			404
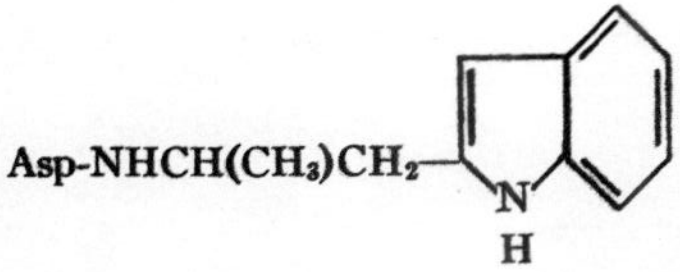									Me	203–205	−22	Me			404

(*continued*)

Table 1 Amino Acid Derivatives (continued)

Compound	Method	Solvent	% Yield	Chromatography Col	Tl	Ion	Paper	CCD	Recrys. Sol.	M.P. °C	$[\alpha]_D$	Solvent Rotation	Elect.	Other	Ref.
Asp-NHCHCHC$_6$H$_5$ (cyclopropyl)										175–178	+5	Me			404
N-(2-Bz-1-methylvinyl)-β-Bzl-Asp									MeCl-Eth	111–115	+17	Alc			589
α-OBzl-Asp-NCA									Chf-Pe	101–102 (dec)	+20	Di			425
									Chf-Pe	85–86 (dec)	+19	Di			425
β-Bzl-N-Z-OBu^t-*threo*-β-hydroxy-Asp											−6.0	EtAc			580
β-Bzl-N-Z-OBu^t-*threo*-β-hydroxy-DL-Asp									Me-Aq	63					580
β-Bzl-OBu^t-*threo*-β-hydroxy-Asp-NCA											−73	EtAc		IR	580
β-Bzl-OBu^t-*threo*-β-hydroxy-DL-Asp-NCA										75					580
β-OBzl-Z-Asp-NHCH$_2$CH$_2$— 									*i*-Pr	113–114	−10	Me			404
β-OBzl-Z-Asp-NHCHCH$_2$— (CH$_2$—) 									*i*-Pr	122–126	−5	Me			404

β-OBzl-Z-Asp-NHCH(CH_3)- 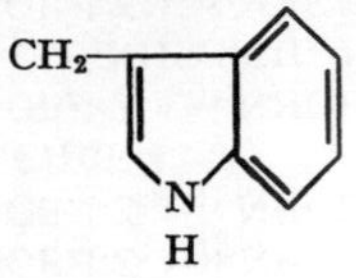									*i*-Pr	102–108	−15	Me			404
β-OBzl-Z-Asp-NHCH(CH_3)CH_2-CH_2CH(CH_3)CH_3										90–103	−8	Me			404
β-OBzl-Z-Asp-NHCH(CH_3)$(CH_2)_3$-CH_3										70–75	−7	Me			404
β-OBzl-Z-Asp-NH$(CH_2)_6$$CH_3$									*i*-Pr	92–93	−6	Me			404
β-OBzl-Z-Asp-NHCH(CH_3)$(CH_2)_4$-CH_3										65–69	−6	Me			404
β-OBzl-Z-Asp-NHCH(CH_3)CH_2-CH(CH_3)C_2H_5										61–71	+11	Me			404
β-OBzl-Z-D-Asp-NHCH(CH_3)CH_2-CH_2CH(CH_3)CH_3									*i*-Pr-Aq	117–119	−4	Me			404
β-OBzl-Z-Asp-NHCH(CH_3)CH_2-*c*-C_6H_{11}									Pe	65–70	−18	Me			404
β-OBzl-Z-D-Asp-NHCH(CH_3)CH_2-*c*-C_6H_{11}									Pe	69–81	+2	Me			404
β-OBzl-Z-Asp-NH-*c*-C_3H_4-CHCHC_6H_5									*i*-Pr	80–92	−13	Me			404
β-OBzl-Z-Asp-NHCH_2CH(CH_3)C_6H_5										72–81	−16	Me			404
β-OBzl-Z-Asp-NHCH(CH_3)CH_2-$CH_2C_6H_5$									Eth-Pen	97–115	−6	Me			404
β-OBzl-Z-Asp-NHCH(CH_3)CH_2-OC_6H_5										76–88	+14	Chf			404
β-OBzl-Z-Asp-NHCH_2CH_2OC_6H_5									*i*-PrAc	101–103	−6	Me			404
β-OBzl-Z-Asp-NH$(CH_2)_5$$CH_3$									*i*-Pr-Aq	72–75	−10	Me			404

(*continued*)

Table 1 Amino Acid Derivatives (continued)

Compound	Method	Solvent	% Yield	Chromatography Col	Tl	Ion	Paper	CCD	Recrys. Sol.	M.P. °C	$[\alpha]_D$	Solvent Rotation	Elect.	Other	Ref.
β-OBzl-Z-Asp-NHCH(C_2H_5)CH_2-CH_2CH_3									*i*-Pr-Aq	96–98	−6	Me			404
β-OBzl-Z-Asp-NHCH(CH_3)CH_2-CH_2CH(CH_3)CH_3									*i*-Pr-Aq	117–119	+3	Me			404
									i-Pr-Aq	104–106	−13	Me			404
β-OBzl-Z-D-Asp-NHCH(CH_3)-(CH_2)$_4CH_3$									*i*-Pr-Aq	95–97	+12	Me			404
									i-Pr-Aq	98–100	−4	Me			404
β-OBzl-Z-Asp-NHCH(CH_3)-(CH_2)$_4CH_3$									*i*-Pr-Aq	99–101	+3	Me			404
									i-Pr-Aq	96–99	−11	Me			404
β-OBzl-Z-Asp-NHCH(CH_3)-(CH_2)$_5CH_3$									*i*-Pr-Aq	80–86	−5	Me			404
β-OBzl-Z-Asp-NHCH(CH_3)CH_2-$CH_2OC_2H_5$										64–81	−8	Me			404
β-OBzl-Z-D-Asp-NHCH(CH_3)CH_2-CH_2CH(CH_3)CH_3									*i*-Pr-Aq	106–107	+13	Me			404
β-OBzl-Z-Asp-NH$CH_2CH_2C_6H_4$F-4									Tol	106–108	−9	Me			404
β-OBzl-Z-Asp-NHCH(CH_3)CH_2-C_6H_4F-4									EtAc-Cy	83–87	−13	Me			404
β-OBzl-Z-Asp-NHCH(CH_3)CH_2-$C_6H_4NHSO_2CH_3$-4									*i*-Pr-Aq	144–155	−24	Me			404
β-OBzl-Z-Asp-NHCH(CH_2OH)CH_2-C_6H_4OH-4									MeCl-Eth	73–77	−33	Me			404

β-OBzl-Z-Asp-$NHCH_2CH_2C_6H_4OH$-4									*i*-PrAc-Chf	148–149	−12	Me			404
β-OBzl-Z-Asp-$NHCH(CH_3)CH_2$-C_6H_4OH-4										106–117	−10	Me			404
β-OBzl-Z-D-Asp-$NHCH(CH_3)CH_2$-$CH(CH_3)CH_3$									*i*-Pr-Aq	94–96	+47	Me			404
β-OBzl-Z-Asp-$NHCH(CH_3)CH_2$-$CH(CH_3)CH_3$									*i*-Pr-Aq	104–105	−16	Me			404
β-OBzl-Z-Asp-$NH(CH_2)_3CH(CH_3)$-CH_3									*i*-Pr-Aq	78–80					404
β-OBzl-Z-Asp-$NHCH(CH_3)CH_2$-CH_2CH_3									*i*-Pr-Aq	91–99	−4	Me			404
β-OBzl-Z-Asp-NH-*c*-C_6H_{11}									*i*-Pr	140–142	−7	Me			404
β-OBzl-Z-Asp-$NHCH_2CH_2$-*c*-C_6H_{11}									*i*-PrAc	129–130	−3	Me			404
β-OBzl-Z-Asp-$NHC(CH_3)_2CH_2C_6H_5$									Cy	96–99	−1	Me			404
β-OBzl-Z-Asp-$NHCH(C_2H_5)CH_2$-C_6H_5									*i*-PrAc-Cy	105–115	−11	Me			404
β-OBzl-Z-Asp-$NHCH_2CH_2C_6H_5$									Cte	112–114	−11	Me			404
β-OBzl-Z-Asp-$NHCH(CH_3)CH_2C_6H_5$									*i*-Pr	133–141	−11	Me			404
β-OBzl-Z-D-Asp-$NHCH(CH_3)CH_2$-C_6H_5									*i*-PrAc	110–111	+4	Me			404
β-OBzl-Z-Asp-$NHCH(CH_3)CH_2C_6H_5$									*i*-PrAc	120–121	−24	Me			404
β-OBzl-Z-Asp-$NHCH(CH_3)$-$CH(OH)C_6H_5$									*i*-Pr-Aq	72–84	−9	Me			404
β-OBzl-Z-Asp-$NHCH(CH_2OH)CH_2$-C_6H_5									*i*-Pr	140–142	−36	Me			404
β-OBzl-Z-Asp-$NHCH(CH_3)CH_2$-$C_6H_3(OCH_2O)$-3,4									*i*-PrAc	119–122	+10	Chf			404
β-OBzl-Z-Asp-$NHCH(C_2H_5)$-$(CH_2)_3CH_3$										91–98	−9	Me			404
Z-Asp-OBu^t															314
Z-Asp-β-OMe										96	−17	Alc			456

(continued)

Table 1 Amino Acid Derivatives (continued)

Compound	Method	Solvent	% Yield	Chromatography Col	Tl	Ion	Paper	CCD	Recrys. Sol.	M.P. °C	$[\alpha]_D$	Solvent Rotation	Elect.	Other	Ref.
Z-Asp-β-OMe ephedrine salt										141–142	+31	Alc			456
Z-D-Asp-β-OMe										98–99	+18	Alc			456
Z-Asp[β-bis(Dmb)-amide]-α-ONp									Pe		+11	DMF			96
Z-Asp-β-semialdehyde										129–131	−21	EtAc			665
Z-Asp-OBu^t										Oil					225
Z-Asp-β-OPic									EtAc	158–160	−20	DMF			168
Z-β-ONp-Asp-OBu^t										80					225
Z-*threo*-β-hydroxy-Asp									EtAc-Pe	172	+8	Alc			580
N-(δ-Bromovaleryl)-Asp									Aq	136–138					88
N^β-Bu^t-N^α-Z-Asp Boc-hydrazide					35 27				di-*i*-PrEth-Pe	53–56	−23	Me			690
β-OBu^t-Asp-Boc-hydrazide acetate hydrate					109 110				Me-Eth		−4.9	Me			690
γ-OBu^t-Boc-Asp															544
Boc-Asp									EtAc-Pe	114–116	−6	Me			544
Boc-Asp-β-OPic									EtAc	133–135	−23	DMF			168
Di-OBzl-N-Z-OBu^t-*threo*-β-hydroxy-Asp									EtAc-Pe	42	−9.8	Alc			580
Di-OBzl-N-Z-*threo*-β-hydroxy-Asp									EtAc-Pe	47–50	−9.1	Alc			580
N-Dca-Asp		Aq								118	+34	EtAc			363
N-Dca-Asp anhydride										150	−37	EtAc			363
DNT-Asp										111–112					227
Di-OPic-Asp Tri-HBr									Alc-Aq	161–162	+1.2	Aq			168

α-OEt-Asp-NCA									Chf-Pe	85–86 (dec)	+19	Di			425
Z-(OMe)-Asp-β-OBzl									Eth	90–92					713
Z-(OMe)-β-OBzl-Asp hydrazide									Eth-Pe	110–115					713
Z-(OMe)-β-OBzl-Asp-TrOC hydrazide										Oil					713
N-α-Methyl-N-Dca-Asp		Me		D-50W						123	+18	EtAc			363
N-α-Methyl-N-Dca-Asp Di-Bzl-ammonium salt										118	+16	EtAc			363
β-OPic-Z-Asp-OTcp									EtAc-Pe	106	−27	DMF			168
α-Tritio-DL-Asp							×								615
δ-Aminovaleric acid															
Z-δ-Ave										102–103					156
Citrullene															
Boc-Cit Dcha										112–116	+11	DMF			641
Boc-Cit-OPcp									Alc	136–137	−20	DMF			641
Cysteine															
S-Acm-Cys-OMe HCl															292
N-Ac-Cys-OMe					2				Eth-Pe	81–82					341
N-Acyl-DL-homo-Cys-thiolactone															327
S-(3-Aminopropyl)-homo-Cys									Aq-Alc	252–254 (dec)					105 514
S-Benzhydryl-N-(2-Bz-1-methyl-binyl)-Cys									MeCl-Eth	135–137	−139	Alc			589
S-Bzl-N-Z-Cys-OCH_2CN										64.5–66	−22	Chf			351
S-Bzl-Z-DL-Cys-NH_2										131–131.5					194
S-Bzl-Boc-Cys									Eth-Pe	63–65	−41	Ac			544
S-Bzl-N-Boc-Cys-OPy									EtAc-Pe	60–61	−16	DMF			134
S-Bzl-N-Boc-Cys-SPy										110–111	−96	DMF			365
S-Bzl-Cys									Ac-Eth		+26	1 *N* NaOH			380 457

(continued)

Table 1 Amino Acid Derivatives (continued)

Compound	Method	Solvent	% Yield	Chromatography Col	Tl	Ion	Paper	CCD	Recrys. Sol.	M.P. °C	$[\alpha]_D$	Solvent Rotation	Elect.	Other	Ref.
S-Bzl-Cys-OBzl(OMe) HCl					10				EtAc	120.5–121.5	−20	DMF			380
S-Bzl-Cys-ONBzl Tos									Ac	167–169	−17	DMF			243
S-Bzl-DL-N-Bz-Cys-seleno-Ph										148–149					264
S-Bzl-Cys-OTMB HCl									Alc-EtAc	170–172	−25				348
S-Bzl-N-Z(OMe)-Cys-ONp									Alc-Pe	86–88					714
S-Bzl-thio-N-Tfa-Cys-OMe										38–40	+40	Chf		IR	217
S-Bzl-N-Tos-Cys															457
Z-Cys(CH_2COOMe) Dcha		Me-Eth								127–128	−5	DMF			188
Z-Cys(CH_2COOMe)-OSu									Pr	63–64	−78	DMF			188
Z-Cys(CH_2CONH_2)		Me								165–167	−42	DMF			188
N,S-Bis-Z-seleno-Cys-OBzh															652
N,N-Bis-Boc-Cys									EtAc-Pe	147–148	−141	Me			431
N,S-Bis-Boc-Cys									Pe	128–130	−67	DMF			546
N,S-Bis-Boc-Cys									Eth-He	127	−27	Alc		NMR IR	431
N,S-Bis-Boc-Cys-OSu									Alc-Dipr-Eth	122–124	−70	DMF			546
S-Bis-OMe-phenylmethyl-Cys									Aq	201	+9.5	0.1 *N* NaOH			489
N,S-Bis-SNp-Cys Dcha									Me-EtAc	144–145	−38	Me			557
S-Bu^t-N-Boc-Cys										118–119	−148	Alc			707
S-Bu^t-Cys											−84				707

S-Bu[t]-N-Nps-Cys Dcha										153–154 (dec)	−66	80% Me			707
S-Boc-Cys HCl									Me-EtAc-He	191–192 (dec)	−29	Me		IR	431
S-(γ-Boc-propyl)Cys									Aq	206–208	−15	Aq			285
S–(γ-Boc-propyl)-N-Nbs-Cys Dcha									Aq-Alc	156–158	−44	DMF			285
S-β-Carboxyethyl-Cys					64 71 125	A-IRC-50				219–221	−10	1 *N* HCl			287
S-β-Carboxyethyl-homo-Cys				D-50					Ac-Aq-Eth	242–245	+17	1 *N* HCl			289
S-γ-Carboxypropyl-Cys									Aq	235–237	−19 −1.6	Aq 1 *N* HCl			285
S-γ-Carboxypropyl-homo-Cys					64 76	DO-50			Aq	229–232	−18 −5.2	1 *N* HCl Aq		Mol. wt	502
S-(Cyclopentylmethyl)-homo-Cys									Alc	187–188 (dec)					105
Cys															41 176
Cys S-monoxide											+29	1 *N* H_2SO_4			655
D-Cys S-monoxide									0.5 *N*-Ac-An		−57	1 *N* H_2SO_4			655
Cys(CH_2COOMe)-OMe HCl									Me	96–98	−9	DMF			188
Cys(CH_2CH_2NHZ)-OMe HCl															188
DL-α-(S-Cys)-β-(5-imidazolyl)-propionic acid				G-10	×	DO-1									725
Decanoyl-Cys(CH_2CONH_2)									Me	142–144	−32	DMF			188
Decanoyl-Cys(CH_2CH_2NH-decanoyl)									EtAc	117–118	−16	DMF			188
Decanoyl-Cys(CH_2CH_2NH-decanoyl)OMe									EtAc	87.5–88.5	−19	DMF			188
Decanoyl-Cys(CH_2CH_2NH-decanoyl)-OSu										Oil					188

(*continued*)

Table 1 Amino Acid Derivatives (continued)

Compound	Method	Solvent	% Yield	Chromatography Col	Tl	Ion	Paper	CCD	Recrys. Sol.	M.P. °C	$[\alpha]_D$	Solvent Rotation	Elect.	Other	Ref.
Decanoyl-Cys(CH_2COOMe)									Eth	72.5–73.5	−40	DMF			188
Decanoyl-Cys(CH_2COOMe)-OMe									Eth-He	44–45	−43	DMF			188
Decanoyl-Cys(CH_2COOMe)-OSu									*i*-Pr-Dipr-Eth	71–72	−62	DMF			188
Decanoyl-Cys(CH_2CONH_2)-OMe									Me	117–117.5	−36	DMF			188
S-β,β′-Dicarboethoxyethyl-Cys					112					167–170					582
S-β,β′-Dicarboethoxyethyl-Cys HCl									Me-Eth	148–150	−14	Aq			582
S-β,β′-Dicarboethoxyethyl-N-For-Cys					96				Eth-He	55–56	+8.2	Alc			582
Di-*p*-hydroxylmethyl-Trt-Cys										143–145 (dec)					731
S-2,4-Dnp-N-Ac-Cys					54 94				Alc-Aq	142–145					623
S-Diphenylmethyl-N-Boc-Cys Dcha										158–159	+6.3	MeCl			234
S-Diphenylmethyl-N-Boc-Cys-OSu															234
S-Diphenylmethyl-Cys										196–198	+17	0.1 *N* HCl-Alc			489
N,S-Di-Trt-Cys diethylamine salt									Chf-Alc	205–207					731
Elephantopin-mono-Cys adduct															341
Gallium mono-*p*-carboxyl-Trt-Cys										224–225 (dec)					731

S-β-OMe-carbonylethyl-N-Z-Cys-5-chloro-OHq									EtAc-Pe	112–115	−39	DMF			595
S-β-OMe-carbonylethyl-N-Z-Cys Dcha									EtAc	121–122	+36	DMF			595
S-β-OMe-carbonylethyl-Cys			51 52 26						Aq-Me	223–225	−23	Aq			595
S-β-OMe-carbonylethyl-homo-N-Z-Cys-5-chloro-OHq										101–102	−31	DMF			289
S-β-OMe-carbonylethyl-homo-N-Z-Cys Dcha									Aq-Alc	156–158	+5.6	Me			289
S-β-OMe-carbonylethyl-homo-Cys						DO- 50				230–233	−6.0	Aq			289
S-(γ-OMe-carbonylpropyl)-N-Z-Cys-5-chloro-OHq									EtAc	115–116	−37	DMF			285
S-(γ-OMe-carbonylpropyl)-N-Z-Cys Dcha									EtAc	118–119	−2.0	DMF			285
S-(γ-OMe-carbonylpropyl)-Cys						A-IR- 4B Z-225			Aq	217–219	−16	Aq	×		285
S-(γ-OMe-carbonylpropyl)-Cys HCl									Me-Eth	155–156	0 −0.4	1 *N* HCl DMF			285
S-(γ-OMe-carbonylpropyl)-homo-N-Z-Cys Dcha					64 125				EtAc	119–121	+6.1	DMF			502
S-(γ-OMe-carbonylpropyl)-homo-Cys					64 76	DO- 50 A-IR- 4B			Aq	220–225	−6.2	Aq		Mol. wt.	502
S-(γ-OMe-carbonylpropyl)-N-Nps-Cys Dcha									Et-Ac-Eth	159–161	−40	DMF			285
S-Bzl(OMe)-N-Boc-Cys										Oil					10
S-Bzl(OMe)-N-Boc-Cys Dcha									EtAc-Pe	132	−15	Chf			10

(*continued*)

Table 1 Amino Acid Derivatives (continued)

Compound	Method	Solvent	% Yield	Chromatography: Col	Tl	Ion	Paper	CCD	Recrys. Sol.	M.P. °C	$[\alpha]_D$	Solvent Rotation	Elect.	Other	Ref.
S-Bzl(OMe)-N-Nps-Cys Dcha					×					158–160	−37	DMF			197
S-Bzl(OMe)-N-Nps-Cys-ONp					×				Oil						197
Z(OMe)-Cys									DMF-Aq	180–182 (dec)	−47	Ac			488
S-Z(OMe)-N-Nps-Cys Dcha									Alc	154–155	−31	DMF			488
α-Methyl-$\Delta^{\alpha,\beta}$-butenolide-Cys adduct										188–190				IR NMR	431
α-Methyl-$\Delta^{\alpha,\beta}$-butenolide-Cys adduct															341
β-Methyl-$\Delta^{\alpha,\beta}$-butenolide-Cys adduct															341
				Ce						203–205 (dec)				Mass spec. IR PMR	341
γ-Methyl-$\Delta^{\alpha,\beta}$-butenolide-N-Ac-Cys-OMe adduct											+46	Chf		IR PMR	341
γ-Methyl-$\Delta^{\alpha,\beta}$-butenolide-Cys adduct										179–180 (dec)					341
(+)-S-Methyl-Cys-sulfoxide										182	+130	Aq			228
(−)-S-Methyl-Cys-sulfoxide										173	−133	Aq			228
Mono-p-carboxyl-Trt-Cys										175–177 (dec)					731

Mono-*p*-hydroxylmethyl-Trt-Cys										145–147 (dec)					731
Mono-3,4-methylenedioxy-Trt-Cys										187–189 (dec)					731
Mono-*p*-phenyl-Trt-Cys										177–178 (dec)					731
Nps-Cys(CH_2CH_2NHZ) Dcha									Me	164.5–165	−73	DMF			188
S-3-Nitro-2-pyridyl-N-Ac-Cys					54 94				Alc-Aq	156–159					623
S-3-Nitro-2-pyridyl-N-Ac-Cys-OMe					54 94				Alc-Pe	152–153					623
S-5-Nitro-2-pyridyl-N-Ac-Cys					54 94					167–169					623
S-(2-Propynyl)-Cys S-dioxide									Aq-Alc	152 (dec)	−0.98	Aq		IR	85
S-(2-Propynyl)-Cys S-oxide									Aq-Alc	189 (dec)	+73	Aq		IR	85
S-*β*-(2-Pyridylethyl)-Cys complex with $HgCl_2$															145
Se-Diphenylmethyl-seleno-N-Z-Cys-OBzh				Si	36 1				EtAc-He	78.5–79.5				IR NMR	650 652
Sodium mono-*p*-carboxyl-Trt-Cys										195–197 (dec)					731
N-Tfa-S-glutathionyl-Cys-OMe					12 84					173 (dec)	−103	1 *N* HCl		IR	217
S-Phthalimido-N-Tfa-Cys-OMe									Me-Aq	125–126	+54	Cte		IR	217
S-Trt-N-Z-Cys									Be-He	114–115	+18	Me			235
S-Trt-N-Boc-Cys										210–211 (dec)	+24	Me			234
S-Trt-Cys										177–179	+16	1 *N* NaOH			489

(*continued*)

Table 1 Amino Acid Derivatives (continued)

Compound	Method	Solvent	% Yield	Chromatography Col	Tl	Ion	Paper	CCD	Recrys. Sol.	M.P. °C	$[\alpha]_D$	Solvent Rotation	Elect.	Other	Ref.
S-Trt-N-(2-Bz-1-methylvinyl)-Cys									MeCl-Eth	186–188	−151	Alc			589
Cystine															
N,N′-Bz-DL-homo-Cyst-dihydroxamic acid										145					327
N,N^{α}-Z-Cyst-bis-OSu									EtAc	112–113	−137				188
N,N′-Z-DL-homo-Cyst-dihydroxamic acid									Me-Aq	135 (dec)					327
N,N′-Z-Seleno-Cyst-OBzh															650
N,N′-Bis-Tfa-Cyst-Di-OMe															218
Boc-Cyst[OBzl(OMe)] Dcha										132					8
N,N′-Boc-Cyst										143–145	−122	Me			431
N,N′-Decanoyl-Cyst-bis-OMe									EtAc	96–97	−89				188
N,N′-Di-Z-Cyst-Di-OEt															218
DL-Homo-Cyst dihydroxamic acid										190–195				IR	327
α,γ-Diaminobutyric acid															
N^{γ}-Z-Dbu										243					174
N^{γ}-Z-Dbu-NCA										65	−26	An			174
Dbu															614
Dbu Di-HCl									Aq-Alc	196–197	+14	Aq			174
N^{γ}-Pht-Dbu-OBzh Tos									DMF-EtAc	159–160 (dec)	+14	Me			221
N^{γ}-Pht-Dbu hydrate									Aq	205–206 (dec)	+14	1 *N* HCl			221

N^{γ}-Pht-Dbu Tos										176–177 (dec)	+9.2	Me			221
Threo-α,β-diaminobutyric acid															
N^{α}-Z-α,β-Dbu-OMe HCl										197–198	−14	DMF			27
α,β-Dbu															27
N^{α}-Tos-α,β-Dbu										233–235	+28	6 *N* HCl			27
Glutamic acid															
N-Ac-γ-fluoro-Glu						DO-50									632
γ-Aminomethyl-DL-Glu									Aq-An						6
γ-Bzl-N^{α}-Boc-Glu-OSu									Pe-Eth	104.5–105.5	−21	Di			333
N^{γ}-Bzl-N^{α}-Nps-Glu-ONp									EtAc-Pe	78–80	−122	Chf			197
γ-OBzl-N-Bz-Glu									Alc	122–124	+6.9	Chf			628
γ-OBzl-N-Bz-Glu-OPcp									Be	141–142.5	0				628
γ-OBzl-Boc-Glu-ONp				D-50W × 8					Alc	118–119	−33	Me			249
γ-OBzl-Boc-Glu-OPcp									*i*-Pr	139–140					628
γ-OBzl-Boc-Glu-OPy									Eth-Pe	48–49	−24	DMF			134
γ-OBzl-Boc-Glu-SPy										82–84	−54	EtAc			365
γ-OBzl-Glu-NCA										93–94	−18	EtAc			185
γ-OBzl-Z(OMe)-Glu									EtAc-Pe	75–76					713
γ-OBzl-Z(OMe)-Glu hydrazide										85–90					713
γ-OBzl-Z(OMe)-Glu trichloroethyloxycarbonylhydrazide															713
Z-Glu-OBu^{t}										85–86	−16	Me			643
Z-Glu-γ-OBu^{t}										79–80	−13				393
Z-α-Glu-OMe-α-tryptamine										128–130	+9.2	THF			442
Z-Glu-α-methyltryptamine									Aq-Alc	126–127	−7.5	Me-THF			442

(*continued*)

Table 1 Amino Acid Derivatives (continued)

Compound	Method	Solvent	% Yield	Chromatography Col	Tl	Ion	Paper	CCD	Recrys. Sol.	M.P. °C	[α]D	Solvent Rotation	Elect.	Other	Ref.
Z-α-Glu-α-methyltryptamine-γ-OBzl									Be-He	124–126	+9.8	THF			442
N-(δ-Bromovaleryl)-Glu									Aq	126–128					88
N-(δ-Bromovaleryl)-Glu-Di-OEt															88
γ-OBu^t-Z-Glu-OSu									EtAc-Pe	101–103	−30	Alc			39
γ-OBu^t-Z-Glu-OPcp										94–95	−12	Chf			643
γ-OBu^t-α-OBu^t-Glu HCl									EtAc-He	112–113	+22	Me			293
γ-OBu^t-Boc-Glu									EtAc-Pe	110–112	−11	Me			544
γ-OBu^t-Boc-Glu-OSu															644
γ-OBu^t-Boc-Glu-OSu									*i*-Pr	146–147	−31	DMF			238
γ-OBu^t-Boc-Glu-OPcp									Me	150–151	−19	Chf		UV	643
														IR	
Boc-Glu										116–118	−16	Me			518
									EtAc-Pe	110–112	−16	Me			544
											−7.3	Alc			
N-Boc-D-Glu-γ-OBzl										65–66	−13	Chf			216
Boc-Glu-α-OBzl Dcha										139–140					8
N-Boc-Glu-α-OBzl Dcha									Alc	174–175					491
Boc-Glu-γ-OBzl Dcha									EtAc	140–142	+13	Me			249
Boc-Glu-OBu^t									Alc-Pen	101–102	−11	Me			643
Boc-Glu-α-OMe Dcha									Chf-Eth	165–	−12	Me			431
										166.5					
Boc-pyro-Glu															578
N-Deuterio-γ-OBzl-Glu-NCA									He					IR	185
N-Dca-Glu									Eth-Pe	128–132	−4.9	EtAc			363

N-Dca-Glu anhydride									MeCl-Pe	112	−34	EtAc			363
N^{γ}-Dca-α-methyl-Glu		Me		D-50W						Oil	−15	Me			363
N^{γ}-Dca-α-methyl-Glu Di-Bzl ammonium salt										126	−7	Me			363
4-Di(2-chlorethyl)amino-DL-Glu 2 HCl									Alc-Eth	60–62				IR	5
γ-Diethylamino-DL-Glu															4
γ-Diethylaminoethyl-DL-Glu									Alc	Oil					6
4-Di(2-hydroxyethyl)amino-DL-Glu									Aq						5
DNT-Glu										191–192					227
Di-4-picolyl-Glu Tri-HBr									Alc	140	+8.0	Aq			168
γ-Fluoro-Glu						DO-1									632
γ-Fluoro-D-Glu						DO-50			Aq		−22	HCl			632
(*R*)(−)-Glu											−0.9	5 *N* HCl			65
Glu-γ-OBzl										194	+20	Ac			185
Glu-α-OBzl HCl									Alc	100–102	−9	Chf			628
Glu-γ-OBu^t Cha									Alc	174–175 (dec)	+15				573
Glu diamide HCl										193–196	+20	Aq			413
Glu methylthiohydantoin derivative														CD	622
α-Glu-α-methyltryptamine									Alc-EtAc	125–126	+21	Me			442
γ-Glu-α-methyltryptamine									Alc-Eth	181–183	+8.3	Me			442
Glu-α-methyltryptamine hydrazide									Alc-Eth						442
α-Glu-α-methyltryptamine hydrazide ditartrate									Alc-Eth	149–151	+26	Me			442
α-Glu-α-methyltryptamine tartrate									Alc-Eth	75–77	+26	Me			442
Morpholino-DL-Glu															4
N[4-(2-Naphthyl)methylamino-Bz]-DL-Glu									Aq	93–97					199
γ-ONp-N-Z-Glu-α-OBzl									Alc-Pe	64–66	−21	Me			669

(*continued*)

Table 1 Amino Acid Derivatives (continued)

Compound	Method	Solvent	% Yield	Chromatography Col	Tl	Ion	Paper	CCD	Recrys. Sol.	M.P. °C	$[\alpha]_D$	Solvent Rotation	Elect.	Other	Ref.
Nps-Glu di-amine										198–202	+7.5	DMF			413
Nps-Glu-OSu										142–146	−61	DMF			413
Piperidino-DL-Glu															4
Pteroyl-Glu															184
Pyro-Glu-OTcp									Me-Pe	155–159	+19	DMF			49
N[4-(2-thienyl)methylamino-Bz]-DL-Glu															199
Tos-D-Glu										143–146	+15	Aq			200
γ-OTMB-Glu-ONBzl HCl									Me-Eth	138–139	+9.9	Me			347
γ-OTMB-Glu-OPcp HCl										163–166					628
γ-OTMB-Nps-Glu-OPcp									Chf-Alc	156–157	$[\alpha]_{546}$ −60				628
γ-OTMB-Nps-Glu-OPcp HCl										163–166					628
α-Tritio-DL-Glu							×								615
Glutamine															
N-Ac-N^{γ}-methyl-Gln															647
N-(2-Bz-1-methylvinyl)-Gln									MeCl-Eth	163–165	+48	Alc			589
N-(2-Bz-1-methylvinyl)-N-xanthyl-Gln									DMF-Eth	198–200	+8.1	Alc			589
Z-Gln Boc-hydrazide									Me-EtAc	193–194	−33	Me			39
Z-Gln-OSu									EtAc	134–136	−31	Alc			39
Boc-Gln									EtAc	116–118	−3.0	Alc			544
									Eth	117	−3	Alc			546
										158	−3.0	Alc			518

Boc-Gln-OPcp										151–152	−27	DMF		IR	643
Boc-Gln-OPy									EtAc-Pe	83–84	−27	DMF			134
Boc-Gln-SPy										105–107	−57	EtAc			365
γ-Bu^t-Boc-Gln									EtAc-Pe	155–156					544
N-Bu^t-Gln										223.5	+7.4	Aq			320
N-Desyl-DL-Gln										220					320
N^γ-(Dmb)-Gln									Me	240–241					491
N^γ-(Dmb)-N^α-Z-Gln									EtAc	136–137					491
N^γ-(Dmb)-N^α-Z-Gln Dcha									EtAc	110	+5.5	DMF			491
N^γ-(Dmb)-N^α-Z-Gln-OMe									EtAc-Pe	125	−12	Me			491
N^γ-(Dmb)-N^α-Z-Gln-ONp									Alc	149–150	−16	Me			491
N^γ-(Dmb)-N^α-Boc-Gln									EtAc-Pe	110–111					491
N^γ-(Dmb)-N^α-Boc-Gln-OBzl									EtAc-Pe	111–112	−6.5	Me			491
N^γ-(Dmb)-N^α-Boc-Gln-ONp										134–135	−18	Me			491
Bpoc-Gln-ONp									EtAc-Pe	124–125	−64	Me			722
N-Ethyl-Gln										215	+8.2	Aq			320
N-Ethyl-DL-Gln										219					320
Gln															491
Gln Boc-hydrazide															492
Gln Boc-hydrazide											+15	Me			39
Gln-OMe HCl															492
Gln-OBzl(OMe) HCl									Alc-Eth	144	+6.9	DMF			380
Gln methylthiohydantoin derivative														CD	622
N-Hexyl-DL-Gln										224					320
N^γ-(4-Hydroxyphenyl)-Gln															659 660
Z(OMe)-iso-Gln									THF-Pe	127.5– 130.5 (dec)	−5.5	Me			511
Z(OMe)-iso-Gln-OMe									Me-Aq	119– 120.5	−5.5	Me			511
N-Methyl-Gln										201	+8.3	Aq			320

(*continued*)

Table 1 Amino Acid Derivatives (continued)

Compound	Method	Solvent	% Yield	Chromatography Col	Tl	Ion	Paper	CCD	Recrys. Sol.	M.P. °C	$[\alpha]_D$	Solvent Rotation	Elect.	Other	Ref.
N-Pentyl-DL-Gln										227					320
Pht-Gln															492
Pht-Gln Boc-hydrazide															492
N-Pr-Gln										223	+6.6	Aq			320
Tos-D-Gln-OiPr									*i*-Pr-Pe	118–120 132–134	+17	DMF			200
Glycine															
α-Acetamidoallyl-Gly-OEt										122–123 (dec)				NMR	665
Ac-N-2-cyanoethyl-Gly-OTMB									Be-Pe	134–135.5				PMR	599
N-Ac-Gly															647
N-Ac-Gly-OCH_2CN									Alc	83–84					600
N-Ac-DL-phenyl-Gly amide									Alc-Aq	233–234					400
N-[2-amino-3-OBzl-4-methyl-Bz]-Gly-OMe									Alc-Aq	111					344
3-Amino-3-methoxy-carbonylpropionyl-Gly-OBzl HBr									Alc-Aq	158–160					223
3-Amino-2-Tos-aminopropionyl-Gly-OBzl HBr									Alc-Eth	167–169					223
N-(2-Aminoethyl)-Gly HBr									Aq-Alc	188–190					26
N-(2-Aminoethyl)-Gly, sodium															531
N-(3-Aminopropyl)-Gly, sodium															531
Bz-Gly-OBhae									THF	157–158					371

Bz-Gly-NHBzl										157.5–158					371
N-(2-Bz-1-methylvinyl)-Gly									MeCl-Eth	171–173					589
Bzl-Gly										196–198 (dec)					531
N-Z-N-(2-Aminoethyl)-Gly									Acn	155–160					26
3-Z-Amino-2-Tos-aminopropionyl-Gly-OBzl									EtAc-Pe	155–157	−15	Chf			223
N-Z-N-(2-Boc-aminoethyl)-Gly									Eth-Pe	90–91					26
N-Z-N-(2-Boc-aminoethyl)-Gly-OSu									EtAc-Pe	109					26
N-Z-N-2-Carboxamidoethyl-Gly									Aq	147.5–148.5					597
N-Z-N-2-Carboxamidoethyl-Gly-OCH_2CN									EtAc-Cy	60–61					597
N-Z-N-2-Carboxamidoethyl-Gly-ONp									Alc	159–159.5					597
N-Z-N-2-Carboxyethyl-Gly									EtAc-Cy	110–110.5					599
Z-DL-(C-Ph)-Gly-OBhae									EtAc	141–143					371
N-Z-N-2-Cyanoethyl-Gly Dcha									Alc	182–183					599
N-Z-N-2-Cyanoethyl-Gly-ONp									Alc	108–108.5					599
Z-Gly															494
Z-Gly-OBhae									EtAc	147					371
N-Z-Gly-α-amino-crotonic acid-OEt									He	86–88				NMR	437
DL-3-(Z-Gly-amino)-1-Tos-2-Piperidone									Di-Alc-Aq	162–163					201
Z-Gly-NHC_6H_5										141.5–143					371
Z-Gly-Bzl amide					8					106–108					354
									Me	114–115					677
Z-Gly-4-carboxyphenylhydrazide									Me	194–195					677

(*continued*)

Table 1 Amino Acid Derivatives (continued)

Compound	Method	Solvent	% Yield	Chromatography Col	Tl	Ion	Paper	CCD	Recrys. Sol.	M.P. °C	[α]D	Solvent Rotation	Elect.	Other	Ref.
Z-Gly-OCH_2CN										65–67					351
Z-Gly-$NHCH_2CN$															195
Z-Gly-NHC_6H_{11}										103–104					371
									Alc	109–110					134
Z-Gly-2,2-diphenyl				Seph-adex LH-20					Cte	63–65					677
Z-Gly-N,N′-diphenylhydrazide									EtAc-Pe	104–106					677
Z-Gly-OBzl(OMe)									Eth-He	61–62.5					380
Z-Gly-*m*-fluorophenylhydrazide										132–133					1
Z-Gly-*o*-fluorophenylhydrazide										94–95					1
Z-Gly-*p*-fluorophenylhydrazide										142–144					1
Z-Gly-isopropylhydrazide									Bz	81–82					354
Z-Gly-*o*-methoxyphenylhydrazide										76–78					1
Z-Gly-*p*-methoxyphenylhydrazide										140–142					1
Z-Gly-N^1-methyl-N^1-phenyl-hydrazide										120–122					1
Z-Gly-α-methyltryptamine									Aq-An	119–121					442
Z-Gly-*m*-nitrophenylhydrazide										149–150					1
Z-Gly-*o*-nitrophenylhydrazide										163–165					1
4-[Z-Gly-oxy]-phthalazone									EtAc	167–168					677
4-[Z-Gly-oxy]-phthalazone									Me	118–120					677
N-Z-Gly-OPh									EtAc	67–68					414

Z-Gly-phenylhydrazide									Me-Aq	140–141					354
										142–144					1
Z-Gly-OPn									EtAc-Cy	101–101.5					595
Z-Gly-OPy															134
Z-Gly-SPy										65–68					365
Z-Gly-*o*-thioisobutyl					×					Oil					354
Z-Gly-*m*-tolylhydrazide										124–125					1
Z-Gly-*o*-tolylhydrazide										66–68					1
Z-Gly-*p*-tolylhydrazide										133–135					1
N-[3,5-Bis-(N-ethyl-Ac-amino)-2,4,6,-triiodo-Bz]-Gly															606
N-[3,5-Bis-(N-ethyl-Ac-amino)-2,4,6-triiodo-Bz]-N-Bzl-Gly															606
N-[3,5-Bis-(N-ethyl-Ac-amino)-2,4,6-triiodo-Bz]-N-ethyl-Gly															606
N-[3,5-Bis-(N-ethyl-Ac-amino)-2,4,6-triiodo-Bz]-N-methyl-Gly															606
N-[3,5-Bis-(N-ethyl-propionyl-amino)-2,4,6-triiodo-Bz]-Gly															606
N-[3,5-Bis-(N-ethyl-propionyl-amino)-2,4,6-triiodo-Bz]-N-ethyl-Gly															606
N-[3,5-Bis-(N-ethyl-propionyl-amino)-2,4,6-triiodo-Bz]-N-methyl-Gly															606
N-[3,5-Bis-(N-methyl-propionyl-amino)-2,4,6-triiodo-Bz]-N-methyl-Gly															606
N-(δ-Bromovaleryl)-Gly									Aq	68–70					88
N-(δ-Bromovaleryl)-Gly-OEt									Eth-He	57–59					88
N-Boc-α-cyclohexyl-Gly					×				Hep	139–141	+13	Alc			283

(*continued*)

Table 1 Amino Acid Derivatives (continued)

Compound	Method	Solvent	% Yield	Chromatography Col	Tl	Ion	Paper	CCD	Recrys. Sol.	M.P. °C	$[\alpha]_D$	Solvent Rotation	Elect.	Other	Ref.
N-Boc-α-cyclopentyl-Gly					×				Hep	90–91	−3.0	Alc			283
Boc-Gly										94–95					544
Boc-Gly-amide															476
Boc-Gly-NHC_6H_5										138–140					371
Boc-Gly-OBhae									Me-Aq	110–112					371
Boc-Gly-4-carboxyphenylhydrazide										184–186					677
Boc-Gly-NHC_6H_{11}									Ac-Aq	113–114					134
O-Boc-Gly-glycolic-OEt										Oil					365
Boc-Gly-D-Hyv-OSu									EtAc-Pe	126–127	+37	Di		IR	368
O-Boc-Gly-DL-lactic-OEt										Oil					365
Boc-Gly-OPy									Be-Pe	71–72					134
Boc-Gly-SPy										95					365
Boc-C-phenyl-Gly									EtAc-Pe	99–100	+139	Alc			544
N-2-Carboxamidoethyl-Gly									Aq-Alc	225–226 (dec)					599
N-2-Carboxamidoethyl-Gly-ONp HCl									Alc-Aq	105–105.5					597
N-2-Carboxamidoethyl-Gly-ONp Tos									Me-Eth	175–176					599
N-2-Carboxamidoethyl-Gly-OTMB HCl									Me-EtAc	173.5–175					597
N-2-Chloroethyl-N-2-fluoroethyl-Gly-OEt										Oil				n_D 1.4465	55
N-2-Chloroethyl-N-2-fluoroethyl-Gly-OEt HCl										92					55

N-2-Chloroethyl-N-2-fluoroethyl-Gly HCl										Oil					55
α-Cyano-Gly						A-CG-120				124 (dec)	−0.05	Ac		IR Mass spec.	511
[Cyano-Boc]-Gly									Eth-Pe	147–148.5					706
N-2-Cyanoethyl-Gly-OTMB									Alc	103–103.5					599
N-2-Cyanoethyl-Gly-OTMB HCl									Me-Eth	158.5–159.5					599
Cyclohexa-Gly															332
N-[Bpoc]-[2,6-dichlor-phenyl]-DL-Gly										136–138					545
Bpoc-Gly Dcha									EtAc	192–193 (dec)					573
DNPS-Gly-OC_7H_7										81–82					415
DNPS-Gly-OCH_3										88–90					415
DNT-Gly										219–220					227
N-Enanthyl-Gly										88					155
N-Enanthyl-Gly-OMe										Oil					155
1-Ethoxyvinyl phthaloyl-Gly									Be-Pe	102–107					564
Ethyl-[2-amino-6-(2,4-dioxo-1,2,3,4-tetrahydro-1-pyrimidinyl)-caproyl]-Gly														NMR	330
Ethyl-Boc-Gly										126–128					544
N-(2-Furacyl)-Gly															362
Gly-NCA															13
															127
Gly-OEt															718
Gly-OEt-2,2-diphenyl					Seph-adex LH-20										677

(*continued*)

Table 1 Amino Acid Derivatives (continued)

Compound	Method	Solvent	% Yield	Chromatography Col	Tl	Ion	Paper	CCD	Recrys. Sol.	M.P. °C	$[\alpha]_D$	Solvent Rotation	Elect.	Other	Ref.
Gly-glycolic acid									Aq-Alc	166–167					601
Gly-glycolic acid HBr									Me-Eth	198.5–200 (dec)					601
Gly-OBzl(OMe) HCl							×		Alc	156 and 172					380
Gly-α-methyltryptamine acetate									Alc	151–152					442
Gly-ONp HBr															367
Gly-SPh										71–72					444
Gly-OPn HCl									Me-Eth	156.5–157					595
Gly-OSu HBr										166–168					367
Gly-THIONCA															127
Hydantoyl-Gly					57				Aq	188–190 (dec)					479
N-Lauroyl-Gly										116–119				IR	687
Methyl-3-Bz-2-oxothiazolidine-4-carbonyl-Gly									Alc	202.5–203.0	−37	DMF			250
DAZ-Gly-OEt															468
MAZ-Gly-OEt															468
2-Mercaptopropionyl-Gly mercury(II) complex															163
2-Mercaptopropionyl-Gly-OEt mercury(II) complex															163

NBzloc-Gly										120–122					478
Z(NO_2)-N-2-carboxamidoethyl-Gly									Aq	165–166					597
N^{α}-Nps-N^{β}-Z-α,β-diaminopropionyl-Gly-OEt									EtAc-Pe	161–163	−19	Chf			730
o-Nps-N-2-carboxamidoethyl-Gly									Alc	166.5–168.5 (dec)					597
o-Nps-N-2-carboxamidoethyl-Gly	DCHA								Me	164–165					597
o-Nps-N-2-carboxamidoethyl-Gly-ONp									DMF-Alc	179–180					597
Nps-N-2-cyanoethyl-Gly									EtAc-Cy	151–152					599
Nps-N-2-cyanoethyl-Gly Dcha									Alc	163.5–164.5					599
Nps-N-2-cyanoethyl-Gly-ONp									DMF-Alc	177.5–178					599
N-(*o*-NPs)-Gly-OBzl(OMe)									EtAc-He	79–79.5					380
N-Pht-Gly									Aq	187–189 191–192				IR	687
Pht-Gly-NHC_6H_{11}									EtAc-Pe	232					365
Pht-Gly-OBhae									Ac	152–154					371
N-Pht-Gly-OBzl(OMe)									Alc or AqAn	123–124.5					380
Pht-Gly-O-3′,5′-dimethoxybenzoin									Be-Me	181–182				Mol. wt.	567
Pht-Gly-O-3,3′,4,4′-dimethylene-dioxybenzoin									MeCl-Me	189.5–190.5				Mol. wt.	567
Pht-Gly-OPh(OH)										192–194					365
Pht-Gly-OPh(OMe)									Alc	119–123					365
Pht-Gly-OP(SMe)										181–185					365
Pht-Gly-NHBzl										214					371
Pht-Gly-ONp										184–188					365

(*continued*)

Table 1 Amino Acid Derivatives (continued)

Compound	Method	Solvent	% Yield	Chromatography				CCD	Recrys. Sol.	M.P. °C	$[\alpha]_D$	Solvent Rotation	Elect.	Other	Ref.
				Col	Tl	Ion	Paper								
Pht-Gly-OPy									EtAc-Pe	118–120 146–161				Mol. wt.	365
Pht-Gly-SPy										122–123					365
Pht-Gly-O-2,2′,3,3′-tetramethoxy-benzoin									Be-Me	139.5–141.5				Mol. wt.	567
Pht-Gly-OTcp										191–193					365
N-Pht-DL-phenyl-Gly OEt									Eth-He	75.5					400
N-Propyl-Gly										191–193 (dec)					531
N-Stearoyl-Gly										119–125					687
N-Tos-azylyl-Gly-OEt									Chf-Eth	147–148	−59	Di		NMR	437
N-Tos-azylyl-Gly-OMe									Chf-Eth	116–117	−56	Di			437
Tos-N-2-carboxamidoethyl-Gly									Aq	209–210					597
N-Tos-3-methyl-azylyl-Gly									EtAc-He	138–139	−52	Me		NMR	437
N-Tos-3-methyl-azylyl-Gly-OEt															437
N-Tos-N-trideuteromethyl-Gly										147–150				NMR	343
N-Tca-Gly									Be	116–121					687
N-Trideuteromethyl-Gly-OBzl HCl									Me-An-Eth	176–179				NMR	343
α-Tritio-Gly															615
N-Trt-Gly-OBzl(OMe)									Alc	110–110.5					380
N-*n*-Valeryl-Gly										81–81.5					155

N-*n*-Valeryl-Gly-OMe										120 (b.p.)					155
Histidine															
N-Ac-DL-His-OEt					56										205
N-Ac-DL-His-OEt HCl									Alc-Eth	153–155				IR NMR	205
N,N′-Adipoylbis[-1-Bzl-His] hydrazide										Oil					445
N$^{\alpha}$-Bz-1-ethyl-His-OMe										Oil					51
N$^{\alpha}$-Bz-His-OMe															51
N$^{\alpha}$-Bz-1-methyl-His-OMe									EtAc	125–126	−53	Alc			51
N-(2-Bz-1-methylvinyl)-His									DMF-Eth	225–226					589
N^{im}-Bzl-N$^{\alpha}$-(2-Bz-1-methylvinyl)-His									MeCl-Eth	165–167	−147	Me			589
N^{im}-Bzl-N$^{\alpha}$-Boc-His									Py-EtAc	180–182	$[\alpha]_{578}$	DMSO			544
									Py-EtAc-Eth	178–180	+54				
N^{im}-Bzl-His										200–201 (dec)	+19	2 *N* HCl			208
N^{im}-Bzl-His-NCA									NMe						483
N^{im}-Bzl-N$^{\alpha}$-Z-His										214–215 (dec)	+6.4	Ac			208
N^{im}-Bzl-N$^{\alpha}$-Z-His-OTcp										108–109					208
N^{im}-Bzl-N$^{\alpha}$-Boc-His										182–183	−0.5	Ac			49
										189–190	+23	Me			208
N^{im}-Bzl-N$^{\alpha}$-Boc-His-OTcp									EtAc-DiPr	123.5–124.5	−9.9	Chf			208
Z-N^{im}-Adoc-His									Me-Aq	125				IR	617
N^{im}-Z-N$^{\alpha}$-Boc-His					36b				Alc-Pe	148–150	+49	Alc			370
N^{im}-Z-TF-N$^{\alpha}$-Z-His-OBu^{t}									EtAc-Pe	50–58	+1.1	Py			96
N$^{\alpha}$,N^{im}-Bis-Boc-His										75–79	−1.6	Ac			546
N$^{\alpha}$,N^{im}-Bis-Boc-His Dcha															546

(*continued*)

Table 1 Amino Acid Derivatives (continued)

Compound	Method	Solvent	% Yield	Chromatography Col	Tl	Ion	Paper	CCD	Recrys. Sol.	M.P. °C	$[\alpha]_D$	Solvent Rotation	Elect.	Other	Ref.
N^{im}-Boc-N^{α}-Z-His										80	+17	Me			152
N^{im}-Boc-N^{α}-Boc-His Dcha									MeCl-Pe	157–159	+18	Chf			722
N^{im}-Boc-N^{α}-Bpoc-His									Alc-Aq	133–134	+21	Me			722
N^{im}-Dpm-N^{α}-Boc-His					36b				MeCl-Pe	60–62	+13	Alc			370
N^{im}-Dpm-N^{α}-Boc-His															369
N^{im}-Dpm-N^{α}-Boc-His-ODpm					36b										370
N^{α}-Boc-His									Alc	199–200 (dec)	+14	Aq			546
									Me-Acn	191–191.5	−11	DMF			208
					36b				Alc-Pe	188–190	+18	Alc			370
N^{im}-Boc-His-NCA HCl									An-Eth	260–265 (dec)	−34	DMF		Mol. wt.	152
N^{im}-Boc-His Di-HCl													×	Mol. wt.	152
N^{α}-Boc-His-OMe										125.5–126	−12	Me			208
											−13	Py			
Decanoyl-His hydrazide									Alc-Aq	215–216	−24	HCl			640
N^{α}-(3-Di-Bz)-His-OMe									Alc-Aq	105–107	−22	Alc			51
N^{α},N^{im}-Di-Boc-His										70	+14	Me			152
N^{α},N^{im}-Di-Boc-His-OMe										ca. 90					208
N^{im}-DNPN-N^{α}-Boc-His										94	+55	EtAc			518
N^{im}-Dnp-His-NCA HCl									An-Eth	215–220	−20	An			153
N^{α}-Bpoc-His										146–149	−13	DMF			722

N^{im}-Dpm-His									EtAc	97					370
N^{im}-Dpm-His-OMe HCl					36b				Me-Eth	238	+8.6	Aq			370
N^{im}-Dpm-His picrate															370
1-Ethyl-His Di-HCl										216–219 (dec)	+17	1 *N* HCl			51
His-NCA HBr									Methyl carbitol-EtAc		−7.4	Methyl carbitol		IR	232
His-hydroxamide acid									Me	159–161					586
His-OMe Di-HCl										205–206 (dec)	+11	Aq			208
His-OMe Di-HCl									Me-Eth	198–199	+3.5	Aq			156
His methylthiohydantoin derivative														CD	622
His-thio-NCA HBr											−7.0	Methyl carbitol			127
N,N′-Malonyl-Di-His-OMe									Alc						445
1-Methyl-His Di-HCl									Alc-Eth	216–219 (dec)	+14	1 *N* HCl			51
1-Methyl-His-OMe Di-HCl									Me-Eth	201				IR NMR Mass spec. Mol. wt.	520
3-Methyl-His-OMe Di-HCl									Me-Eth	208–209				IR NMR Mass spec. Mol. wt.	520
N,N′-Oxalyl-Di-His-OMe									Aq-Alc						445

(*continued*)

Table 1 Amino Acid Derivatives (continued)

Compound	Method	Solvent	% Yield	Chromatography: Col	Tl	Ion	Paper	CCD	Recrys. Sol.	M.P. °C	$[\alpha]_D$	Solvent Rotation	Elect.	Other	Ref.
N,N′-Succinyl-Di-His				Seph-adex G-15	114		×							Mass spec.	445
N^{im}-Trt-N^{α}-Boc-His					36b				MeCl-Pe	113	+5.9	Alc			370
															369
N^{im}-Trt-N^{α}-Boc-His-OMe HCl					36b				MeCl-Pe	210	+3.8	Alc			370
N^{im}-Trt-N^{α}-Boc-His-OTrt					36b										370
N^{im}-Trt-His									EtAc	139					370
N^{im}-Trt-His-OMe HCl					36b					145–148	−12	Aq			370
Pht-δ-amino-valeryl-His									Aq-Me	220–222	+15	Aq			156
Isoleucine															
N-Ac-Ile															553
Aoc-Ile-OSu									Eth-Pe	82–83	−24	DMF			219
N-(2-Bz-1-methylvinyl)-Ile									MeCl-Eth	149–151	+145	Alc			589
Z-Ile										Oil					456
Z-D-Ile										Oil					456
Z-D-Ile (+) ephedrine salt										139	+16	Alc			456
Z-Ile-OPht															306
Z-Ile-SPy										Oil	−37	DMF			365
N-Boc-allo-Ile									Eth	137–139	+6.3	Alc			283
Boc-Ile									Pe	66–68	+3	Ac			544
Boc-Ile-SPy										99–100	−79	EtAc			365
Ile (−) ephedrine salt										138	−17	Alc			456

Ile-OEt															484
Ile methylthiohydantoin derivative														CD	622
Ile-OTMB HCl									Alc-Eth	150–150.5	+20				347
Z(OMe)-Ile-OSu										Oil	−16	Ac			325
Nbs-Ile-OSu									Alc	110–112	−104	Di			285
N-Tfa-Ile-OBu^t															671
Leucine															
Aceto-Ac-Leu										124–125	−16				42
Aceto-Ac-Leu-NHBzl										112–113	−44	Alc			42
N-Ac-Leu															553
N-Ac-Leu-OEt										b.p. 116–117 (1 mm)	−43	Alc		IR	439
Ac-Leu-*p*-nitroanilide									Alc	192–194	−11	An		UV	508
Bz-Leu															191
Bz-Leu Anilide									Alc	214–215.5	+9	Ac			563
N-Bz-Leu anilide										220	−23	Py			534
N-Bz-N-methyl-Leu										135–137	−55	DMF			101
N-(2-Bz-1-methylvinyl)-Leu									MeCl-Pen	117–119	+118	Me			589
Z-Erythro-β-methyl-Leu										Oil					456
Z-Erythro-β-methyl-D-Leu										Oil					456
Z-Erythro-β-methyl-D-Leu (+)-ephedrine salt										124–125	+18	Alc			456
Z-Leu										Oil					456
Z-D-Leu										Oil					456
Z-Leu amide										125–126	−12	Alc			365
Z-DL-Leu amide										129–131					194
Z-Leu-NHBzl										102–104					371
Z-Leu-$NHCH_2CH_2OH$									EtAc-Eth	125–126	−18				384
Z-Leu(+) ephedrine salt										91–93	+12	Alc			456

(*continued*)

Table 1 Amino Acid Derivatives (continued)

Compound	Method	Solvent	% Yield	Chromatography Col	Tl	Ion	Paper	CCD	Recrys. Sol.	M.P. °C	$[\alpha]_D$	Solvent Rotation	Elect.	Other	Ref.
Z-Leu-SPy										89–90	−58	DMF			365
										89–91					
N-Z-N-Methyl-Leu										73–74	−26	DMF			101
Z-*Threo*-N,β-dimethyl-Leu										98–99	−76	Alc			456
Z-*Threo*-N,β-dimethyl-D-Leu										98–99	+75	Alc			456
Z-*Threo*-N,β-dimethyl-D-Leu(+) ephedrine salt										159–160	+73	Alc			456
N-(δ-Bromovaleryl)-Leu-OEt															88
Boc-Leu															470
									Pe	78–81	−24	Ac			544
										Oil					8
Boc-Leu-NHC_6H_{11}									EtAc	135–136					134
Boc-Leu-OPy										Oil	−33	DMF			134
Boc-Leu-SPy										93–95	−90	DMF			365
D-$CH_3(CH_2)_{12}CH(OH)CH_2CO$-Leu hydrazide									Be-Pe	100–105	−28	Fm			230
D-$CH_3(CH_2)_{10}CH(OH)CH_2CO$-Leu hydrazide									Me-Eth	121–123	−28	Ac			230
D-$CH_3(CH_2)_9CH(OH)CH_2CO$-Leu hydrazide									Me-Be	126–130	−31	Ac			230
D-$CH_3(CH_2)_8CH(OH)CH_2CO$-Leu hydrazide									Me-Eth	127.5–130	−37	Ac			230
D-$CH_3(CH_2)_7CH(OH)CH_2CO$-Leu hydrazide									Me-Be	130–131.5	−41	Ac			230

D-$CH_3(CH_2)_6CH(OH)CH_2CO$-Leu hydrazide									Me-Eth	146–147	−34	Ac			230
D-$CH_3(CH_2)_5CH(OH)CH_2CO$-Leu hydrazide									Me-Eth	145–147	−46	Ac			230
D-$CH_3(CH_2)_4CH(OH)CH_2CO$-Leu hydrazide									Me-Eth	145–150	−42	Ac			230
$CH_3(CH_2)_{14}CH(OH)CH_2CO$-Leu hydrazide									Alc	144–147	−14	Alc			230
$CH_3(CH_2)_{12}CH(OH)CH_2CO$-Leu hydrazide									Alc	151–152	−17	Fm			230
$CH_3(CH_2)_{10}CH(OH)CH_2CO$-Leu hydrazide									Me	159–161	−16	Ac			230
$CH_3(CH_2)_9CH(OH)CH_2CO$-Leu hydrazide									Me	157–159	−20	Ac			230
$CH_3(CH_2)_8CH(OH)CH_2CO$-Leu hydrazide									Me	160–162	−17	Ac			230
$CH_3(CH_2)_7CH(OH)CH_2CO$-Leu hydrazide									Me	159–161	−21	Ac			230
$CH_3(CH_2)_6CH(OH)CH_2CO$-Leu hydrazide									Me	165–167	−17	Ac			230
$CH_3(CH_2)_5CH(OH)CH_2CO$-Leu hydrazide									Me	168–169	−24	Ac			230
$CH_3(CH_2)_4CH(OH)CH_2CO$-Leu hydrazide									Me	180–181	−16	Ac			230
$CH_3(CH_2)_2CH(OH)CH_2CO$-Leu hydrazide									*i*-Pr	188–189	−13	Ac			230
N-(3,5-Dinitro-2-pyridyl)-DL-Leu															66
DNT-Leu										140–141					227
Bpoc-Leu									Eth	227–230 (dec)	−12				573
Bpoc-Leu									Eth-Pe	224–226	−12	Me			545
N-Bpoc-Leu									Eth-Pe	216–218	−11	Me			545

(*continued*)

Table 1 Amino Acid Derivatives (continued)

Compound	Method	Solvent	% Yield	Chromatography				CCD	Recrys. Sol.	M.P. °C	$[\alpha]_D$	Solvent Rotation	Elect.	Other	Ref.
				Col	Tl	Ion	Paper								
Erythro-β-hydroxy-Leu									Aq	249–250 (dec)				IR	20
Erythro-β-methyl-Leu(−) ephedrine salt										124–125	−18	Alc			456
OEt-thiocarbonyl-Leu											−28	MeCl			127
β-Hydroxy-DL-Leu										225 (dec)					245
β-Hydroxy-DL-Leu															20
D-Leu															387
Leu-$NHCH_2CH_2OH$ HCl									Me-EtAc-Eth	153–154	+23				384
Leu-OBzl-benzenesulfonate									Chf-Cy	171–172					643
S-Leu-O-*i*-Bu Tos										137–138	+10	Alc			210
D-Leu(−) ephedrine salt										92–94	−12	Alc			456
Leu-OMe															494
DL-Leu-OMe															494
Leu methylthiohydantoin derivative														CD	622
DL-Leu methylthiohydantoin derivative														CD	622
Leu-thio-NCA					27				EtAc-He	77–78	−57	MeCl			127
									Cy	74–75	−35	MeCl			
									EtAc-He	79.5–81	−27	MeCl			
Leu-OTMB HCl									Alc-Eth	177–179	+21				347
Methyl-DL-Beu									Eth-Pe	90–92					544

DAZ-Leu-OEt															468
N-Methyl-Leu-OMe HBr										105–106	+18	DMF			101
MAZ-Leu-OEt															468
Pht-Leu									Eth-He	118.5–119.5	−24	Alc			563
Pht-Leu anilide									Be-Pe	154.5–156	−21	Ac			563
Pht-Leu-OCH_2CN										72–73	−35	An			351
N-[δ-(6-Purinylthio)valeryl]-Leu-OEt									An-He	144–145					88
Threo-N,β-dimethyl-Leu(−) ephedrine salt										160–161	−75	Alc			456
Threo-β-hydroxy-Leu									Aq-Alc	239–240 (dec)				IR	20
Lysine															
N^{ϵ}-Ac-homo-Lys										248–249	+18	HCl			61
(+)-N^{ϵ}-Ac-thia-Lys sulfoxide										163	+4.3	Aq			228
(−)-N^{ϵ}-Ac-thia-Lys sulfoxide										163	−4.4	Aq			228
N^{ϵ}-Z-N-(2-Bz-1-methylvinyl)-Lys									MeCl-Eth	135–137	+49	Alc			589
N^{ϵ}-Z-N^{α}-Z-Lys-SPy										Oil	−28	DMF			365
N^{ϵ}-Z-N^{α}-Boc-Lys					126				Di-Aq	254–256					498
N^{ϵ}-Z-N^{α}-Boc-Lys									EtAc	110–111					544
N^{ϵ}-Z-N^{α}-Boc-Lys-NH-$(CH_2)_4$-NH-Boc									*i*-Pr-Eth	83–85					171
Z-DNS-Lys										115–117					550
N^{ϵ}-Z-Lys										189	+13	HCl			537
															457
N^{ϵ}-Z-Lys-NCA															483
Z-Lys methylthiohydantoin derivative														CD	622
N^{ϵ}-Z-N^{α}-SNp-Lys Dcha									EtAc	185–187	−30	DMF			537
N^{ϵ}-Z-N^{α}-Tos-Lys															457
N^{ϵ}-Z-Lys-OTMB HCl									Chf-Eth	125–126	+6.7	DMF			600

(*continued*)

Table 1 Amino Acid Derivatives (continued)

Compound	Method	Solvent	% Yield	Chromatography Col	Tl	Ion	Paper	CCD	Recrys. Sol.	M.P. °C	$[\alpha]_D$	Solvent Rotation	Elect.	Other	Ref.
N^{ϵ}-Z-Lys-OTMB Tos									Alc-Aq	166–167	+2.6	DMF			600
N^{α},N^{ϵ}-Bis-Boc-Lys-OSu															645
N^{α},N^{ϵ}-Bis-Nps-Lys															537
$N^{\alpha}N^{\epsilon}$-Bis-Nps-Lys															537
$N^{\alpha}N^{\epsilon}$-Bis-Nps-Lys Dcha									Alc	156–158					537
N^{ϵ}-Boc-N^{α}-Z-Lys										Oil					10
									EtAc-Pe	76–78					544
N^{ϵ}-Boc-N^{α}-Z-Lys Dcha									*i*-Pr	154–155					10
N^{ϵ}-Boc-N^{α}-Bpoc-Lys Dcha											+10				573
N^{ϵ}-Boc-Lys									EtAc-Pe	136–138					544
N^{ϵ}-Boc-D-Lys									Aq	227–230 (dec)	−6.5	1 *N* NaOH			546
											−17	Ac			
N^{ϵ}-Boc-Lys-NCA									EtAc-He	123 (dec)	−17	An			232
N^{ϵ}-Boc-Lys-NH-$(CH_2)_4$-NH-Boc Tos															171
N^{ϵ}-Boc-N^{α}-nonanoyl-Lys															423
N^{ϵ}-Boc-Lys-OSu										84–86	−25	Me			645
N^{ϵ}-2-Chloroethyl-N^{ϵ}-2-fluoroethyl-Lys Di-HCl															55
N^{α},N^{ϵ}-Di-Boc-Lys-OSu									EtAc-Pe	92–96	−31	Alc			569
Di-decanoyl-DL-Lys-OMe					11					Oil					527
N^{ϵ}-Hexanoyl-Lys										227–229	+12	Ac			640
N^{α}-Hexanoyl-Lys-OMe HCl											+11	Alc			640

Homo-Lys HCl										263–264	+19	HCl			61
DL-Homo-Lys HCl						DO-50			Alc	267–268					61
D-Lys															265
Lys methylthiohydantoin derivative														CD	622
N^{α}-Methyl-Lys															112
N^{ϵ}-(2-Methyl-2-ONp-propionyl)-Lys															112
N^{α}-Nps-Lys									Py-Aq	186–188	−19				537
N^{ϵ}-Pht-N^{α}-Z-Lys										129–130.5	−1.8	Me			219
N^{ϵ}-Pht-N^{α}-Z-Lys Dcha									Me-Eth	158–159	+6.1	Me			219
N^{ϵ}-Pht-N^{α}-Z-Lys-ONp									Alc	99–100	−16	DMF			219
N^{ϵ}-Tos-N^{α}-Boc-Lys															544
N^{ϵ}-Tos-N^{α}-Boc-Lys-OSu										114–117	−19	Me			656
N^{ϵ}-Tos-Lys										241 (dec)	+13	HCl			513
N^{ϵ}-Tos-Lys-ONp HBr									Me	182	+18	Alc			559
N^{ϵ}-TrOC-N^{α}-Z-Lys Dcha									Me-Eth	162–165	+2.9	Me			716
N^{ϵ}-TrOC-N^{α}-Boc-Lys										94–98	−9.8	Me			716
N^{ϵ}-TrOC-Lys									Ac-Aq	218–220	+14	Ac-Aq			716
N^{ϵ}-TrOC-Z(OMe)-Lys Dcha									Me-Eth	138–140	+42	Me			716
N^{ϵ}-Tfa-N^{α}-Boc-Lys-OSu									EtAc-Pe	118	−23	DMF			689
β-Lysine															
N^{β}-Ac-β-Lys-OBu^{t}										Oil					526
N^{ϵ}-Z-N^{β}-Ac-β-Lys-OBu^{t}					20										526
N^{ϵ}-Z-N^{β}-Decanoyl-β-Lys															528
N^{ϵ}-Z-N^{β}-Decanoyl-β-Lys-OBu^{t}										Oil					528
N^{ϵ}-Z-β-Lys										Oil					528
N^{ϵ}-Z-β-Lys-OBu^{t}										Oil					528
N^{β}-Decanoyl-β-Lys					115										528
N^{β}-Decanoyl-β-Lys-OMe															528
N^{ϵ}-Decanoyl-β-Lys															528

(*continued*)

Table 1 Amino Acid Derivatives (continued)

Compound	Method	Solvent	% Yield	Chromatography Col	Tl	Ion	Paper	CCD	Recrys. Sol.	M.P. °C	$[\alpha]_D$	Solvent Rotation	Elect.	Other	Ref.
N^{ϵ}-Decanoyl-β-Lys-OMe															528
N^{ϵ}-Decanoyl-N^{β}-Pht-β-Lys															528
Di-Ac-β-Lys					85					Oil					526
Di-Ac-β-Lys-OBu^t					85					Oil					526
Di-Z-β-Lys										152–153	−9.0	Me		IR	610
														UV	
Di-Z-β-Lys-OBzl														IR	610
									An-Eth-Pe	113–114					610
Di-Z-β-Lys-OBzl										113–114				IR	610
														UV	
Di-decanoyl-β-Lys															528
Di-decanoyl-β-Lys-OMe															527
β-Lys															610
N^{β}-Pht-β-Lys															528
Methionine															
N-Ac-Met															553
															647
N-(2-Bz-1-methylvinyl)-Met									Chf-Eth	168–170	+21	Alc			589
Z-Met										69–70	−19	Alc			456
Z-D-Met										70–71	+17	Alc			456
Z-Met (+) ephedrine salt										130–131	+28	Alc			456
Z-Met-SPy										85–86	−60	DMF			365
N-Z-DL-Seleno-Met									EtAc-He	114–116				NMR	650
N-Z-DL-Seleno-Met-OBzh				Si										NMR	650

Boc-Met															8
									EtAc	138–139					544
N-Boc-Met										Oil					10
Boc-Met Dcha									EtAc	141	−18	Me			249
Boc-Met-O·CH_2·CO_2Et	OPy		80						Pe	46–48					134
Boc-Met-NHC_6H_{11}	OPy		95						Alc-Eth	135–136					134
Boc-Met-ONp				D- 50W × 8					Alc	96–97	−48	Me			249
N-Boc-Met-OPcp									Alc	138–139	−43	Alc			196
Boc-Met-OPh															314
Boc-Met-OPy									Eth	74–75	−34	DMF			134
Boc-Met-SPy										91–92	−84	EtAc			365
DMPOC-Met Dcha										117–119					590
DNT-Met										114–117					227
Bpoc-Met Dcha									EtAc-Pe	143–145	+14				573
Met															500
D-Met (−) ephedrine salt										129–130	−28	Alc			456
Met methylthiohydantoin derivative														CD	622
Met sulfoxide															500
Z(OMe)-Met-OTcp									Alc	127–128	−23	DMF			325
Nvoc-Met										150–151	−26				478
N-Pyridoyl-DL-Met															262
N-Methylisoleucine															
N-Ac-N-MeIle										121.5– 122.5	−135	DMF			101
N-Z-N-MeIle										Oil					101
N-MeIle-OMe HBr										126.5– 127.5	+24	DMF			101
N-Methylvaline															
N-Ac-N-MeVal										112–114	−144	DMF			101

(continued)

Table 1 Amino Acid Derivatives (continued)

Compound	Method	Solvent	% Yield	Chromatography Col	Tl	Ion	Paper	CCD	Recrys. Sol.	M.P. °C	$[\alpha]_D$	Solvent Rotation	Elect.	Other	Ref.
N-Z-N-MeVal										Semi-solid					101
N-MeVal-OMe HBr										127–128	+17	DMF			101
Norleucine															
N-Ac-Nle-OEt										117–118	−30	Alc		IR	439
Z-Nle-OSu									EtAc-Pe	69–71	−22	An			238
Boc-Nle-OSu									*i*-Pr-Pe	107–108	−31	An			238
6-Diazo-N-Dca-5-oxo-Nle OMe									EtAc-Pe	73	+1.4	EtAc			363
Erythro-β-hydroxy-DL-Nle						DO-50				253–254 (dec)					20
Erythro-β-hydroxy-DL-Nle tetra-chlorophthalic acid salt										129.5–130 (dec)				IR	20
β-Hydroxy-DL-Nle										229 (dec)					245
Threo-β-hydroxy-DL-Nle										232–233 (dec)					20
Norvaline															
N-Ac-Nva-OEt										33–35				IR	439
β-Hydroxy-DL-Nva						DO-50	×		Aq-Alc	247–248 (dec)					20
						A-IR-120B				219 (dec)					245

β-Hydroxy-DL-Nva-α-naphthyl-phosphoric acid salt									Aq-Me	172.5–173.5 (dec)				IR	20
Methyl 5-diazo-N-Dca-4-oxo-Nva									EtAc-Pe	82	+63	EtAc			363
Ornithine															
N^{δ}-Ac-N^{α}-Tos-Orn Dcha									Alc	182–183					200
N^{δ}-Ac-N^{α}-Tos-Orn-OEt									Alc-Eth	131–132					200
N^{δ}-Ac-N^{α}-Tos-Orn-O-*i*-Pr															200
N^{δ}-Ac-N^{α}-Tos-Orn-O-*i*-Pr									*i*-Pr	158–160	−15	DMF			200
N^{δ}-Ac-N^{α}-Tos-D-Orn-O-*i*-Pr									*i*-Pr	159–161	+14	DMF			200
N^{α}-Bz-Orn				Sephadex					Alc-Aq	227–228				IR	18
N^{α}-Bzl-N^{α}-methyl-N^{ϵ}-Tos-Orn															583
N^{δ}-Z-N^{α}-Boc-Orn															544
N^{δ}-Z-Orn															174
N^{δ}-Z-Orn-OMe HCl										140–142	+2.2	Ac			261
N-α-Bromo-Ac-Orn										260	+18	5 *N* HCl			459
N-α-Bromo-Ac-DL-Orn										260					459
N^{δ}-Boc-N^{α}-Z-Orn									Eth-Pe	97–99	−3.2	Me			544
N^{δ}-Boc-Z-Orn-OMe					9				Eth-Pe	71	−14	90% Ac		Mol. wt.	613
					24				Di-Pr-Eth						
N-α-Chloro-Ac-Orn										254	+20	5 *N* HCl			459
N-α-Chloro-Ac-DL-Orn										240					459
N^{δ}-Cyano-N^{α}-Bz-Orn-OEt										85–85.5					18
N^{α},N^{δ}-Ditos-Orn									Alc-Aq	186.5–187.5	+26	Alc-Aq			201
N-α-Iodo-Ac-Orn										250	+16	5 *N* HCl			459
N-α-Iodo-Ac-DL-Orn										254					459
β-Methyl-Orn HCl										238 (dec)					583
N^{α}-Methyl-Orn HCl										238–240					583

(*continued*)

Table 1 Amino Acid Derivatives (continued)

Compound	Method	Solvent	% Yield	Chromatography Col	Tl	Ion	Paper	CCD	Recrys. Sol.	M.P. °C	$[\alpha]_D$	Solvent Rotation	Elect.	Other	Ref.
N^δ-Z-(NO_2)-Orn									Aq	231–232 (dec)	+14	6 *N* HCl			453
N^δ-Z(NO_2)-N^α-Z(OCH_3)-Orn									Me-Eth-Pe	122–124	−4	Me			453
N^δ-Z(NO_2)-N^α-Z(OCH_3)-Orn Dcha									Alc-Eth	128–130	+3.2	Me			453
D-Orn HBr									Aq-An	232–234 (dec)	−1.0	Aq			200
D-Orn HCl										240–241.5 (dec)	−25	HCl			200
N^δ-Pht-D-N^α-Tos-Orn									Alc-Aq	145–146	−0.4				730
N^δ-Phenyl-DL-Orn										244–245					583
N^δ-Tos-N^α-Z-Orn amide									Me	150–151	0	Me			201
N^δ-Tos-N^α-Z-DL-Orn									EtAc-Pe	138–139.5					201
N^δ-Tos-N^α-Z-DL-Orn amide									Me-Aq	134–135.5					201
N^δ-Tos-N^α-Z-DL-Orn hydrazide									Alc-Eth	146					201
N^δ-Tos-N^α-Z-Orn-OPcp									EtAc or THF-Eth	153–154	−4.3	Chf			201
N^δ-Tos-Orn										236–238 (dec)	+22	6 *N* HCl			201 494 495 496

N^{α}-Tos-D-Orn									Aq	190–194	+5.6	5 *N* HCl			200
N^{δ}-Tos-DL-Orn					20				Aq	223–229					201
										(dec)					494
															495
N^{δ}-Tos-Orn amide HBr									Me-Eth	141–143	+14	Me			201
N^{δ}-Tos Orn-OMe															496
N^{δ}-Tos-DL-Orn-OMe															495
N^{δ}-Tos-Orn-OMe HCl															494
															496
Phenylalanine															
N-Aceto-Ac-Phe										109–110	+68				42
N-Ac-N-methyl-Phe															101
N-Ac-Phe															553
															647
(*R*)-N-Ac-Phe															117
N′-Ac-D-Phe-3,5-dimethylpyrazole									Be-Pe	147.5–149	$[\alpha]_{436}$ −296	Alc			508
N-Ac-Phe-OEt										b.p. 141–142					439
Ac-D-Phe hydrazide										166–168	−22	Alc			508
N-Ac-Phe-*p*-nitroanilide									Alc	252–253	+117	An			508
											+216				
N-Ac-DL-Phe-*p*-nitroanilide									Alc	270–272					508
									Alc	261–263					
m-Amino-Phe															123
4-Aminobutyryl-Phe-OMe acetate									Chf-Pe	100–102	+2.4	Ac			584
Aoc-Phe-OSu									*i*-Pr	130–131	−52	DMF			219
N-Bz-2,5-Di-OMe-4,α-dimethyl-Phe									Alc	175–176					237
N-Bz-2,5-Di-OMe-4-methyl-Phe									Alc	187–188					237
N-(2-Bz-1-methylvinyl)-Phe									MeCl-Eth	152–154	−277	Alc			589
3-Bz-2-oxothiazolidine-4-carbonyl-Phe-OMe									EtAc	216–217	+34	DMF			250

(*continued*)

Table 1 Amino Acid Derivatives (continued)

Compound	Method	Solvent	% Yield	Chromatography Col	Tl	Ion	Paper	CCD	Recrys. Sol.	M.P. °C	$[\alpha]_D$	Solvent Rotation	Elect.	Other	Ref.
N-Bz-Phe anilide										224–225	+25	Py			534
N-Bz-Phe-α-methyl-Bzl amide									Alc	181–183					562
N-Bz-DL-Phe-SePh										155–157					264
Z-4-Aminobutyryl-Phe-OMe									Chf-Aq	101–103	−11	DMF			584
N-Z-N-Methyl-Phe										Oil					101
Z-Phe										86–89	+8.4	Alc			456
Z-D-Phe										86–87	−8.3	Alc			456
Z-DL-Phe									EtAc	137–139					371
N-Z-Phe															494
N-Z-D-Phe															494
Z-Phe-OBu^t										79–80					671
Z-Phe Boc-hydrazide									Eth	81–83	−27	Me			83
Z-DL-Phe-$NHCH_2CN$															195
Z-Phe-NHC_6H_{11}									Alc	163–165					134
Z-DL-Phe-OCH_2CN															195
Z-Phe-O-2,2-diphenylethyl									Cte	58–62					677
Z-Phe-N,N′-diphenyl hydrazide															677
Z-D-Phe (+) ephedrine salt										139–140	−11	Alc			456
N^α-Z-Phe-*p*-nitroanilide									Alc-Aq	158.5–159.5	+141	An			508
Z-DL-Phe-ONp															195
4-[Z-Phe-oxy]-phthalazone										186–188					677
Z-Phe-SPy										Oil	−63	DMF			365
Z-DL-Phe-OTcp															195

N-Boc-N-methyl-Phe										Oil				IR	317
Boc-Phe										85–86					8
									Eth-Pe	84–86	−0.8	Ac			544
Boc-D-Phe															470
Boc-Phe amide															476
Boc-Phe-4-carboxy-phenylhydrazide										204–206					677
Boc-Phe-O-2,2′-dinitrodiphenyl-methyl										124–126	−38				478
Boc-Phe hydrazide									EtAc-Pe	93–95					677
Boc-Phe-nitro-agmatine										84–89	−3.3	DMF			279
Boc-Phe-OPy									Eth	111–112	−32	DMF			134
Boc-Phe-SPy										123–125	−86	EtAc			365
N-Chlor-Ac-(3*S*)-3-deuterio-DL-Phe									Aq	130					246
α-Deuterio-DL-Phe							×							NMR	615
(3*R*)-3-Deuterio-DL-Phe									Aq	258–262					246
(3*S*)-3-Deuterio-DL-Phe									Aq	260–265					246
3,4-Dihydroxy-Phe(L-DOPA)															132
DL-3,4-Dihydroxy-Phe hydroxamic acid										177–179					586
3,4-Dihydroxy-Phe-OMe				Si										UV	132
														IR	
N-Z-5,6-Dioxo-Phe-OBzl														IR	132
2,5-Di-OMe-4,α-dimethyl-Phe									Aq	234 (dec)					237
2,5-Di-OMe-4,α-dimethyl-Phe-OMe HCl									Me-Eth	183–184					237
2,5-Di-OMe-4-methyl-Phe									Aq	235 (dec)					237
2,5-Di-OMe-4-methyl-Phe-OMe									Me-Eth	191–192					237
Dmpoc-Phe Dcha										204–206					590
DNbzloc-Phe										196–200	−21				478
DNT-Phe										182–185					227

(*continued*)

Table 1 Amino Acid Derivatives (continued)

Compound	Method	Solvent	% Yield	Chromatography				CCD	Recrys. Sol.	M.P. °C	$[\alpha]_D$	Solvent Rotation	Elect.	Other	Ref.
				Col	Tl	Ion	Paper								
Bpoc-Phe Dcha									EtAc	116–119 (dec)	+33				573
Ethoxythiocarbonyl-Phe									Be-He	82–85	+78	MeCl			127
For-Phe-OBhae									EtAc	111–114					371
Z(OMe)-Phe-ONp									EtAc-Pe	92–93	−16	Me			325
Z(OMe)-Phe-OTcp									EtAc-Pe	131–132	−35	DMF			325
Methyl-3-(N-Z-Phe)-2-oxothiazolidine-4-carboxylate									EtAc	112–112.5	−76	Me			250
DAZ-Phe-OMe															468
MAZ-Phe-OMe															468
N-Methyl-Phe-OMe-HBr										132–133	+37	DMF			101
(*S*)(−)-α-Methyl-Phe									Aq-An	318 (dec)					662
(*S*)(−)-α-Methyl-β-Phe															662
Nbzloc-Phe										112–114	−374				478
Nps-Phe-OPhOBzl									EtAc-Pe	129–131	−59	EtAc			110
N-Nps-Phe-ONp									Me	105–113	−114	DMF			537
Nvoc-Phe										167	−28				478
(*R*)(+)-Phe											+1.3	Aq			65
Phe amide						×				94.5–95.5	+21	Aq			466
Phe anilide									Alc-Bu_2O	223–224					563
D-Phe anilide HCl									Alc-Bu_2O	226–227	−105	Abs. Alc			563
Phe-OPhOBzl HCl									Alc-Eth	200–201	−4.1	Alc			110

Phe-OBu^t										Oil					671
Phe Boc-hydrazide									EtAc-Pe	117–119	+49	1 *N* HCl			83
D-Phe-OEt Tos										153–154	−20	Alc			401
Phe hydroxamic acid									Me	186–188					586
(*S*)-Phe-O-*i*-Bu Tos										165–166	+25	Alc			210
Phe (−) ephedrine salt										139–140	+11	Alc			456
Phe-OMe										90–91					671
Phe methylthiohydantoin derivative														CD	622
Phe-OP(SMe) HBr									Me-Eth	226	+35	Me			275
Phe-*p*-nitroanilide									Alc	156.5–157.5	$[\alpha]_{436}$ −314	An			508
Phe-OPh HBr										232–233					314
Phe thiocarbamate, dipotassium									Alc-*i*-Pr				×		127
Phe-thio-NCA									EtAc-He		−155	MeCl			127
Phe-ONBzl Tos									Aq	117–178	−9	DMF			480
Phe-OTMB HCl									Me-Eth	160–161.5 177–181	−46	Aq			347
Pht-Phe									Me-Aq	183–185	−212	Alc			563
Pht-D-Phe										180–180.5	+207	Alc			563
Pht-Phe anilide									Alc	203–205	−104	Chf			563
Pht-D-Phe anilide									Me	207–208	+105	Chf			563
Pht-Phe chloride									Be-Pe	83–84	−197	Be			563
Pht-Phe-OCH$_2$CN										138–139.5	−191	An			351
Pht-Phe-N-cyanomethylamide									EtAc-Pe	199.5–200.5	−199	DMF			365
Pht-Phe-OPy									EtAc-Pe	96–98	−204	DMF			365
Pht-Phe-4-pyridyl thiolester									EtAc-Pe	135–136.5	−184	DMF			365

(*continued*)

Table 1 Amino Acid Derivatives (continued)

Compound	Method	Solvent	% Yield	Chromatography Col	Tl	Ion	Paper	CCD	Recrys. Sol.	M.P. °C	$[\alpha]_D$	Solvent Rotation	Elect.	Other	Ref.
Pht-Phe pyrimidin-2-yl thiolester									EtAc-Pe	94–97	−229	DMF			365
Pht-Phe-SPy										103–105	−237	DMF			365
N-Tca-Phe-OBu^t															671
Tfa-Phe-OBu^t															671
Phenylglycine															
N-Ac-DL-phenyl-Gly OEt										63					400
Z-DL-Phenyl-Gly										123–124					400
R(−)N,N-Dimethyl-phenyl-Gly									Alc-An	257 (dec)	−133 ±0.6	Aq		IR	114
R(−)-N,N-Dimethyl-phenyl-Gly-OMe															114
N-For-DL-phenyl-Gly										174					400
Phenyl-Gly										223–224	+144	Py			534
(*R*)-Phenyl-Gly-O-*i*-Bu-Tos										185–186	−62	Alc			210
Proline															
N-(2-Bz-1-methylvinyl)-Pro									MeCl-Eth	186–187	+94	Alc			589
Z-Pro-NH$(CH_2)_{15}CH_3$					45 14				Me	87–88	−19	DMF		IR	77
Z-Pro-NH$(CH_2)_9CH_3$					14 113					82.5–84	−22	DMF		IR	77
Z-Pro amide									EtAc-Pe	90–91	−35	Alc			248
Z-Pro-OBu^t										44					671
Z-Pro-SPy										Oil	−75	DMF			365

Boc-Pro										134–135					8
										137–138	−105	Chf			530
									EtAc-Pe	134–136	−60	Ac			544
										134	−60	Ac			518
Boc-Pro amide															476
Boc-Pro-OPy										Oil	−47	DMF			134
Boc-Pro-SPy										68–70	−92	DMF			365
1-Cyanoethyl-Pro									An	155–156	−63	Aq			435
1-Cyanoethyl-Pro-1-oxide									Me-Eth	109–111 (dec)	+26	Aq			435
2-Deutero-DL-Pro-OBzl HCl									Alc-Eth	120–128	0	Me		NMR	343
Bpoc-Pro Dcha									EtAc-Pe	173–175 (dec)	−12				573
Tritio-ethoxythiocarbonyl-Pro									EtAc-He						127
Imaa-Pro									Eth-Pe	103–105					704
Imaa-Pro-OMe															704
Imaa-Pro-OSu										88–91 (dec)					704
Z(OMe)-Pro-OSu										Oil	−47	DMF			325
cis-3,4-Methylene-Pro						DO-1 A-IR-120				235–245	−144	Aq		IR NMR	158
cis-3,4-Methylene-Pro HCl									Alc-Eth	219–221					158
trans-3,4-Methylene-Pro										245–250	−94	Aq		IR NMR	158
6-Nvoc-Pro										111–112	−11				478
N-Pz-Pro-ONp					88				Alc	107–108.5					705
Pro					105						−87	Aq			530
D-Pro															387

(*continued*)

Table 1 Amino Acid Derivatives (continued)

Compound	Method	Solvent	% Yield	Chromatography Col	Tl	Ion	Paper	CCD	Recrys. Sol.	M.P. °C	$[\alpha]_D$	Solvent Rotation	Elect.	Other	Ref.
Pro-O$(CH_2)_{13}CH_3$ HCl					43 14 134					79.5– 80.5	−22	DMF			77
Pro-O$(CH_2)_9CH_3$ HCl					27 14 134					Oil					77
Pro-O$(CH_2)_{17}CH_3$ HCl					27 14 134					89–90	−20	DMF			77
Pro-O-CH[$(CH_2)_{16}CH_3$]$_2$					27 14 134				Alc	39.5– 40.5	−14	Ac			77
Pro-O$(CH_2)_{23}CH_3$ HCl					43 14 134					95–96.5				IR	77
Pro-NH$(CH_2)_9CH_3$ HCl					134 14 24										77
Pro-NH$(CH_2)_{15}CH_3$										56.8– 57.3					77
Pro-NH$(CH_2)_{15}CH_3$ HCl					14 135					111–115	−30	DMF			77
Pro-OBzl Tos					124					Oil					263
Pro-OMe															494

Pro-OBzl(OMe) HCl									Alc-Eth	124.5–125	−35	DMF			596
Pro methylthiohydantoin derivative													CD		622
Pro-thio-NCA											−157	Chf			127
Pro-OTMB HCl									Me-Eth	153.5–154	−21				347
									Alc-Pe	66.5–67	−55	DMF			596
N-Tfa-3,4-dehydro-Pro-OMe										Oil				IR NMR	158
3-Hydroxyproline															
3-Hyp									Me	157–160 (dec)	−80	Aq		NMR	435
DL-3-Hyp									An	158–161 (dec)	+78	Aq			435
DL-3-Hyp									Me	151–155 (dec)					435
Boc-3-Hyp									EtAc-Pe	194–195					544
Decanoyl-3-Hyp hydrazide									EtAc	150–151	−40	Alc			640
4-Hydroxyproline															
4-Hyp amide HCl									Me-EtAc	219–220	−61	Aq			598
4-Hyp-OBzl(OMe) HCl									Alc-Eth	141.5–142.5	−33	DMF			596
4-Hyp-ONp HCl									Me-Eth	180.5–181.5	−84	DMF			598
4-Hyp-ONp HCl									Me	201–201.5	−35	DMF			598
4-Hyp-OTMB HCl									Me-EtAc	201–202	−24	DMF			598
N-Nitroso-4-Hyp Dcha									Alc-Eth	196–197 (dec)	−61	DMF			598
N-Nitroso-4-Hyp-ONp									Alc	150–150.5	−80	DMF			598
N-Nitroso-4-Hyp-OPcp									Alc	180–181	−65	DMF			598

(*continued*)

Table 1 Amino Acid Derivatives (continued)

Compound	Method	Solvent	% Yield	Chromatography				CCD	Recrys. Sol.	M.P. °C	$[\alpha]_D$	Solvent Rotation	Elect.	Other	Ref.
				Col	Tl	Ion	Paper								
N-Nitroso-4-Hyp-OTMB									Alc-Cy	126.5–127	−63	DMF			598
Sarcosine															
Boc-Sar									Eth-Pe	84					544
N-Hydroxy-Sar									Me	128–130 (dec)					435
Sar-OBzl(OMe) HCl									Alc	135–135.5					596
Sar-OTMB HCl									Alc-Eth	158.5–159					596
N-Stearoyl-Sar									Acn	64–67					687
Tos α-phenyl-Sar									EtAc-Pe	144–145					441
Tos D(−)-α-phenyl-Sar									EtAc-Pe	115–117	−61	Alc			441
Serine															
OAc-N^{α}-Z-Ser										Oil					208
OAc-N^{α}-Z-Ser-ONp									Alc	98–99					208
OAc-N^{α}-Z-Ser-OTcp									Be	88–89					208
OAc-Ser HCl										160–161 (dec)	+18	Alc			208
α-N-Bz-DL-homo-Ser-N-methyl-hydroxamic acid															327
N-(2-Bz-1-methylvinyl)-OBu^t-Ser										165–166	−1.1	Alc			589
N-(2-Bz-1-methylvinyl)-Ser									Alc	192–193	+99	Me			589

OBzl-α-N-Bz-DL-homo-Ser-hydroxamic acid															327
OBzl-N-Boc-Ser Cha										156–157	+30	Me			261
OBzl-N-Boc-Ser-SPy										94.5–95.5	−75	EtAc			365
OBzl-N-Z(OMe)-Ser-OTcp									Eth-Pe	67–68	−5.4	Alc			325
O,N-Z-Ser										83–84					456
O,N-Z-D-Ser										83–84					456
O,N-Z-Ser (+) ephedrine salt										132	+33	Alc			456
O-Z-D-Ser (−) ephedrine salt										133–134	−34	Alc			456
N-Z-DL-Ser-OEt										Oil				IR NMR	205
Boc-Ser(C_4H_9)										Oil					8
O,N-Boc-Ser									Eth-Pe	132–133	+8.8	Me			546
O,N-Boc-Ser Dcha									Eth	157–159	+1	DMF			546
O-Bu^t-Boc-Ser										161–163					544
OBu^t-N-Bpoc-Ser Cha									EtAc	180–181 (dec)	+21				573
N-Boc-Ser									Eth-Pe	75–78					544
									Eth-Pe	88–89	−4.5	Ac			546
β-(*p*-Chlorophenyl)-DL-Ser										186 (dec)					245
N-Bpoc-Ser										Oil					545
N-Bpoc-Ser Dcha										154–156	+2.5	Me			545
DNPS-Ser-OMe										98					415
DL-Homo-Ser-hydroxamic acid															327
N-Z(OMe)-Ser-2,4-DNPO									EtAc-Pe	112–114	−23	DMF			325
N-Z(OMe)-Ser-OPcp									Acn	170–171	−18	DMF			325
β-(*o*-Nitrophenyl)-DL-Ser										179 (dec)					245
β-(*p*-Nitrophenyl)-DL-Ser									Aq	182 (dec)					245

(*continued*)

Table 1 Amino Acid Derivatives (continued)

Compound	Method	Solvent	% Yield	Chromatography				CCD	Recrys. Sol.	M.P. °C	$[\alpha]_D$	Solvent Rotation	Elect.	Other	Ref.
				Col	Tl	Ion	Paper								
DL-Phenyl-Ser										194.5 (dec)					245
O-Pyrophospho-D,N-Bz-Ser-methyl-amide															29
D-Ser															387
DL-Ser						A-IR-120B									245
Ser-NCA										82 (dec)				IR	232
DL-Ser-OEt HCl										103–104					205
Ser-OMe															540
Ser methylthiohydantoin derivative														CD	622
β-(*p*-Tolyl)-DL-Ser										186 (dec)					245
N-Tos-Ser-OEt										103–104	−8.2	Alc			437
O,N-Tos-Ser-OEt									EtAc-He	117–118	+22	Me			437
Tca-Ser-NCA															232
Tfa-Ser-NCA										Oil				IR	232
O-Trimethylsilyl-Ser-NCA										(dec)	−47	Chf		IR	232
Threonine															
N-(2-Bz-1-methylvinyl)-OBu^t-Thr									Eth-Pe	174–177	−11	Alc			589
N-(2-Bz-1-methylvinyl)-Thr									Chf-Eth	203–204	+114	Me			589
N-Bz-Thr-OEt										Oil				IR NMR	205

Z-Thr Boc-hydrazide					4 3				EtAc-Pe	143–144	−11	DMF			691
N-Z-DL-Thr-OEt										Oil				IR NMR	205
N,O-Bis-Boc-Thr									Eth-Pe	138–140	+33	Me			546
N-(4-Bromo-2-nitro-3-hydroxy-Bz)-Thr									EtAc-Cy	192–193	−52	Alc			552
OBu^t-Boc-Thr									Pe	73–75					544
Boc-Thr									EtAc-Pe	74–77	−2.5	Me			544
N-Boc-Thr									Eth-Pe	73–75 80–81					546
Decanoyl-Thr											+19	DMF			640
N,O-Di-Tos-Thr-OEt									EtAc-He	118–119	+20	Me			437
N,O-Di-Tos-Thr-OMe															27
N-[2-Nitro-3-OBzl-4-methyl-6-deutero-Bz]-2-deutero-Thr									Aq-Alc		−22	Aq		NMR	343
N-[2-Nitro-3-OBzl-4-methyl-6-deutero-Bz]-2-deutero-D-allo-Thr										133–152	+2.8	Me		NMR	343
N-[2-Nitro-3-OBzl-4-methyl-6-deutero-Bz]-Thr									Ac-Bz-Cy		+58 ±2	Me			344
N-Nps-OBu^t-Thr-OSu									Alc	138–140	−56	DMF			591
Thr															21 386
DL-Thr						A-IR-120B				216 (dec)					245
Thr Boc-hydrazide acetate hemi-hydrate									Me-Eth	68–70	−15	DMF			691
Thr-NCA										94–98 (dec)	−63	Ac			232
Thr, silver															232
Thr methylthiohydantoin derivative														CD	622

(*continued*)

Table 1 Amino Acid Derivatives (continued)

Compound	Method	Solvent	% Yield	Chromatography Col	Tl	Ion	Paper	CCD	Recrys. Sol.	M.P. °C	$[\alpha]_D$	Solvent Rotation	Elect.	Other	Ref.
Thr-OTMB HCl									Alc-Eth	153.5–154.5	−6.1				347
N-Tos-Thr-OEt										106	−4.4	Me			437
O-Trimethylsilyl-Thr-NCA															232
Tryptophan															
N-Ac-Trp															536
N-(2-Bz-1-methylvinyl)-Trp									Alc-Pe	195–197	−344	Me			589
Z-Trp-NHC_6H_{11}									Alc	169–170					134
N-Z-Trp-OP(SMe)									EtAc-He	111	+9.3	Chf			275
Z-Trp-SPy										152–154	−61	DMF			365
Boc-Trp									EtAc-Pe	135–137	−18	Ac			544
Boc-Trp-NHC_6H_{11}									Cy	73–74					134
Boc-Trp-OPy									Eth	136–137	−24	DMF			134
Boc-Trp-SPy										160–162	−66	DMF			365
Cyano-Boc-Trp										161–163	+13	Alc			706
N,N-Diethyl-Trp-OEt									Alc-Eth	162–165				NMR	205
2-Methyl-Trp															536
6-Nvoc-Trp										194–195	−23				478
2-Thiol-Trp															680
Trp hydroxamic acid									Me	166–168					586
D-Trp-OMe HCl										212 (dec)	−18	Me			342
Trp methylthiohydantoin derivative														CD	622

Tyrosine															
N-Ac-Tyr hydrazide									Aq	235–236	+15	DMF			612
OAc-Tyr-NCA										122 (dec)	−37	CH_3CN			232
N-Bz-Tyr-OMe									Me	151–153	+103	Chf			646
OBz-N-Ac-Tyr-OMe									Me-EtAc	144–146	+83	Chf			646
OBzl-N-Ac-Tyr-OEt										125–128	+3.5	DMF			612
OBzl-N-Ac-Tyr hydrazide										182–184	+15	DMF			612
OBzh-Boc-Tyr									Eth-Pe	108–110					544
OBzl-N-Boc-Tyr-ONBzl									Me	109–110	−13	DMF			124
OBzl-N-Boc-Tyr-ONp										140–141	−0.3	EtAc			249
O-Bzl-N-Boc-Tyr-SPy										103–105	−72	EtAc			365
OBzl-N-Boc-Tyr-OSu									EtAc-Pe	137	−33	DMF			689
OBzl-Tyr									Ac	254 (dec)	−9.7	Ac			513 678
OBzl-Tyr-ONBZl HCl									Me-Eth	202–204 (dec)	−8	Me			124
N-Z-OBz-Tyr-OMe									Me	98–100	+42	Chf			646
N-Z-Tyr-OBzl									Be	121					132
N-Z-Tyr-OMe										94–95	+52	Chf			646
O,N-Bis-Z-Tyr-OPht															306
OBu^t-N-Aoc-Tyr									He	108–109.5	−15	DMF			219
OBu^t-N-Aoc-Tyr Dcha									He	114–116	+35	DMF			219
OBu^t-N-Aoc-Tyr-OSu									He	112–113	−43	DMF			219
OBu^t-N-(2-Bz-1-methylvinyl)-Tyr									MeCl-Eth	156–159	−339	Alc			589
OBu^t-N-Z-Tyr Dcha										159–162	+32	Alc			285
OBu^t-N-Boc-Tyr Dcha										134–136	+24	DMF			285
OBu^t-N-Boc-Tyr-OMe									Pe	98–100	−17	DMF			285
OBu^t-N-Boc-Tyr-OSu									Alc	140–142	−40	DMF			285
OBu^t-Tyr						D-50 DO-50					−26	Aq			285

(*continued*)

Table 1 Amino Acid Derivatives (continued)

Compound	Method	Solvent	% Yield	Chromatography Col	Tl	Ion	Paper	CCD	Recrys. Sol.	M.P. °C	[α]D	Solvent Rotation	Elect.	Other	Ref.
Boc-Bzl-Tyr									Pe	104–106					544
									EtAc-Pe	104–105					
										108–109					
Boc-OBzl Tyr amide															476
Boc-Tyr									EtAc-Pe	96–98	+3.9	Ac			544
Boc-Tyr-OSu									EtAc	192–193	−43	DMF			693
Decanoyl-Tyr									EtAc	114–116	+28	Alc			640
Decanoyl-Tyr-OMe									EtAc-He	65–66	+11	Alc			640
Decanoyl-Tyr-OSu									*i*-Pr	150–152	−27	Alc			640
N,O-Di-Bz-Tyr-OMe									Me	143–145	+101	Chf			646
DNPS-Tyr-OCH_3										50					415
Bpoc-Tyr Cha									An	248–252 (dec)	+39				573
[S]-(−)-α-Methyl-Tyr									Aq	320	−4.4	1 *N* HCl			662
Tyr hydrazide									Me	195–196	+78	Ac			194
Tyr hydrazide N-Z-pipecolinic acid salt															31
Tyr hydrazide N-Z-pipecolinic acid salt										187–188	+71	Aq			31
Tyr hydroxamic acid									Me	178–180					586
Tyr methylthiohydantoin derivative														CD	622
Tyr-OTMB HCl									Alc-EtAc	191.5–193	−16				347

Valine															
Aceto-Ac-Val										124–125	+7.4				42
N-Ac-Val															647
															553
N-Ac-Val-OEt										33–35	−49	Aq		IR	439
N-Ac-Val-α-methylbenzylamide									Alc	188–189					562
2-Benzamido-5-(2-nitroso-formamide)-Val										183–184					18
N-(2-Bz-1-methylvinyl)-Val									MeCl-Eth	161–162	+189	Alc			589
N-Bz-Val anilide										220	−37	Py			534
N-Bz-Val-α-methylbenzylamide									Alc	202–204	−63	Alc			562
N-Bz-Val-D-α-methylbenzylamide									Alc	205–207					562
N-Z-δ-hydroxy-Val					Dowex 50w × 8				Aq-Alc	119–122				IR	253
Z-Val										66–67	+0.7	Alc			456
Z-D-Val										66–67	+0.53	Alc			456
Z-Val amide						108				212 (dec)	+23				710
										210–211					194
Z-DL-Val amide										187–189					194
3-(Z-Val-amino)-1-Tos-2-piperidone									Alc-DiPr-	144–145	−14	Alc-Aq			201
									Eth	144–146	−13	Alc-Aq			
Z-DL-Val-OBhae									EtAc	110–112					371
Z-D-Val (+) ephedrine salt										145	+19	Alc			456
N-Z-Val 4-(methyl-sulfonyl)phenyl ester									EtAc-He	100–101	+9.0	Chf			275
Z-Val-NHC_6H_{11}									EtAc	201–202					134
Z-Val-SPy										Oil	−28	DMF			365
N-(δ-Bromovaleryl)-DL-Val									Be	121–123					302
N-(δ-Bromovaleryl)-DL-Val-OEt															302
N-Boc-β-ethyl-Val									Pe	91–93	+11	Alc			283

(*continued*)

Table 1 Amino Acid Derivatives (continued)

Compound	Method	Solvent	% Yield	Chromatography Col	Tl	Ion	Paper	CCD	Recrys. Sol.	M.P. °C	$[\alpha]_D$	Solvent Rotation	Elect.	Other	Ref.
Boc-methyl-DL-Val									Eth-Pe	86					544
Boc-Val									Eth-Pe	72–73	−5.8	Ac			544
Boc-Val amide															476
Boc-Val-D-Hyv									EtAc-Pe	142–143	+2.0	Bz			368
Boc-Val-D-Hyv-OBzl										Oil	−8.5	Di			368
Boc-D-Val-Lac									Eth-Pe	95–96	−0.5	Bz			368
Boc-D-Val-Lac-OBzl											−9.5	Bz			368
Boc-Val-OPy										Oil	−47	DMF			134
Boc-Val-SPy										120–122	−85	EtAc			365
2-Deutero-DL-Val									Aq-Alc					NMR	343
DNT-Val										170.5–171					227
Bpoc-Val Cha									*i*-Pr	178–180 (dec)	−8				573
For-2-deutero-Val										108–110	+10	Alc		NMR	343
N-For-2-deutero-D-Val									Aq	143–146				NMR	343
N-For-2-deutero-D-Val brucine salt									Aq	151–153	−14	Alc		NMR	343
											−8.4	Me			
Gallium(III) chloride Val complex															23
Gallum(III) chloride Val complex dihydrate															23
Gallium(III) chloride Val complex hydrate HCl															23
Gallium(III) dichloride valinate															23

β-Hydroxy-Val															46
Indium(III) chloride Val complex															23
Indium(III) chloride Val complex dihydrate															23
Z(OMe)-Val-OSu									EtAc-Pe	74–75	−26	Alc			325
Z(OMe)-Val-OTcp									Eth-Pe	83–84	−14	Alc			325
Z(OMe)-Val-ONp									Alc-Pe	76–78					714
N-Methyl-2-deutero-DL-Val-OMe HCl									An	133–135				NMR	343
DAZ-Val-OMe															468
MAZ-Val-OMe															468
4-(Nitroguanidino)-Val									Aq	200–203					279
Nvoc-Val										148	−7.4				478
Tos-N-trideuteromethyl-Val										77–81	−25	Me		NMR	343
N-Trideuteromethyl-Val-OMe HCl									An-Eth	136–139	+29	Me		NMR	343
Val						DO-50W					−11	HCl			562
											−24	Ac			
Val amide HBr					54				Eth	242 (dec)	+22	Aq			710
Val (−) ephedrine salt										145	−20	Alc			456
Val-D-Hyv											+25	Me			368
(*R*)-Val-O-*i*-Bu tos										146–147	−9.1	Alc			210
Val methylthiohydantoin derivative														CD	622
Val-OTMB HCl									Alc-Eth	173–174	+14				347
DL-Val-OTMB HCl									Me-Eth	148–150					347

Chapter 3

Dipeptides

Some of the more innovative and creative aspects of peptide chemistry have centered around the peptide bond-forming reaction. Because of the asymmetric center adjacent to the carboxylic acid carbonyl group, the superficially simple operation of amide formation has required study, unparalleled in detail, to avoid racemization. While racemization can usually be limited to less than 1% there is still no universally infallible reaction technique and the advantages of having several good methods at one's disposal is well known to every peptide organic chemist. Presently, in our laboratory, we prefer to first use the MCA method employing isobutyl chloroformate with one equivalent of a relatively weak base such as N-methylmorpholine in tetrahydrofuran. If this proves unsatisfactory we then resort to a method from one of the series A, DCCI-HOBT, EDCI, WRK, or one of the active ester methods such as OSU, OPCP, or ONP. Soon, we may find it appropriate to add the BBDQ, NCA, SPY, and DPPA methods to this list for reasons outlined below.

The best known peptide bond-forming reagent, albeit not now necessarily the one of choice, is DCCI and a great many dipeptides have been prepared by this means over approximately the last 10 years. Reaction periods in these examples have varied from several hours to several days and monitoring by chromatographic techniques was generally not employed. For these reasons it seemed appropriate to first outline an example of this older peptide bond-forming method which was used in the context of modern thin-layer and column chromatographic procedures. Such an illustration appears in a synthesis of the β-Lys-β-Lys unit of racemomycin B as outlined in Chart 3.[610] The racemomycins A → D vary from one to four β-Lys units. All are active against both gram-negative and gram-positive bacteria and their antimicrobial activity, antiviral activity, and toxicity increase with increasing β-Lys content. While the β-Lys-β-Lys synthesized as shown in Chart 3 was inactive against *Escherichia coli* and *Salmonella aureus*, the DCCI reaction (12 hr required) and chromatographic procedures reflect some useful contemporary techniques. Also it should be noted that recent kinetic work has indicated (see Chapter 14) that DCCI promoted coupling with most common amino acids will be essentially complete in 1 to 2 min.[103]

In this general area of experimental methods should be mentioned the recent application of 5% aqueous potassium hydrogen sulfate in place of aqueous citric acid for washing amino acids or peptides protected with acid-labile groups.[591] Use of trichloroethyloxycarbonylhydrazine has been suggested for applications where removal of the protecting group by treatment with zinc powder and acetic acid at room temperature (1 hr) would be advantageous.[713] Addition of carbon dioxide in aqueous solution to amino acids undergoing coupling by the ONP or OSU method has been found to about half the reaction time.[257]

Improved methods for synthesis of amino acid N-carboxyanhydrides (NCA) and their application in peptide syntheses have appeared recently. The Leuchs' anhydrides of Asp and Glu were among those receiving additional study and the advantages of using both bearing unprotected β- and γ-carboxylic acid groups was emphasized.[232] Here it should be noted that the saponification of peptides containing Asp β-ester groups can lead to the formation of β- as well as the normal α-Asp peptides. The method outlined in Chart 3 for synthesis of Asp-NCA and subsequent conversion to Asp-Phe-NH_2 is illustrative. The more specialized synthetic methods for obtaining the NCA derivatives of Arg and Lys are also presented in Chart 3. The NCA of Gly has proved troublesome in aqueous reaction mixtures and can produce more than 20% hydantoic acid as a side product. This shortcoming has been overcome by using a heterogeneous solvent system composed of acetonitrile-aqueous sodium hydroxide as summarized in Chart 3.[296]

The NCA coupling method has been made more workable in recent years by Denkewalter and his colleagues at Merck by conducting the reaction in an aqueous sodium borate buffer in the pH range 10.2–10.5.[232] The pH of such buffer solutions is adjusted by adding 0.2 *N* sodium hydroxide to 50 ml of a mixture composed of 0.2 *M* sodium borate and 0.2 *M* potassium chloride. In the case of water-insoluble products, isolation is easily achieved by filtering the solution. If the product is soluble in water, but less soluble in, say alcohol, the product can be crystallized from the mixture by adding the particular solvent. With products soluble in organic solvents, extraction of the reaction mixture can be employed. Where these approaches are unworkable, the reaction mixture must be separated by chromatography. To circumvent the latter inconvenience, a modified technique has recently been proposed employing triethylamine as base (pH 10.3). The new reaction conditions were successfully applied to preparation of Met-Asp (53%), Met-Glu (65%), and Ala-Gly (80%) in the yields specified.[164] Apparently, such condensations proceed by nucleophilic attack of an amino group on the unprotonated form of the Leuchs' anhydride, and as long as the pH of the reaction mixture is kept above that of the corresponding amino group pKa, the reaction should proceed.

The analogous N-thiocarboxyanhydrides have now been investigated.[127] It was hoped that these thiocarbamates would be more stable to side reactions below pH 11 (decarboxylation) and those above pH 11 (NCA hydrolysis and hydantoic acid formation). While the His-NTA and those of Gly and Ala did prove more useful than the corresponding NCA in each case, the usefulness of the NTA derivatives was found limited by epimer formation of up to 20%.[127]

From the observation that isobutyl chloroformate can lead to higher yields as compared to ethyl chloroformate in the MCA coupling procedure, 2-isobutoxy-1-isobutoxycarbonyl-1,2-dihydroquinoline (BBDQ) was prepared and found a useful alternative to the ethoxy derivative of the EEDQ procedure.[323] In the Bodanszky racemization test[59] (quantitative amino acid analysis of D-Ile from Ac-Ile plus Gly-OMe). The BBDQ procedure gave 4% racemate while DCCI led to 27% racemate.

Among peptide bond-forming reactions of little present practicality, but of current theoretical interest and possible future utility, is the trimetaphosphate method.[100] An abbreviation of the newly proposed mechanism for this pH-dependent reaction has been included in Chart 3. The negligible yield of Gly-Gly at pH 8 was raised to 28% at pH 12. Sodium ion may play a catalytic role. The prebiotic origin of peptides by such a route is mentioned in Chung *et al.*[100] and the analogous implication of trimetaphosphate in nucleotide bond production has already been discussed in Volume 1 of "Synthetic Nucleotides."[486]

One of the newer and more promising additions to the expanding list of new peptide coupling methods is the 2-pyridyl thiolester[365] method (SPY). The thiolester starting materials were found stable in the absence of moisture but to decompose slowly in moist air and in aqueous solution. On the other hand, they were stable in acetic acid and in trifluoroacetic acid solutions. Reactions of the thiolesters with amines were found quite rapid and exothermic even with sterically hindered amino acids. Of particular interest was the observation that 2-pyridyl thiolesters react with alcohols, but of course more slowly than with amines. For example, Boc-Gly-SPy condensed with ethyl glycolate in ethyl acetate at room temperature (1 day) to afford the corresponding ester in about 80% yield. After condensation, the excess active ester may be conveniently destroyed by addition of water. In ethanol solution, 2-pyridinethiol was observed nearly 20 years ago by Djerassi and myself[129] to exist in the thione form drawn in Chart 3 but to be easily removed by routine peptide washing procedures to the point of not affecting subsequent hydrogenation reactions.[365] In the Val-Val-OMe reaction sequence of Chart 3, no racemate was discovered using a procedure which is capable of detecting down to 0.15%. Thus, this aminolysis reaction with neighboring group participation may prove to bear little risk of racemization.

More details of the new oxidation-reduction peptide bond-forming sequence employing triphenylphosphine-dipyridyl-2,2′-disulfide (TPP-DPDS) have been reported[429] and an example has been entered in Chart 3. Because the amide-forming reaction is believed to proceed through the pentavalent acyloxyphosphonium salt the possibilities for racemization are greatly reduced. Application of this method in a new synthesis of oxytocin gave a specimen of the hormone with avian depressor activity amounting to 104 u/mg. The reaction was carefully studied in respect to its applicability to solid phase peptide synthesis and found workable. A study has also been made with NPS-protected peptides as the amine component, and led to the discovery that such sulfenamides would react with an N-blocked amino acid copper(II) salt [or free carboxylic acid in the presence of copper(II) chloride] and triethylamine with triphenylphosphine to afford the corresponding peptide thereby eliminating prior removal of the *o*-nitrophenylsulfenyl (NPS) group.[630]

Another new peptide bond-forming method based on oxidation of trivalent phosphorous to the pentavalent state arose from the fairly recent observation that carboxylic acid chlorides can be prepared using carbon tetrachloride and triphenylphosphine. A summary of the new method is presented in Chart 3. The piperazine phosphine also serves as base, and the Young racemization test was quite encouraging.[717] The newest peptide bond-forming reaction based on phosphorus chemistry employs diphenylphosphorylazide. Here the azido group is apparently transferred to the amino acid carboxyl.[571] Results of the Young racemization test have been entered in Chart 3. Due to the well-known propensity of Bz-Leu to give extensive oxazolinone ring formation the results appear satisfactory and the new reaction seems to offer a useful experimental alternative to the usual A technique.

In quite different new approaches to peptide bond synthesis treatment of Gly-OEt with copper(II) chloride in ethanol was found to yield di-, tri-, and tetra-Gly-OEt derivatives,[718] and a four-component condensation approach with computer planning has been summarized.[631]

References received too late to tabulate in the table of this chapter concerned the following miscellaneous subjects. Oxidation of Thr dipeptides with DCCI-DMSO (Moffat oxidation) was shown to yield racemic ketone derivatives and cleavage of the peptide bond.[43] A Dowex 50 column chromatographic procedure has been developed for separating, for example, α-Asp-His and β-Asp-His.[231] An enzymatic synthesis of D-Ala-D-Ala has been reported using alanine racemase and D-Ala-D-Ala ligase.[84] Some peptides containing aminophosphonic acid groups have been prepared,[723] and the relationship of bitterness to peptide composition has been examined.[396] In the latter work the bitterness of some peptides was attributed to the presence of hydrophobic amino acid units and to blocking amino and/or carboxyl groups.

Chart 3

610

Racemomycin B

HN, NH, N, O, OH (streptolidine moiety); CH_2OCONH_2, HO, O, HO

NH—$\overset{O}{\overset{\|}{C}}CH_2\underset{NH_2}{CH}CH_2CH_2CH_2NH\overset{O}{\overset{\|}{C}}CH_2\underset{NH_2}{CH}CH_2CH_2CH_2NH_2$

β-Lys (5 g) —[Cu^{2+}, Aq]→ $NH_2CH_2CH_2CH_2\underset{NH_2}{CH}CH_2C(=O)O$ (Cu complex) —[(1) $C_6H_5CH_2OCOCl$, 2*N* NaOH, Aq, ~0°, 70 min; (2) Ac, H_2S; (3) Cellulose chromat.; (4) Recrys. An]→

N^{ϵ}-Z-β-Lys (1.4 g) —[Me, $SOCl_2$, 12 hr]→ N^{ϵ}-Z-β-Lys-OMe (~100%) —[(1) N^{ϵ}, N^{β}-Z-β-Lys, Py, Et_3N, DCCI, di, ~0°, 12 hr (to neg. ninhydrin); (2) Recrys. Me-Eth]→

N^{ϵ}, N^{β}-Z-β-Lys-N^{ϵ}-Z-Lys-OMe (0.7 g ⟶ 0.4 g) —[(1) 1*N* NaOH, Me, 1 day, room temp.; (2) Ac, Aq, Pd, H_2, 36 hr, room temp.; (3) Sephadex LH-20 (Aq)]→ β-Lys-β-Lys —[Aq]→

↓

HCl salt

d.p. 218°-225°

$\alpha]_D^{26}$ + 6° (Aq)

HO—C_6H_4—N=N—C_6H_4—$SO_3^{\ominus}$ salt

d.p. 230°

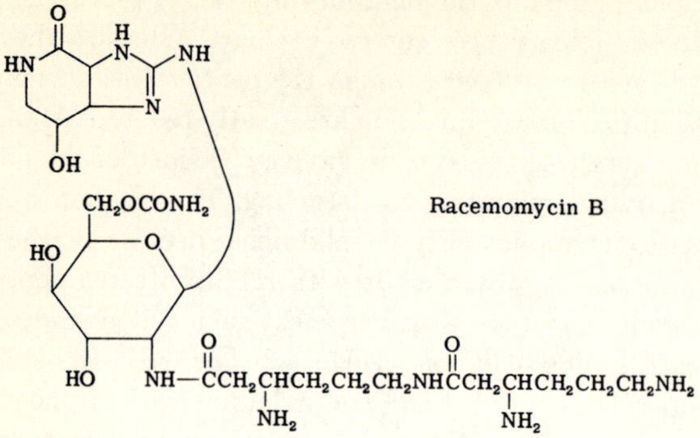

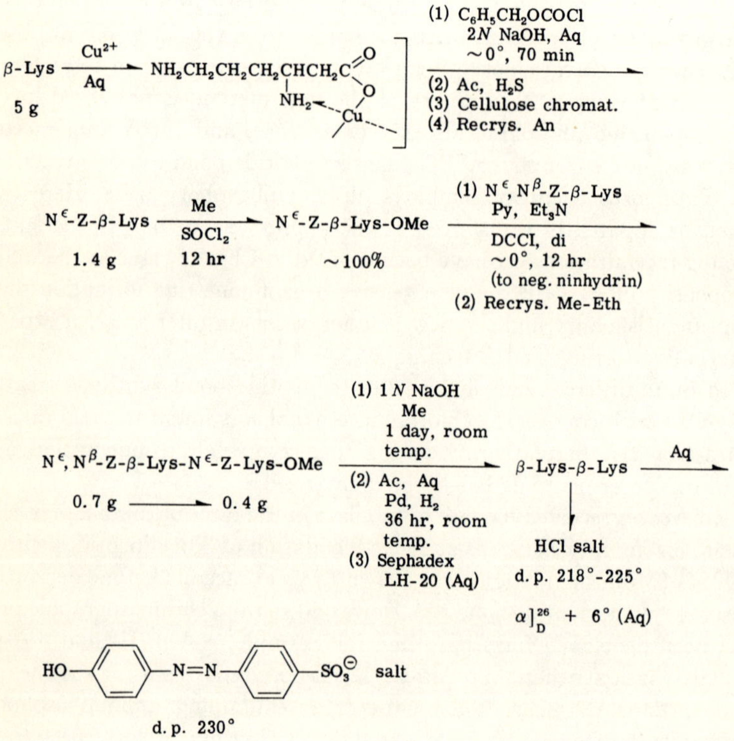

Chart 3 (cont.)

232

Asp —[(1) ClCOCl, THF, ~1 hr, room temp. (2) Ppt. from EtAc with He]→ N-carboxyanhydride of Asp (HOC(=O)CH$_2$C(H)–C(=O)–O–C(=O)–NH) 21-25% d.p. Δ —[Phe-NH$_2$, Na$_3$BO$_3$ buffer, pH 10.2→20.5, rapid mixing, ~0°]→ Asp-Phe-NH$_2$

N$^{\alpha}$-Z-Arg —[PBr$_3$, THF]→ $NH_2\overset{\oplus}{C}(=NH_2Br^{\ominus})NHCH_2CH_2CH_2C(H)$–C(=O)–O–C(=O)–HN (N-carboxyanhydride) ~50%

N$^{\epsilon}$-Boc-Lys —[(1) AgCN (2) ClCOCl]→ OCNHCH$_2$(CH$_2$)$_3$C(H)–C(=O)–O–C(=O)–HN (N-carboxyanhydride)

296

Gly —[(1) ClCOCl, THF, 45°, 2 hr (2) Ppt He (3) Recrys. EtAc]→ Gly-NCA 73% m.p. 100° —[(1) Trp, 1*N* NaOH-Aq, Acn, -10°, 3 hr (2) Extract Acn (3) Recrys. Me]→ Gly-Trp 89% + Gly-Gly-Tyr small amt.

323

Quinoline —[(1) CH$_3$)$_2$CHCH$_2$OCOCl (2) CH$_3$)$_2$CHCH$_2$OH, Et$_3$N]→

Z-Ala + 1-(CO$_2$CH$_2$CH(CH$_3$)$_2$)-2-(OCH$_2$CH(CH$_3$)$_2$)-1,2-dihydroquinoline —[DMF, 1 day, room temp.]→

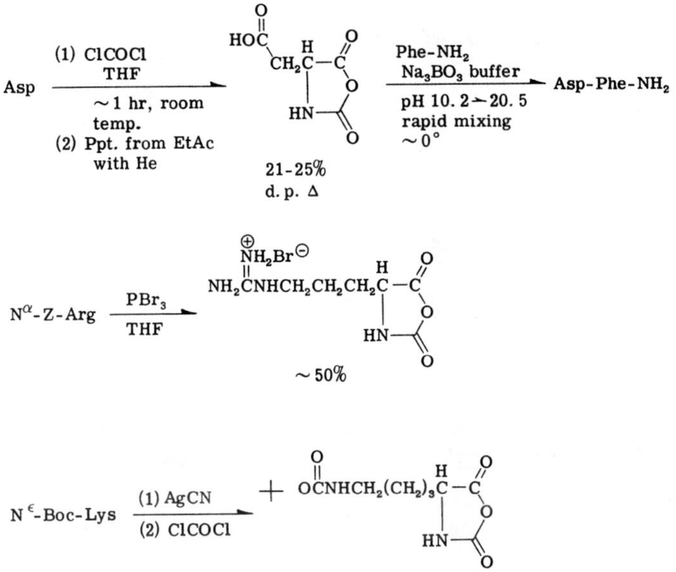

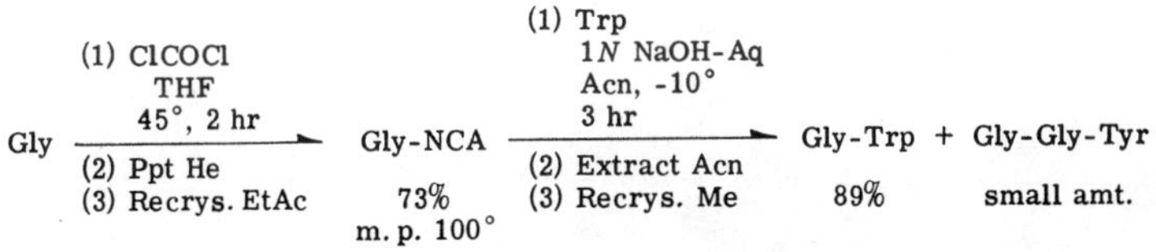

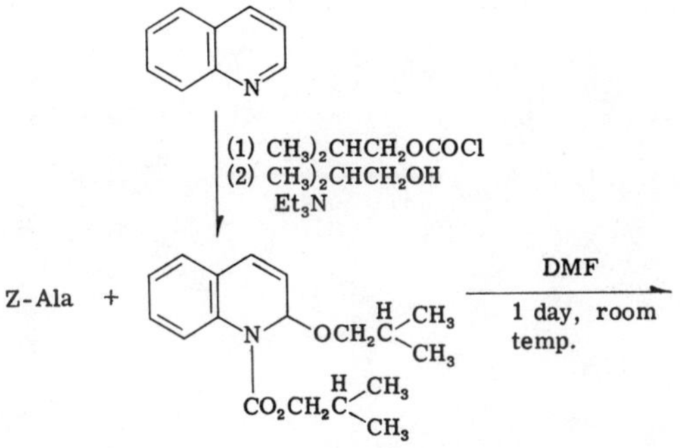

Chart 3 (cont.)

$$\left[CH_3CH(NHZ)C(=O)-O-C(=O)-OCH_2CH(CH_3)_2 \right] \xrightarrow{\text{Tyr-OMe}} \text{Z-Ala-Tyr-OMe}$$

88%

100

$$\text{Gly} + 3\,Na^+ \text{ (cyclic trimetaphosphate)} \xrightarrow[\text{10 days}]{\text{pH 12, NaOH}} \left[\text{cyclic } C(=O)-O-P(=O)(O^{\ominus})-NH-CH_2 \right] \xrightarrow{\text{Gly}}$$

$$\left[{}^{\ominus}O-P(=O)(O^{\ominus})-NHCH_2C(=O)-NHCH_2C(=O)-O^{\ominus} \right] \xrightarrow{H_2O} \text{Gly-Gly} + HPO_4^{2+}$$

28%

365

$$\text{Val} + \text{2-pyridinethione} \underset{H_2O}{\overset{\text{DCCI, EtAc}}{\rightleftharpoons}} (CH_3)_2CHCH(NH_2)C(=O)-S-\text{(2-pyridyl)} \xrightarrow[\text{di, 1 hr}]{\text{Val-OMe}}$$

73%

$$\text{Val-Val-OMe} + < 0.15\%\ \text{DL} + \text{2-pyridinethione (N-H)}$$

86%

429

$$\text{S-Bzl-N-Z-Cys} + (C_6H_5)_3P + \text{(2-pyridyl)}-S-S-\text{(2-pyridyl)} \xrightarrow{\text{MeCl}}$$

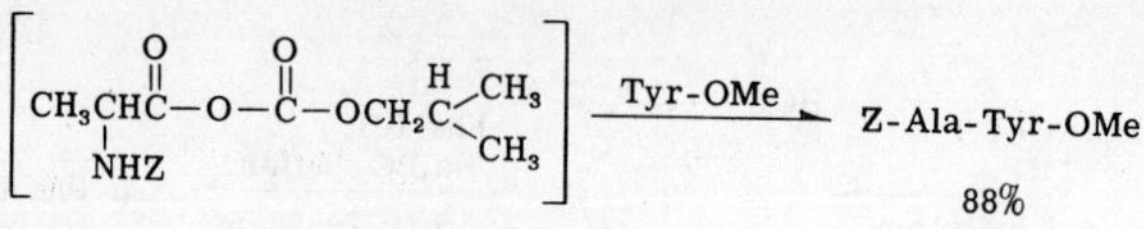

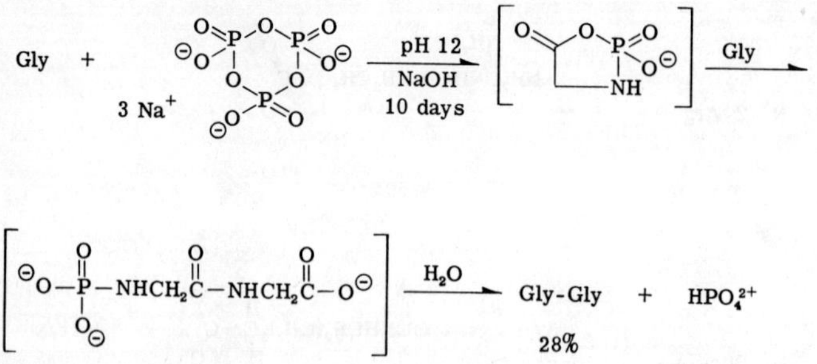

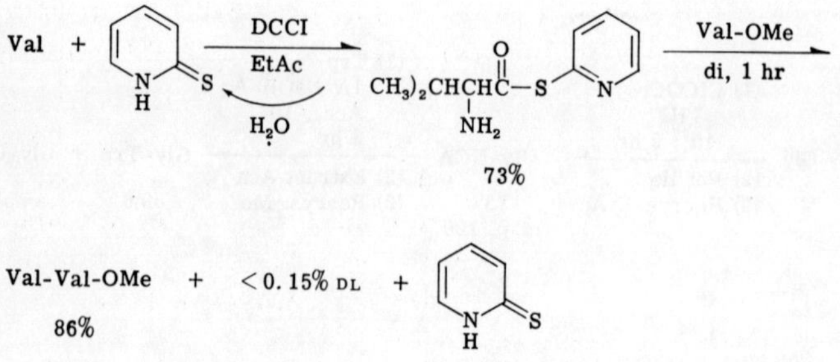

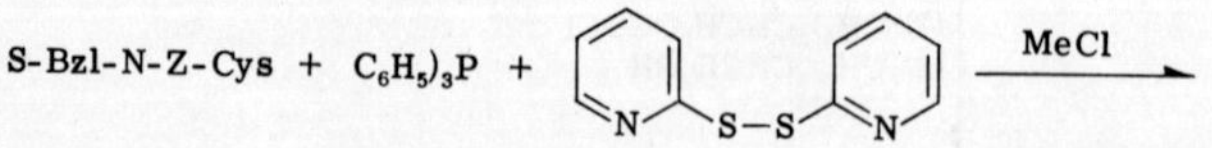

Chart 3 (cont.)

$$\left[Bzl-S-CH_2\underset{NHZ}{CH}-\overset{O}{\overset{\|}{C}}-O-P\begin{matrix}\diagup S-\text{2-pyridyl} \\ \diagdown (C_6H_5)_3\end{matrix} \right] \xrightarrow[\text{1 day, room temp.}]{\text{Gly-OEt}}$$

$$\underset{91\%}{\text{S-Bzl-N-Z-Cys-Gly-OEt}} + (C_6H_5)_3P{=}O + \text{pyridine-2(1H)-thione}$$

717

$$\text{Bz-Leu} + \left(CH_3N\langle\rangle N\right)_{\!3}P + CCl_4 \xrightarrow[\text{1 day, room temp.}]{\text{Gly-OEt, THF, -15°}}$$

$$\left[\left(CH_3N\langle\rangle N\right)_{\!3}\overset{\oplus}{P}-CCl_3Cl^{\ominus} \longrightarrow (CH_3)_2CHCH_2\underset{NHBz}{CH}\overset{O}{\overset{\|}{C}}-O-\overset{\oplus}{P}\left(N\langle\rangle N-CH_3\right)_3 \; Cl^{\ominus} \right]$$

$$\longrightarrow \underset{\substack{\sim 60\text{-}85\% \\ (97\%\ \text{L})}}{\text{Bz-Leu-Gly-OEt}} + \left[\left(CH_3N\langle\rangle N\right)_{\!3}P{=}O\right] HCl + CHCl_3$$

571

$$(C_6H_5O)_2\overset{O}{\overset{\|}{P}}-Cl \xrightarrow[\text{An}]{NaN_3} \underset{>90\%}{(C_6H_5O)_2\overset{O}{\overset{\|}{P}}-N_3} \xrightarrow[\text{DMF, }\sim 0°\text{; 1 day, room temp.}]{\text{Bz-Leu, Gly-OEt, Et}_3\text{N}}$$

$$\underset{\substack{83\% \\ (94.5\%\ \text{L})}}{\text{Bz-Leu-Gly-OEt}} + (C_6H_5O)_2\overset{O}{\overset{\|}{P}}-OH + HN_3$$

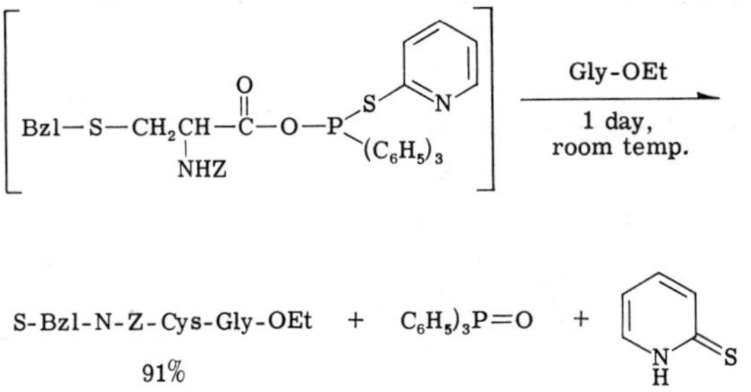

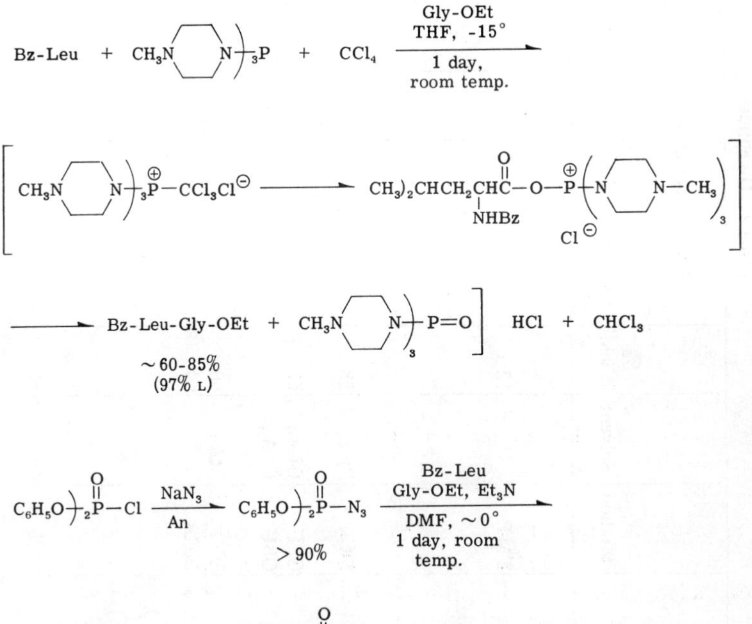

Table 2 Dipeptides

Compound	Method	Solvent	% Yield	Chromatography				CCD	Recrys. Sol.	M.P. °C	$[\alpha]_D$	Solvent Rotation	Elect.	Other	Ref.
				Col	Tl	Ion	Paper								
Alanine															
Ala-Ala	TMP		12												504
Ala-Ala-OEt HCl										Oil					186
Ala-D-Ala methylthiohydantoin derivative														CD	622
Ala-Ala-morpholine amide HCl										230–232 (dec)					186
Bz-DL-Ala-DL-Ala-OEt	$SiCl_4$	Py	58												89
Z-Ala-Ala									Chf	149–151	−50	TFA			186
	OSU	Di-Aq	67							150	−35	Me			703
Z-Ala-D-Ala	OSU	Di-Aq	80							117	−7.0	Me			703
Z-D-Ala-D-Ala Boc-hydrazide															259
N^{α}-Z-Ala-Ala Boc-hydrazide	ONP	EtAc	90						EtAc	172–173	−16	DMF			197
Z-Ala-Ala-OEt	MCA		89						EtAc-He	116–116.5	−51	TFA			186
Z-β-Ala-β-Ala-OMe									Be	110–111				IR	718
Z-Ala-Ala-OSu										152	−74	Me			703
Z-Ala-D-Ala-OSu										154	+24	Me			703
Z-Ala-Ala-morpholine amide									EtAc	135–136	−35	TFA			186
Pht-DL-Ala-DL-Ala-OEt	$SiCl_4$	Py	51							193–195					89
N-Thio-Bz-DL-Ala-Aib									Eth	161–162					32
Ala-D-Arg HCl										92 (dec)					703

Z-Ala-D-Arg	OSU	Di-Aq	74							130–134					703
										dec					702
N-Z-Ala-S-chlorocarbonyl-Cys-OMe									Chf-Pe	94.5–95.5					250
N-Z-Ala-Cys-OMe									EtAc	114–116	−27	Me			250
										116.5–118					
Z-β-Ala-Val-OMe	SPy		86						MeCl-Pe	62–65	+5	DMF			365
N-Z-Ala-S-Phoc-Cys-OMe									Be	158.5–159.5	−30	DMF			250
Ala-Gly									Alc-Aq		+22	0.5 *N* HCl			164
										233					694
Ala-Gly, ammonium															444
Ala-Gly-OPcp HCl										190 (dec)	+2.0			Mol. wt.	71
Ala-Gly-Tfa										136–139	+9.2	Me			694
Z-Ala-N-2-carboxamidoethyl-Gly									EtAc-Eth	89.5–90.5	−8.4	DMF			597
Z-Ala-N-2-carboxamidoethyl-Gly-OTMB	DCCI	Acn	73						Alc-Cy	119.5–120	−10	DMF			597
Z-Ala-Gly	SPh		55							132–133					444
Z-Ala-Gly-OBhae									Ac-Eth	142–144					371
Z-Ala-Gly-OEt										97–98	−24	Alc			717
	OBHAE	THF	80							93–96					371
Z-DL-Ala-Gly-OEt	OBHAE	THF	71							80–81					371
	$SiCl_4$	Py	15												89
Z-D-Ala-Gly-ONp	MCA	DMF	66						Alc						259
Z-DL-Ala-Gly-T										167–169					195
Z-DL-3-(2-Pyridyl)-Ala-Gly										173–174					339
Z-DL-3-(3-Pyridyl)-Ala-Gly										183–185					339
Z-DL-3-(2-Pyridyl)-Ala-Gly-OEt										125–126					339

(*continued*)

Table 2 Dipeptides (continued)

Compound	Method	Solvent	% Yield	Chromatography Col	Tl	Ion	Paper	CCD	Recrys. Sol.	M.P. °C	$[\alpha]_D$	Solvent Rotation	Elect.	Other	Ref.
Z-DL-3-(3-Pyridyl)-Ala-Gly-OEt										103–104					339
Z-DL-3-(4-Pyridyl)-Ala-Gly										186–188					339
									Aq-Alc	Oil					339
Z-DL-3-(4-Pyridyl)-Ala-Gly-OEt									Alc-Eth						339
Z-β-(8-Theobrominyl)-α-Ala-Gly										125–129					637
Z-β-(8-Theobrominyl)-α-Ala-Gly-OBzl									Alc	155–156					637
Z-β-(8-Theobrominyl)-α-Ala-Gly-OBu^t									Be	166–168					637
Z-β-(8-Theobrominyl)-α-Ala-Gly-OEt										177–179					637
Boc-Ala-Gly monohydrate	OSU	Glyme-Aq	75							78–80	−23	Me			694
Z(OMe)-Ala-Gly Dcha	ONP	Di-Aq	70						Me-An	143–144					712
Nps-Ala-Gly									EtAc-Cy	142–143	−43	DMF			71
Nps-Ala-Gly-OPcp										182–184 (dec)	−33	DMF		Mol. wt.	71
Pht-DL-Ala-Gly-T										224–226					195
Nps-D-Ala-Gly-OBu^t	OSU	Chf	85						Eth-He						259
β-(8-Theobrominyl)-α-Ala-Gly										347–350					637
										349–352					637
N-Thio-Bz-DL-Ala-Gly									Eth	171					32
N-Z-Ala-His-OMe	OSU	Di	83						Me-EtAc-Me	164–165	−19	Me-HCl			94

N-Boc-Ala-His-OMe	DCCI	Di-MeCl	36						EtAc	153.5–155	−20	Me			94
	OSU	Di	50						EtAc	154.5–156	−19	Me			
Decanoyl-Ala-His-OMe															321
Z-D-Ala-Leu											−14	DMF			640
N-Thio-Bz-DL-Ala-Leu									Eth	182–184					32
N-Thio-Bz-DL-Ala-D-Leu									Eth	176–178					32
Ala-Met															499
N-Ac-β-(1,3-dioxy-4,4,5,5-tetra-methyldihydroimidazol-2-yl)-DL-Ala-Phe										Oil					664
Ala-Phe	SP														616
														NMR	127
Ala-Phe-OMe HCl									Alc	104–106					679
Z-Ala-Phe									EtAc	124–125	+13	Alc			372
Z-Ala-Phe-OMe										99	−14	Me			717
Boc-Ala-Phe															679
Boc-Ala-Phe-OMe	TPP-CTe	Chf		Seph-adex LH-20						84–86					679
Boc-Ala-Phe amide	OPy		96						Eth	165–166					134
N-Thio-Bz-DL-Ala-Phe										50–55					32
D-Ala-Pro-OEt HCl										Oil					349
Z-Ala-Pro										120–122					676
Z-Ala-Pro amide									Eth	168–169	−94	Me			574
Z-D-Ala-Pro-OEt	MCA	THF-Chf								Oil					349
Boc-Ala-Pro					66					154–156					676
			74						EtAc-He	155–156					367
									Glyme-He	156–158					
Boc-Aa-Pro-OBzl	MCA	Chf	96							Oil					367

(*continued*)

Table 2 Dipeptides (continued)

Compound	Method	Solvent	% Yield	Chromatography Col	Tl	Ion	Paper	CCD	Recrys. Sol.	M.P. °C	$[\alpha]_D$	Solvent Rotation	Elect.	Other	Ref.
Boc-Ala-Pro-OMe	MCA	THF	80							Oil					676
					×					Oil					367
CHla-Ala-Pro															151
Z-β-Cyano-DL-Ala-Sar															360
Z-Ala-Tyr-OMe									EtAc	113–120	−9.0	Me			712
[S]-(−)-α-Methyl-β-[4-hydroxyphenyl]-Ala-α-methyl-Tyr										122–123	+94	Chf			662
Z-Ala-Val-OMe										84	−38	Alc			717
β-Alanine															
β-Ala-Gly															595
Z-β-Ala-Gly-OPn									EtAc	126.5–127.5					595
Z-β-Ala-Gly-T										179–181					195
Z-β-Ala-Gly-OTcp									Alc	134–135					223
Z-β-Ala-Phe-OPhBzl	SPy	Chf	93						Chf-Pe	138.5–140.5	−22	Chf			110
Z-β-Ala-Tyr hydrazide										225–228	+3.9	Ac			249
Z-β-Ala-Tyr-OMe	DCCI	MeCl	83						EtAc-Eth	114–116	+6.2	Me			249
γ-Aminobutyric acid															
γ-Abu-His										242–244 (dec)	+4.0	Aq			156
ε-Aminocaproic acid															
ε-Acp-His										236–237 (dec)	+25	Aq			156

Pht-ε-Acp-His										225–226	+17	Aq			156
α-Aminoisobutyric acid															
N-Bz-Aib-DL-Ala-OEt															523
Pht-Aib-Aib-OMe	SPy		83						MeCl-Pe	153–154					365
Pht-Aib-Ile-OMe	SPy		79						MeCl-Pe	86–87	−4.0	Alc			365
Z-Aib-Tyr hydrazide			92						Aq-Alc	164–165	−31	Me			249
Pht-Aib-Val-OMe	SPy		80						MeCl-Pe	95	+15	Chf			365
Arginine															
Tris-Z-Arg-Gly	MCA								Alc	177					667
N^G-Nitro-$N^α$-Z-Arg-Gly-OC_7H_7															415
N^G-Nitro-$N^α$-Nps-Arg-His-OMe	OPCP	DMF	79		×					110–115					196
Arg-Leu acetate	NCA	Aq	60		68				Aq-An		+9.8	Aq			232
Arg-Leu Tfa salt									Me-Ac	203–204	+9.0	Aq			640
Tris-Z-Arg-Leu	OSU	Di	77						Me Aq	126–128	−5.8				711
N^G Nitro-Arg-Leu-OBz HCl			95												640
N^G-Nitro-Arg-Leu-OMe HBr			95												640
N^G-Nitro-Arg-Leu-OMe HCl			95												640
N^G-Nitro-$N^α$-Z-Arg-Leu									EtAc	160–161	−5.0	Py			640
N^G-Nitro-$N^α$-Z-Arg-Leu-OMe			58						Me-Aq	157–159	−22	Me			640
N^G-Nitro-$N^α$-Boc-Arg-Leu											−9.0	EtAc			640
N^G-Nitro-$N^α$-Boc-Arg-Leu-OBz			50								−9.0	EtAc			640
N^G-Nitro-$N^α$-Boc-Arg-Leu-OMe									EtAc-Eth	129–130	−19	EtAc			640
Arg-Phe									Alc	191					667
				Si											127
N^G-Nitro-$N^α$-Z-Arg-Pro										105–107	−28	DMF			26
N^G-Nitro-$N^α$-Z-Arg-Pro-OMe									*i*-Pr-Pe	155–156	−22	DMF			26
Asparagine															
N-Boc-$N^α$-(Dmb)-iso-Asn-Ala									Alc-Eth	194–195					491
N-Boc-$N^α$-(Dmb)-iso-Asn-Ala-OBzl	DCCI	MeCl	70						EtAc	140–141	−15	DMF			491
$N^β$-(Dmb)-$N^α$-Z-Asn-Ala-OMe	ONP	Py	74						Me-EtAc	181–182					491
Asn-S-Bz-Cys-OMe HBr															558

(*continued*)

Table 2 Dipeptides (continued)

Compound	Method	Solvent	% Yield	Chromatography Col	Tl	Ion	Paper	CCD	Recrys Sol.	M.P. °C	$[\alpha]_D$	Solvent Rotation	Elect.	Other	Ref.
Asn-S-Bzl-Cys amide HCl										176–178			×		308
Nα-Z-Asn-S-Bz-Cys-OMe	MA	Py-Chf	67						Ac-Aq	175–177	−7.5	Ac			558
Nα-Z-Asn-S-Bzl-Cys-OMe	ONP	DMF	80							199–201	−31	Py			62
Nα-Boc-Asn-S-Bzl-Cys-OBzl	ONP	THF-DMF	66						An	145–147	−38	DMF			142
SNp-Asn-S-Bzl-Cys amide									EtAc-He	206–207					308
Asn-Gly									Aq-Alc		+57	Aq			595
Asn-Gly HCl															559
Z-Asn-Gly-OEt										188–189	−4.4	DMF			717
Nα-Z-Asn-Gly-OEt															415
Z-Asn-Gly-OPn									Alc	155–155.5	−4.9				595
Nα-Nps-Asn-Gly										130–135					559
Nα-Nps-Asn-Gly-Cha										174–175	−63	Me			559
Nα-Nps-Asn-Gly-OEt	MA	Chf	54							201–203	−117	DMF			559
α-Bzl-Asn-γ-Bzl-Glu formate							×			109–110	−11	DMF			689
α-Bzl-Boc-Asn-γ-Bzl-Glu	MA	Chf-Py	67		4 3		×			98–99	−15	DMF			689
Nβ-Bzl-Nα-Boc-Asn-Nε-Z-Lys-OBzl	DCCI-NHS	Di-EtAc	82						EtAc-Pe	96–97	−16	Me			333
Nα-Z-Asn-Pro-OBzl(OMe)										98–99	−48				596
Nα-Z-Asn-Pro-OTMB										107–108	−36				596
Nα-Boc-Asn-Pro									EtAc-Pe	108–116	−50	Me			693

Boc-Asn-Pro-OBzl	MA	Chf-Py	85						EtAc-Pe	115–116	−54	Me			693
Boc-Asn-Pro-OSu	DCCI-NHS	Di	84						EtAc-Pe	121	−54	EtAc			693
Aspartic acid															
β-OBzl-Z(OMe)-Asp-Gly															712
β-OBzl-Z(OMe)-Asp-Gly Dcha	MCA	THF-Aq	45						Me-Eth	137–141					712
β-OBl-Nps-Asp-α-OBzl-Glu-ONp	MCA								An	146–148	−40	Di			348
Z-Asp-S-Bzl-Cys-ONBzl	ONP	DMF							EtAc	177–178.5	−30	DMF			243
β-OBu^t-Z-Asp-Gly-OPh															314
β-OMe-Z-Asp-Cys(CH_2COOMe)-OMe										Oil					188
c-Asp-His															733
β-Asp-Lys															24
β-OBzl-NPs-Asp-N^ε-Z-Lys-OBzl	DCCI		84							127–129	−17	DMF			727
Asp-Met-OMe															499
Asp-Phe amide	NCA	Aq	55	Sephadex G-25			×								232
Asp-Phe-OMe															499
α-Asp-Phe-OMe										235–236 (dec)	+32	Ac			19
											±0	Aq			
β-Asp-Phe-OMe										198–199 (dec)	+43	Ac			19
											+4.0	Aq			
α-Asp-Phe-OMe cinnamic acid salt															19
α-Asp-Phe-OMe 3,5-dinitrobenzoic acid salt															19
α-Asp-Phe-OMe-gentisic acid salt															19
γ-OBzl-Boc-Asp-Phe-OMe	OPy		95						Me	115–116					134
β-OBu^t-Asp-Phe amide										122–124					25

(*continued*)

Table 2 Dipeptides (continued)

Compound	Method	Solvent	% Yield	Chromatography Col	Tl	Ion	Paper	CCD	Recrys. Sol.	M.P. °C	$[\alpha]_D$	Solvent Rotation	Elect.	Other	Ref.
β-OBu^t-Asp-Phe-OMe HCl															574
β-OBu^t-Asp-Phe-OPh HCl										153–156					314
β-OBu^t-Z-Asp-Phe amide	OSU	DMF	89						EtAc	156–157					25
β-OBu^t-Z-Asp-Ser	OSU		78						EtAc-Pe	133–135	+6.2	Me			257
δ-Aminovaleric acid															
δ-Ave-His	MCA	MeCl	72			A-CG-120				239–240 (dec)	+24	Aq			156
Cysteine															
S-Bzl-N-Z-Cys-DL-Ala-OPfp	ONP	Di							Alc						309
N,S-Di-Trt-Cys-DL-Ala										130 (dec)					731
N,S-Di-Trt-Cys-DL-Ala-OEt										125 (dec)					731
S-Trt-Cys-DL-Ala										140 (dec)					731
Boc-Cys-Asn-OBu^t					43					Oil					293
S-Trt-Boc-Cys-Asn-OBu^t	MCA	THF	55						EtAc-Pe	155–157	+34	Chf			293
S-Trt-Cys-Asn-OBu^t					43										293
S-Trt-N-Trt-Cys-Asn-OBu^t	DCCI	EtAc	76						He	209–210	+39	Chf			293
Boc-S-Trt-Cys-Asp-OTMB	DCCI-NHS	DMA	61							161	+20	Me			234
N,S-Di-Trt-Cys-Asp										145 (dec)					731

N,S-Di-Trt-Cys-Asp-Di-OMe										140 (dec)					731
S-Trt-Cys-Asp-OTMB									Me-Eth	119–121	+49	Me			234
Boc-Cys(Ame)-Cys(Ame)-OMe			78						EtAc-Pe	96–98	−52	Me			292
Boc-$(Cys)_2$									EtAc	146–147	−138	Me			544
Boc-Cys-Cys-OMe (S—S bridge between the two Cys)									Chf-Pe	185–187 (dec)	−65	Me			292
S,S′-Di-Trt-Cys-Cys									Me-Aq	158 (dec)					731
S,S′-Di-Trt-Cys-Cys-OEt HCl										135 (dec)					731
N-Tfa-S-Cys-Cys-OMe HCl										151–153 (dec)	−142	1 *N* HCl		IR	217
S,S′-Trt-Cys-Cys-OEt	DCCI	MeCl	79						MeCl-Alc	171–173					731
N,S-Di-Trt-Cys-S-Trt-Cys-OEt										171–173					731
S-Trt-Cys-S-Trt-Cys										158 (dec)					731
S-Trt-Cys-S-Trt-Cys-OEt HCl										135 (dec)					731
Z-Sulfonamido-Cys-Cyst										Oil					7
Z-Sulfonamido-Cys-Cyst-OEt									EtAc-Alc	113–116	−82	DMF			7
N-Bz-Cys-Gly-OMe										114–115	−32	Me			250
S-Bzl-N-Z-Cys-Gly-OPcp	MCA	THF	93		51a				THP-Eth	190–191	−32	DMF			336
S-Bzl-Cys-Gly									Aq	157					380
										191–192	75	Aq			
S-Bzl-Cys-Gly-OBzl(OMe) HCl									Me-Eth	135–135.5	8.9				380
S-Bzl-Cys-Gly-5-OMe-tryptamine HCl									EtAc-Me	181–183	−32	Me			557
S-Bzl-Cys-Gly-OPcp HBr					126				Me-Eth	193–194 (dec)	+11	DMF			336

(continued)

Table 2 Dipeptides (continued)

Compound	Method	Solvent	% Yield	Chromatography Col	Tl	Ion	Paper	CCD	Recrys. Sol.	M.P. °C	$[\alpha]_D$	Solvent Rotation	Elect.	Other	Ref.
S-Bzl-N-Tfa-Cys-Gly-5-OMe-tryptamine									Alc-Aq	159–162	−3.7	Me			557
N,N′-Bis-Bz-Cys-Gly-Di-OMe									Alc	224	−206	DMF			250
N,S-Bis-Boc-Cys-Gly									EtAc-Pe	147–149	−21	Alc			546
N,S-Bis-Boc-Cys-Gly-OMe	DCCI	Chf	66						Eth-Pe	100.5–101	−26	Alc			431
	MCA	THF	21						Eth-Pr-Eth	99–100.5	−26	Alc			
S-Boc-Cys-Gly-OMe HCl															431
S-Chlorocarbonyl-N-Bz-Cys-Gly-OMe										109–110 (dec)					250
S-Bzh-N-Boc-Cys-Gly-OBzl				Si											234
S-Bzh-N-Boc-Cys-Gly Dcha										153–154	−24	DMF			234
S-Bzh-N-Boc-Cys-Gly-OMe	OSU	Glyme	89						Alc-Aq	122–124	−19	Alc		Mol. wt.	69
	DCCI-HBT	Glyme	92							122–124	−18	Alc			69
S-Bzh-Cys-Gly										182–185 (dec)	−2.6	HMPA			234
S-Bzh-Cys-Gly-OBzl-toluene-sulfonate salt			68							181–183	+26	Me			234
N,S-Di-Trt-Cys-Gly										127–129 (dec)					731

N,S-Di-Trt-Cys-Gly-OEt										112–114 (dec)					731
S-*p*-Methoxy-Bzl-N$^{\alpha}$-Z-Cys-Gly-OMe	DCCI	MeCl	90						Cte	104–105	−58	DMF			10
S-*p*-Methoxy-Bzl-Cys-Gly-OMe HBr															10
S-Z(OMe)-N-Nps-Cys-Gly-OEt	DCCI	Chf	80						Alc	135–137	−18	THF			488
Nps-S-Bzl-Cys-Gly-OBzl(OMe)									EtAc-He	118.5–122	−21	Alc			380
Nps-Cys(An)-Gly-OBzl(OMe)									EtAc-He	102.5–125	−6.5	DMF			380
S-Sulfonamido-N-Z-Cys-Gly-OEt	DCCI	Chf	46						Alc-EtAc	170–171	−10	DMF			7
S-Sulfonamido-Cys-Gly-OEt									EtAc-Alc	201–203					7
S-Trt-N-Boc-Cys-Gly Dcha	DCCI-NHS	Glyme	91						Eth	130.5–133	+15	Chf			234
S-Trt-Cys-Gly										114–116	−6.0	HMPA			234
										228–230 (dec)					731
N,S-Di-Trt-Cys-Glu-Di-OEt										143–144					731
S-Sulfonamido-N-Z-Cys-His															7
S-Sulfonamide-N-Z-Cys-His-OMe	DCCI	EtAc-DMF	35						EtAc-Alc	147–150					7
S-Bzl-N-Z-Cys-D-Leu-OMe	MCA		74						EtAc-Pe	101	−1.7 −12	Me DMF			54
S-Bzl-Cys-N-Z-Leu-OMe	MCA	Chf	84							Oil					54
	DCCI	Di	51							Oil					
S-Bzl-N-Nps-Cys-Leu Boc-hydrazide	DCCI	Chf	77						EtAc-Pe	152–153	−10	Me			54
S-Bzl-N-Nps-Cys-D-Leu Boc-hydrazide	DCCI	Chf	61						EtAc-Pe	153–155	+28	Me			54
Z-Cys(CH_2CONH_2)-Leu-OMe									An	135–136.5	−29				188

(*continued*)

Table 2 Dipeptides (continued)

Compound	Method	Solvent	% Yield	Chromatography Col	Tl	Ion	Paper	CCD	Recrys. Sol.	M.P. °C	$[\alpha]_D$	Solvent Rotation	Elect.	Other	Ref.
Decanoyl-Cys(CH_2CONH_2)-Leu-OMe			87						EtAc	122–123	−33				188 321
N,S-Di-Trt-Cys-Leu										140–142					731
N,S-Di-Trt-Cys-Leu-OMe										161–163					731
S-Trt-Cys-Leu-OMe HCl										163–165					731
S-Bz-N-Z-Cys-N^{ϵ}-Tos-Lys-ONp	DCCI	Py-DMF	68						Alc	173–174	−1.7	Alc			559
N-Bz-Cys-Phe-OMe									EtAc	154–155	−32	Alc			250
S-Bz-Cys-Phe-OMe HCl									Me-Eth	151–152 (dec)	+1.5	Alc			250
										153.5–154 (dec)	+3.1	Alc			
S-Bzl-N-Z-Cys-DL-fluoro-Phe									Alc-*i*-Pr						309
S-Bzl-N-Z-Cys-Phe	ONP	Di							Alc		+85	Alc	×		309
N-Z-S-Bz-Cys-Phe-OMe	DCCI	Chf	80						Alc	136–137	−31	Alc			250
S-Chlorocarbonyl-N-Bz-Cys-Phe-OMe									Be	137–137.5 (dec)	−40	Di			250
N,S-Di-Trt-Cys-Pro										148–150 (dec)					731
N,S-Di-Trt-Cys-Pro-OBzl										135 (dec)					731

S-AcOMe-N-decanoyl-Cys-Tyr-OMe															460
Z-S-Bzl-Cys-Tyr-OCH_3															415
Decanoyl-Cys(CH_2COOMe)-Tyr-OMe		Eth								107–109	−27	DMF			188
N,S-Di-Trt-Cys-Tyr										150 (dec)					731
N,S-Di-Trt-Cys-Tyr-OEt										183–185					731
S-Sulfonamide-N-Z-Cys-Tyr-OMe	DCCI	DMF-EtAc	43						EtAc-Alc	174–175	−14	DMF			7
Sulfonamide-Cys-Tyr-OMe									EtAc-Alc	189–190					7
N,S-Di-Trt-Cys-Val										135 (dec)					731
N,S-Di-Trt-Cys-Val-OMe										163–165					731
S-Sulfonamide-N-Z-Cys-Val															7
S-Sulfonamide-N-Z-Cys-Val-OMe	DCCI		44						EtAc-Alc	227–228	−13	DMF			7
S-Trt-Cys-Val-OMe HCl										125 (dec)					731
Cystine															
Z-Cyst-sulfonamido-Cys															7
Z-Cyst-sulfonamido-Cys-OEt_2										143–144	−83	DMF			7
Cyst-sulfonamido-Cys						IRA-410			Aq-Alc	157–158					7
NN′-Bz-Cyst-bis-Gly-bis-OMe										226–227	−211	DMF			250
Z-Cyst-Gly-OMe									Me	169–170	−147	DMF			188
N,N′-Z-Cyst-Gly-OMe									Me	172–173	−162	DMA			618
Z-S-Cyst-Gly-OPn									DMF-Alc	160–160.5	−136				595
N,N′-Di-Z-OMe-Cyst-Gly-OEt															218
N,N′-Bis-Boc-Cyst-bis-Gly-OBzl			93						EtAc-Pe	125–127	−112	DMF			375
Cyst-Bis-Gly-OBzl Di-HCl										125–128	−88	Me			375
Decanoyl-Cyst-Gly-OMe									Me	156–157	−86	DMF			188

(*continued*)

Table 2 Dipeptides (continued)

Compound	Method	Solvent	% Yield	Chromatography Col	Tl	Ion	Paper	CCD	Recrys. Sol.	M.P. °C	$[\alpha]_D$	Solvent Rotation	Elect.	Other	Ref.
Glutamic acid															
N-Z-Pyro-Glu-Ala-OBzl	MCA		88						Me	128–130	−26	Chf			378
N-Z-Pyro-Glu-β-(pyrazolyl-3)-Ala Boc-hydrazide	OTCP		60							140–142 (dec)	−49	Me			239
γ-OBu^t-N-Z-Glu-Ala-OMe	A	MeCl	90						Me-Aq	102	−15	DMF			299
γ-OBu^t-Glu-Ala-OMe HCl					43										293
Pyro-Glu-Ala										203–205	−31	Me			578
Pyro-Glu-β-(3-pyrazolyl)-Ala hydrazide									Me-Aq	232–235	−16	Aq			177
Pyro-Glu-β-(3-pyrazoyl)-Ala-OMe	DCCI	DMF			92					74–76	−2.7	Me			177
N-Z-γ-Glu-α-OBzl-Abu-OBzl									Alc-Pe	122–124	−35	Me			669
γ-Glu-Abu					118 55 138					142–145	−15	Aq			669
Pyro-Glu-N^G-nitro-Arg									Aq	198–200	−11	Aq			177
Pyro-Glu-N^G-nitro-Arg-OMe									Me-Aq	195–197	−28	Aq			177
γ-OBzl-N-Z-Glu-Asn-OTMB	ONP	DMF							DMF-Alc	191–192	−1.2	DMF		IR	594
γ-OBu^t-N-Z-Glu-γ-OBu^t-Asp Boc-hydrazide hemihydrate	OSU	THF	95		35 27					Oil	−25	Me			690
γ-OBu^t-Glu-γ-OBu^t-Asp Boc-hydrazide acetate					109 110				Me-Eth	139–140	−12	DMF			690
Pyro-Glu-N^γ-Z-Dbu hydrazide									Me	223–225	−25	Aq			177
Pyro-Glu-N^γ-Z-Dbu-OMe											−19	Me			177

α-OBzl-Glu-α-OBzl-Glu														628
γ-OBzl-Glu-γ-OBzl-Glu-ONp HCl	ONP	Chf	66						*i*-Pr	117–119	−30 −33	DMF Chf			226
γ-OBzl-Glu-γ-OBzl-Glu-OPcp HCl	OPCP	EtAc	70							140 (dec)	−28	DMF			226
α-OBzl-N-Nps-Glu-α-OBzl-Glu														628
α-OBzl-N-Nps-Glu-α-OBzl-Glu Dcha									Alc	143–145	+31	Chf			628
γ-OBzl-N-Nps-Glu-γ-OBzl-Glu-ONp	DCCI	Chf	93						Alc	84–86	+30	EtAc			226
γ-OBzl-N-Nps-Glu-γ-OBzl-Glu-OPcp	OPCP		72							106–107	+13	EtAc			226
Z-$Glu_2(OBu^t)_3$	OPCP		66							Oil	−27				410
	MCA	EtAc	82							83–84	−25				
Z-α-Glu-α-Glu-α-methyltryptamine-γ,γ-Di-OMe									Be-Eth	104–106	−9.5	Me			442
Z-Glu-Glu-α-methyltryptamine-γ,γ′-Di-OMe															442
$N^α$-Boc-Glu-Glu-α-OSu-γ,γ′-Di-OBu^t										Oil					644
$N^α$-Boc-Glu-Glu-α-OSu-γ,γ′-Di-OBu^t Dcha															644
Di-γ-OBu^t-N-Z-Glu-Glu										143–144	−3.8				393
Di-γ-OBu^t-Glu-Glu										180–181	+25				393
Di-γ-OBu^t-D-Glu-Glu										132–134	−22				393
Di-γ-OBu^t-Glu-D-Glu									Alc	132–133	+22				393
Di-γ-OBu^t-α-methyl-N-Z-Glu-Glu										46–47	−19				393
Di-γ-OBu^t-α-methyl-N-Z-D-Glu-Glu										60–61	−5.3				393
Di-γ-OBu^t-α-methyl-N-Z-Glu-D-Glu										64–65	+6.2				393
Di-γ-OBu^t-α-methyl-Glu-Glu HCl										129–130	+6.3				393
Di-γ-OBu^t-α-methyl-Glu-D-Glu HCl									Chf-Pe	62–63	+36				393
Glu-Glu-γ,γ′-Di-OBu^t															644
Glu-Glu-α-methyltryptamine									Alc-Eth	104–106	+20	Me			442

(*continued*)

Table 2 Dipeptides (continued)

Compound	Method	Solvent	% Yield	Chromatography Col	Tl	Ion	Paper	CCD	Recrys. Sol.	M.P. °C	$[\alpha]_D$	Solvent Rotation	Elect.	Other	Ref.
Glu-α-methyltryptamine-Z-Glu-γ-OBzl-α-OBzl									EtAc-Pe	82–84	−15	Me			442
Glu-α-methyltryptamine-Z-Glu-γ-OBzl-γ-OMe									Aq-DMF	178–180	−7.7	DMF			442
α-*p*-Iodo-OBzl-N-Nps-Glu-α-OBzl-Glu-ONp	MCA								EtAc-Eth	132–133	−34	EtAc			348
α-ONBzl-N-Nps-Glu-α-OBzl-Glu-ONp	MCA								EtAc-Eth	60–63	−32	DMF			348
γ-OTMB-Glu-Glu-γ-OTMB-α-OPcp HCl										142–144					628
γ-OTMB-N-Nps-Glu-γ-OTMB-Glu-OPcp										159–160					628
γ-OBzl-Glu-Gly									Aq-Alc					IR	594
γ-OBzl-N-Nps-Glu-Gly-ONp	MCA								EtAc-Eth	97–100	−37	An			348
γ-OBzl-N-Nps-Glu-Gly-OTMB									EtAc-Cy	96–97	−14	DMF			594
γ-OEt-Z-Glu-Gly-OEt										97–99	−16	Alc			717
γ-OTMB-N-Nps-Glu-Gly-OTMB									Alc	139–139.5	−17	DMF			594
Z-Pyro-Glu-1-methyl-His-OMe									Me-Eth	140–141				IR NMR Mass spec. Mol. wt.	520

Z-Pyro-Glu-3-methyl-His-OMe					23									UV NMR Mass spec. Mol. wt.	520
Pyro-Glu-His amide									Me	225–227	−21	Ac			578
Pyro-Glu-His Dcha															64
Pyro-Glu-1-methyl-His hydrazide										242 (dec)				IR NMR Mass spec.	520
Pyro-Glu-3-methyl-His hydrazide										226 (dec)				IR NMR Mass spec.	520
Pyro-Glu-His-OMe	DCCI	Acn	55							210–212 (dec)					64
	MCA	Chf													
Pyro-Glu-1-methyl-His-OMe									Me-Eth	179–180				IR NMR Mass spec. Mol. wt.	520
Pyro-Glu-3-methyl-His-OMe										Oil				NMR Mass spec. Mol. wt.	520
γ-OBzl-N-Boc-D-Glu-Leu									Aq-Alc	53–55	+6.1	Chf			216

(*continued*)

Table 2 Dipeptides (continued)

Compound	Method	Solvent	% Yield	Chromatography Col	Tl	Ion	Paper	CCD	Recrys. Sol.	M.P. °C	[α]D	Solvent Rotation	Elect.	Other	Ref.
γ-OBzl-2-methylsulfonylethyl-N-Boc-D-Glu-Leu										58–60	+7.5	Chf			216
γ-OBzl-2-methylthio-ethyl-N-Boc-D-Glu-Leu									Dipr-Eth	48–50	+4.5	Chf			216
γ-OBzl-N-succinimidyl-N-Boc-D-Glu-Leu									Anhy. Eth	110	−8.8	Chf			216
γ-OBzl-N-succinimidyl-D-Glu-Leu HCl										105–106	−39	Chf			216
Glu-Leu			55		82		×				+7.1	HCl			232
α-OBzl-N-Nps-Glu-N^ε-Z-Lys-ONp	MCA								EtAc-Eth	143–145	−37	EtAc			348
Pyro-Glu-N^ε-Z-Lys hydrazide									Me-Eth	197–198	−18	Aq			177
Pyro-Glu-N^ε-Z-Lys-OMe	DCCI	DMF	73		29					109–112	−9.6	Me			177
α-OTMB-N-Nps-Glu-N^ε-Z-Lys-ONp	MCA								EtAc-Eth	125–131	−36	EtAc			348
γ-OBzl-N-Nps-Glu-Phe-OTMB	DCCI								Alc	127.5–128	+4.6	DMF			594
Z-Pyro-Glu-Phe-OBzl	MCA	THF	88						Alc-Aq	144–146	−0.2	Chf			578
Decanoyl-(Glu-OMe)-Phe-N^G-nitro-OMe			70						Alc	183–185	−19	Ac			640
Pyro-Glu-Phe									EtAc-Me	223–225 (dec)	+19	Me			578
Z-Pyro-Glu-Trp-OBzl	MCA		66						Alc	149–151	+7.5	Chf			578
Pyro-Glu-Trp															578
γ-OBzl-N-Z-Glu-Tyr-OBzl										124–127	−12	DMF			612
Z-Glu-Tyr									Aq	184–185	+6.4	Alc			612

Pyro-Glu-OBzl-Tyr									Chf-EtAc	182–185	+13	Me			578
Pyro-Glu-OBzl-Tyr-OMe	DCCI	Acn	86						Alc-He	124–126	+8.1	Me			578
Pyro-Glu-Tyr amide									Me	275 (dec)	−17	DMF			578
									Me	213–215	+3.1	DMF			578
Glutamine															
N^{γ}-Dmb-N-Z-Gln-Ala									EtAc	193–194	−1.9	DMF			491
N^{γ}-Dmb-N-Z-Gln-Ala-OMe									EtAc	179–180	−6.3	DMF			491
N^{γ}-Dmb-N-Boc-Gln-Ala									Alc-Eth	159–160	+1.2	DMF			491
N^{γ}-Dmb-N-Boc-Gln-Ala-OMe									EtAc-Pe	133–134	+7.0	DMF			491
Gln-Ala									Ac-Aq	174–175	+8.3	1 *N* HCl			491
Boc-Gln-N^{γ}-Pht-Dbu									Me-Eth	170–171 (dec)	+7.5	DMF			221
Boc-Gln-N^{γ}-Pht-Dbu-OBzh	ONP	DMF	51						EtAc	148–150	−12	DMF			221
Z-Gln-Gly					130				Aq	183	−3.1				710
Z-Gln-Gly-OEt															415
Z-Gln-Gly-OMe					48					175	−6.7				710
N^{γ}-Z-Gln-Gly-OMe	WRK	Acn	56						Me-Aq	174–175	−6.6	N,N-DMF			208
N^{γ}-Bu^{t}-N^{α}-Z-Gln-Gly	OSU	THF	81						Eth-Pe	49–54	−13	Me			39
N^{γ}-Bu^{t}-N^{α}-Z-Gln-Gly Boc-hydrazide	DCCI	THF	92							68–73	−8.8	Me			39
N^{γ}-Bu^{t}-Gln-Gly Boc-hydrazide			91						Alc-Aq		+5.6	Me			39
Gln-Gly HCl					130										710
Gln-Gly-OMe HCl															208
N-Z-Glu-N^{im}-Bzl-His-OBzl	ONP	DMF	88						Aq-Me	169–171	−26	Ac			300
N-Z-Gln-His-OMe	ONP	MeCl	84						Me-Eth	195	−36	HCl			300
Z-Gln-Leu hydrazide									Me	239	−15	DMF			141
Gln-Leu	NCA	Aq	40							206–208	+4.4	HCl			232
γ-OBzl-N-Boc-Gln-N^{ϵ}-Tos-Lys-OBzl	DCCI	MeCl	74						He/Alc (6:3)	128.5–129.5	−3.7	DMF			378
γ-OMe-N-Z-Gln-N^{ϵ}-Tos-Lys-OMe	OSU	DMF	91							Oil					378
Nps-Gln-N^{ϵ}-Z-Lys-OMe	OSU	DMF	77						Me-Aq	155–157					569

(*continued*)

Table 2 Dipeptides (continued)

Compound	Method	Solvent	% Yield	Chromatography Col	Tl	Ion	Paper	CCD	Recrys. Sol.	M.P. °C	$[\alpha]_D$	Solvent Rotation	Elect.	Other	Ref.
Z-Gln-N$^{\delta}$-Boc-Orn-OMe	ONP	Py	94						Alc-Eth	161–162	−3.3	DMF			68
Z-Gln-D-Ser hydrazide									Aq-Eth						492
Z-Gln-Ser-OMe															492
Z-Gln-D-Ser-OMe	A	EtAc													492
Z-Gln-Thr-OMe									Me-EtAc	164–167	+1.9	DMF			712
Z-Gln-Val-OMe															415
Glycine															
Z-Gly-Ala										133	−9.8	Alc			263
Z-Gly-DL-Ala									Alc-Eth	182					263
Z-Gly-Ala-OBzl										77	−14	EtAc			263
Z-Gly-DL-Ala-OBzl	MCA	Chf	69						EtAc-Pe	91					263
Z-Gly-Ala-β-OBzl					124										263
Z-Gly-Ala-OEt										60–62					354
Z-Gly-DL-Ala-OEt	$SiCl_4$	Py	54												89
Z-Gly-Ala hydrazide									An-Eth	134	−30	Me			263
N-Z-Gly-dehydro-Ala-OEt									EtAc-Eth	81–81.5				NMR	437
Z-Gly-DL-3-(2-pyridyl)-Ala										141–143					339
Z-Gly-DL-3-(2-pyridyl)-Ala amide										145–147					339
Z-Gly-DL-3-(2-pyridyl)-Ala-OEt										68					339
Z-Gly-DL-3-(3-pyridyl)-Ala										157–158					339
Z-Gly-DL-3-(3-pyridyl)-Ala amide										152–155					339
Z-Gly-DL-3-(3-pyridyl)-Ala-OEt										88–90					339
Z-Gly-DL-3-(4-pyridyl)-Ala										184–185					339
Z-Gly-DL-3-(4-pyridyl)-Ala amide										164–165					339

Z-Gly-DL-3-(4-pyridyl)-Ala-OEt										100–101					339
Z-Gly-β-(8-theobrominyl)-α-Ala	ONP	Aq-Alc							Alc	139–140					637
Gly-Ala	NCA		87								−59	0.5 *N* HCl			296
Gly-Ala-SPh										97–98					444
Gly-DL-3-(2-pyridyl)-Ala amide									Eth	Oil 123–126					339
Gly-DL-3-(3-pyridyl)-Ala amide acetate										152					339
Gly-DL-3-(4-pyridyl)-Ala amide acetate										157					339
Gly-β-(8-theobrominyl)-α-Ala									Aq	352–355					637
Gly-β-(8-theobrominyl)-α-Ala HBr									Alc	216–218					637
Z-(C_2H_4CN) Gly-Asn-OTMB									DMF-Eth	141.5–142.5	−7.8				599
Nps-(C_2H_4CN) Gly-Asn-OTMB									Me	173–173.5	−5.2				599
N-Z-Gly-S-Bzl-Cys-OPfp	MCA	EtAc	44						Eth-Pen	84–85	−31	EtAc			322
N-Z-Gly-Se-Bzl-Cys-SePh									Alc-He	93–95	−10	DMF			264
Z-Gly-Cys(CH_2COOMe)-OMe			94							Oil					188
Z-Gly-sulfonamide-Cys-OEt	DCCI	EtAc	83						Alc	152–153	−5.0	DMF			7
Boc-Gly-S-Bu^t-Cys										146–148 (dec)	−100	80% Me			707
Decanoyl-Gly-Cys(CH_2COOMe)-OMe			42						EtAc	58–60	−24				188
Boc-Gly-S-Bzh-Cys	OSU	Chf	83							174–176	−13	Chf-DMF			234
Gly-S-Bz-Cys-OMe HCl									Me-Eth	187.5–188 (dec)	+5.7	Me			235
Gly-S-Bzl-Cys									Aq	189–191	−16	1 *N* HCl			380
Gly-S-Bzl-Cys-OBzl(OMe) HCl									Me-Eth		−28				380

(*continued*)

Table 2 Dipeptides (continued)

Compound	Method	Solvent	% Yield	Chromatography Col	Tl	Ion	Paper	CCD	Recrys. Sol.	M.P. °C	$[\alpha]_D$	Solvent Rotation	Elect.	Other	Ref.
Gly-S-Bu^t-Cys											−43	1 *N* HCl			707
Nps-Gly-S-Bzl-Cys-OBzl(OMe)									EtAc-He	118.5–119	−35	DMF			380
Z-Gly-Glu(OBu^t)$_2$	MCA	THF	96						EtAc-He	71–74	−17	Me			293
Gly-Glu(OBu^t)$_2$					44										293
Z-(C_2H_4CN) Gly-Gln-OTMB									DMF-Eth	129–130	−11				599
N-($C_2H_4CONH_2$) Gly-Gln									Aq-Alc	200–201.5	−4.2	Aq			599
Nps-Gly-Gln-OBzl(OMe)									Di-He	174–175	−8.8	DMF			380
N-Ac-Gly-Gly-OCH_2CN									Alc	135–135.5					600
[^{14}C]Ac-Gly-Gly-OEt									Alc	146–148				IR	440
Bz-Gly-Gly-OEt	OBHAE	THF	53							117–118.5					371
Z-(C_2H_4CN) Gly-Gly-OTMB									EtAc-Cy	113–113.5					599
Z-N-2-Cyanoethyl-Gly-Gly									EtAc-Cy	131–131.5					599
N-Gly-Gly	MA	Be	99							105–107					414
Z-Gly-Gly-OBzl									Alc	178–179					480 415
Z-Gly$_2$-OBzl									Chf-Eth	109–110				IR	718
Z-Gly-Gly OBu^t	OSU	Di		Al_2O_3						Oil					259

Z-Gly-Gly-OEt															415
	$SiCl_4$	Py	70												89
										82					717
	DCCI	Acn								80–81				IR	718
Z-Gly_2-OMe									Chf-Eth	63–64				IR	718
Z-Gly-Gly-ONp	MA	Di	56							106–107					414
	ONP	Di	60							106–107					414
Z-Gly-Gly-OPn									EtAc-Cy	127.5–128					595
Z-Gly-Gly-SPy										124–126					365
Z-Gly-Gly-SePh										109–110					264
Z-Gly-Gly-OSu										166					703
									EtAc-Eth	171–173					480
Z-Gly-Gly-T										188–190					195
N-Z-Gly-3-methyl-azlyl-Gly-OBzl									An-PrEth	114–116	−36	THF		NMR	437
N-(δ-Bromovaleryl)-Gly-Gly									An	135–136					302
N-(δ-Bromovaleryl)-Gly-Gly-OEt									An	161–162					302
N-($C_2H_4CONH_2$) Gly-Gly									Aq-Alc	204.5–205					599
Cu-(Gly-OEt)$_2Cl_2$									Chf-Alc	165–167				IR	718
										105–106 (dec)				IR	718
N-2-Cyanoethyl-Gly-Gly-OTMB HCl									Me-Eth	147.5–148.5					599
DNS-Gly (1-dimethylaminonaphthalene-5-sulfonyl-Gly)										157–157.5					120
Gly-Gly	TMP		40												504
															395
	NCA		90												296
															503
Gly-Gly-OPcp HBr									DMF-EtAc	225 (dec)				Mol. wt.	71

(*continued*)

Table 2 Dipeptides (continued)

Compound	Method	Solvent	% Yield	Chromatography Col	Chromatography Tl	Chromatography Ion	Chromatography Paper	CCD	Recrys. Sol.	M.P. °C	$[\alpha]_D$	Solvent Rotation	Elect.	Other	Ref.
Gly-Gly-OPh	TP-TEA									137					414
Di-Gly N-triphosphate															100
Nps-(C_2H_4CN)Gly-Gly-OTMB									EtAc	88–89					599
N-Nps-N-2-carboxamidoethyl-Gly-Gly-OTMB	ONP	DMF	92						Me	187–187.5					597
N-Nps-N-2-cyanoethyl-Gly-Gly-ONp	ONP	DMF	68						Me	134–135					599
Nps-Gly-Gly-OPn									DMF-Alc	165.5–166.6					595
Pht-Gly-Gly									Alc	230–231					563
Pht-Gly-Gly-OEt									Alc	191–193					564
	$SiCl_4$	Py	45												89
										189–191					365
	OBHAE	THF	59							192					371
Pht-Gly-Gly-OMe	AC	Chf	90							203–204					563
Z-Gly-Hyp-OBzl(OMe)										74–74.5	−42				596
Z-Gly-Leu hydrazide									Eth-Me	Oil					494
Z-Gly-D-Leu hydrazide															495
															496
Z-Gly-Leu-OMe									Eth	Oil					494
	MCA	Chf							Me-Eth						496
															495

Z-Gly-D-Leu-OMe															495
															496
Z-Gly-DL-Leu-OMe	MCA	Chf							Me-Eth	Oil					494
2,5-Di-(N-Gly-Leu-OEt)-1,4-benzoquinone									Be	154–156 (dec)					111
Gly-Leu	NCA		92								−36	Aq			296
Pht-Gly-Leu-OMe	$SiCl_4$	Py	48												89
N^{α}-(Z-Gly)-N^{δ}-Tos-DL-Orn hydrazide			92							156–158					201
Pht-Gly-N^{δ}-Boc-Orn-OMe					24				EtAc		−7.5	Me		Mol. wt.	613
Pht-Gly-Orn-OMe HCl					24 9				Me-Eth	225–230 (dec)	−16	Me		Mol. wt.	613
N-Z-N-(2-Boc-aminoethyl)-Gly-Phe-OMe	DCCI	DMF	67						EtAc-Pe	115–116	−3.6	DMF			26
Z-(C_2H_4CN)Gly-Phe-OTMB									EtAc-Cy	106.5–107	−5.8				599
Z-Gly-Phe															337
	TPP-DDD									125–126	36	An			397
Z-Gly-Phe-OBhae															371
Z-Gly-Phe-Bzl amide				Seph-adex LH-20					Me	152–153					677
Z-Gly-Phe-OBu^t										60–62					354
Z-Gly-Phe-OCH_2CN															351
Z-Gly-Phe-$OCOOCH_2$-$CH(CH_3)_2$															16
Z-Gly-Phe-OEt	TPP-DDD									Oil					397
Z-Gly-DL-Phe-OEt	OBHAE	THF	91							89–91					371
N-Z-Gly-Phe-OPfp									Alc-Aq	96–98	−11	Chf			322
N-Z-Gly-Phe-SePh										95–96	−60	DMF			264

(*continued*)

Table 2 Dipeptides (continued)

Compound	Method	Solvent	% Yield	Chromatography Col	Tl	Ion	Paper	CCD	Recrys. Sol.	M.P. °C	[α]D	Solvent Rotation	Elect.	Other	Ref.
N-Z-Gly-DL-Phe-SePh									EtAc-Pe	123					264
Z-Gly-DL-Phe-T										69–71					195
Boc-Gly-Phe amide	OPy		85						EtAc-Pe	162–164					134
Boc-Gly-Phe-OEt	OBHAE	THF	89							Oil					371
N-($C_2H_4CONH_2$) Gly-Phe									Aq-Alc	211.5–212	+18	Aq		IR	599
Gly-Phe	NCA		89								41	Aq			296
Z-Gly-Pro										154–155					676
	DCCI	Chf								152–155	−72	Alc			372
										155.5–157	−77	Chf			
Z-Gly-Pro-diketopiperazine									EtAc-Pe	107–109	−107	EtAc			372
Z-Gy-Pro-ONp									EtAc-Pe	101–103	−99	EtAc			372
									Alc	104–105	−103	EtAc			
Boc-Gly-Pro									EtAc-Eth	126–135				IR NMR	121
Boc-Gly-Pro-OBzl	MCA	Chf	100						EtAc-He	76–77				IR NMR	121
Boc-Gly-4-Hyp-ONp	ONP	DMF	60						EtAc-Cy	120–121	−67	DMF			598
Boc-Gly-4-Hyp amide									Alc-EtAc	197–198.5	−42	DMF			598
Z-Gly-4-Hyp-OTMB	ONP	DMF	68						EtAc-Cyclo-He	119–120	−29	DMF			598
N-Z-Gly-O-Boc-Ser-OMe										Oil	−6.9	Alc			365

N-Z-Gly-Ser-OEt	A	EtAc	76						EtAc	112.5	+17	Di			437
Z-Gly-Ser-OMe															415
N-Z-Gly-Ser-OMe															540
	DCCI	MeCl	77						Me-Eth	94–96					300
N-Z-Gly-O-Tos-Ser-OEt									EtAc-Eth	109–110	+23	Di			437
N-Z-Gly-Thr hydrazide									Hot Me	172–173					437
N-Z-Gly-O-Tos-Thr-OEt									Eth	79–80	+21	Alc			437
Cyano-Boc-Gly-Trp Dcha	DCCI-NHS	THF	85								+19	Alc			706
Gly-Trp					67						+30	HCl			706
Z-Gly-Tyr hydrazide										181–183	−10	DMF			331
Z-Gly-Val										Oil					528
Z-Gly-Val-OBu^t	MCA	Chf	55							Oil					527
Z-Gly-D-Val-OMe			59							Oil	+14	DMF			188
Z-Gly-Val-OSu									*i*-Pr	87–89					528
Z-Gly-DL-Val-T										125–127					195
Decanoyl-Gly-Val										Oil					528
Decanoyl-Gly-Val-OBu^t										Oil					528
Gly-Val	NCA		91								−20	Aq			296
Gly-Val-OBu^t															527
Nps-(C_2H_4CN)-Gly-Val-OTMB									EtAc-Cy	81.5–83	−10				599
Histidine															
N^{im}-Bzl-His-Glu															684
N^{im}-Bzl-D-His-Glu															684
N$^{\alpha}$,N$^{\alpha'}$-Adipoyl-Di-His hydrazide										186–188					445
N$^{\alpha}$,N$^{\alpha'}$-Adipoyl-Di-His-OMe															445
N^{im}-Bzl-N$^{\alpha}$,N$^{\alpha'}$-adipoyl-Di-His hydrazide										194–196					445
N$^{\alpha}$,N$^{\alpha'}$-Glutaroyl-Di-His-OMe										156–157	−61	DMF			445
N$^{\alpha}$,N$^{\alpha'}$-Malonoyl-Di-His-OMe										172–174	+5.0	DMF			445
N$^{\alpha}$,N$^{\alpha'}$-Oxaloyl-Di-His-OMe										183–184					445
N$^{\alpha}$,N$^{\alpha'}$-Succinoyl-Di-His										66–67					445

(*continued*)

Table 2 Dipeptides (continued)

Compound	Method	Solvent	% Yield	Chromatography: Col	Tl	Ion	Paper	CCD	Recrys Sol.	M.P. °C	$[\alpha]_D$	Solvent Rotation	Elect.	Other	Ref.
N^{α},$N^{\alpha'}$-Succinoyl-Di-His hydrazide										246–248					445
N^{α},$N^{\alpha'}$-Succinoyl-Di-His-OMe										209–210					445
N^{im}-Bzl-N^{α}-Z-His-Leu hydrazide									Me-Eth	110–111	−19	Ac			299
N^{im}-Bzl-N^{α}-Boc-His-Leu-OBzl	DCCI	DMF	72						Dipr Eth-Pe	67–68	−24	Me			643
N^{im}-Bzl-His-Leu-OBzl Tfa					×					115–116					643
N^{im}-Bzl-N^{α}-Boc-His-Pro amide	DCCI	Acn	92	Si							−19	Chf		NMR	90
N^{im}-Bzl-N^{α}-Boc-His-Pro-OBzl	DCCI-NHS	DMF-MeCl	83						EtAc-He	127–127.5	−20	DMF			377
N^{im}-Bzl-His-Pro amide 2 HBr									Me-Eth	196–198 (dec)	−27	Me			90
Z-His-Pro amide	A	EtAc	85						Aq	102–104	−41	Me			248
N^{im}-Bzl-N^{α}-Z-His-Ser										181–182					208
N^{im}-Bzl-N^{α}-Z-His-Ser hydrazide									DMF-Aq	147.5–149					208
N^{im}-Bzl-N^{α}-Z-His-Ser-OMe	OTCP	MeCl	83							173.5–175					208
Isoleucine															
Z-Ile-Gly-OPn									Alc	170.5–171	−7.8				595
Boc-Ile-Ile-OMe	SPy		83						Eth-Pe	155	−7	EtAc			365
Ile-Leu					×										671
Ile-Leu-OBu^t			95								−37	Me			671
N-Tfa-Ile-Leu-OBu^t			98							107–108	−95	Me			671

Z-Ile-Phe	ONP		74						EtAc-Pe	177–179	−19	Me			257
Boc-Ile-Val-OMe	SPy		85						Eth-Pe	166–167	−15	EtAc			365
Leucine															
Z-Leu-D-Aa-OEt	DCCI	Chf	83						EtAc-Pe	93–94	+3.3	Me			349
Z-Leu-D-Ala hydrazide									DMF-Aq	171–174	−14	DMF			349
N-Thio-Bz-DL-Leu-Ala										55–59					32
N-Thio-Bz-DL-Leu-Aib									Eth	177–178					32
Boc-Leu-Arg	OSU	Di-Aq	72		54					198					705
Z-Leu-N$^{\delta}$,N$^{\omega}$-Adoc$_2$-Arg					27				Me-Aq	93–96	−7.2	Me			375
Z-Leu-Glu-Di-OBzl									Aq-An	206–209	+10	1 *N* HCl			365
Boc-Leu-Gln	OSU	Di	89		130				EtAc-Pr-Eth	187–188 (dec)	−22	Alc			710
Leu-Gln-HCl					130					202–203 (dec)					710
N-Aceto-Ac-Leu-Gly-OEt	DCCI	Di	84	Si						87–88	−48	Aq-Alc			42
Bz-Leu-Gly-OEt	TPP		99												630
	TPP-DDD		81							144–147	−7.9	Alc			397
	MCA	THF	93		23a				Alc-Aq	157–158	−33	Alc			16
	TPP-DDD		92								−33	Alc			397
										148–152	−33				397
	SiCl$_4$	Py	65												89
Z-Leu-Gly-OBzl									Aq-Me	235–241 (dec)	+88	Aq			365
Z-Leu-Gly-OEt									Alc-Aq	104–105.5	−27	Alc			62
	TP	EtAc	74						EtAc-Pe	104–105					415
Boc-Leu-Gly-OMe	OPy	Chf	95						EtAc-Pe	131	−21	DMF			134
Boc-Leu-Gly-OPic	SPy		97							88–91	−24	EtAc			365
Leu-Gly	SP									246	+80				381
Leu-Gly-OEt HCl	SP														381

(*continued*)

Table 2 Dipeptides (continued)

Compound	Method	Solvent	% Yield	Chromatography Col	Chromatography Tl	Chromatography Ion	Chromatography Paper	CCD	Recrys. Sol.	M.P. °C	[α]D	Solvent Rotation	Elect.	Other	Ref.
Leu-Gly-OBzl(OMe) HCl									Me-Eth	117.5–119.5	+17				380
Nps-Leu-Gly-OBzl(OMe)									EtAc-He	96–99	−41	DMF			380
Pht-Leu-Gly-OEt															415
N-Thio-Bz-DL-Leu-Gly									Eth	139					32
N-Thio-Bz-DL-Leu-Gly-OBzl									Eth-Pe	103–104					32
N-Boc-Leu-His-OMe	OSU	Di-MeCl	72							140–141.5	−25	Me			333
(Bz-Leu-O)$_2$Cu															630
Boc-Leu-Leu-OMe										132–133	−50				448
Boc-Leu-D-Leu-OMe										108	−2.2				448
										95–96	−16				448
Leu-Leu															499
N-Thio-Bz-DL-Leu-Leu										56–60					32
Boc-Leu-N$^\epsilon$-Z-Lys					27						−12	Me			91
Boc-Leu-N$^\epsilon$-Z-Lys-OMe			42						EtAc	72–74	−17	DMF			188
	MCA	THF			27				Pe	84–87	−21	Me			91
Z-Leu-D-Phe-OEt	DCCI		84		124 25					106–107	−7.8	Me			401
Z-Leu-D-Phe hydrazide										163					401
Boc-Leu-Phe amide	OPy		80						EtAc-Pe	164–165					134
Boc-Leu-Phe-OMe										78–79	−28				448
Bpoc-Leu-Sar-OMe	DCCI-HBT	DMF	70		93 100					Oil				NMR	290

Z-Leu-OBu^t-Ser					86				Eth-Pe	128–129	+12				708
Z-Leu-OBu^t-Ser-OMe	DCCI-NHS	EtAc								Oil					708
Leu-OBu^t-Ser					130				Me		+33				708
Boc-Leu-N-Z-Tyr-OMe	SPy	DMF	62						EtAc-Pe	104–109	+5	Chf			365
Boc-Leu-Tyr-OMe	MCA	EtAc	81						Eth	106–108	−20	Me			293
Leu-Tyr-OMe HCl					27										293
Boc-Leu-Val			89						Be-He	109–111	−32	Me			293
Z-Leu-Val amide	DCCI-NHS	MeCl	90		48				An-Aq	231–232 (dec)	−24				710
Boc-Leu-Val amide	OPy		90						EtAc-Pe	195–196					134
Boc-Leu-Val-OMe	MCA	Acn	81						EtAc-Pe	143–144	−44	Me			293
Leu-Val amide HBr					54					210 (dec)	+24				710
Leu-Val-OMe HCl			97						EtAc-Pe	173–175	+3	Me			293
N-Tfa-Leu-Val-OBu^t															671
N-*t*-Tfa-Leu-Val-OBu^t															671
Trt-Leu-Val			83						EtAc-Pe	148–151	−40	Me			293
Trt-Leu-Val-OMe			86						EtAc-He	170–172	−51	Chf			293
Lysine															
N$^\epsilon$-Z-Lys-Ala	NCA	Aq	82	Sephadex G-10	×						+3.2	Aq			582
N$^\epsilon$-Z-Lys-Ala-OBzl HCl				Si	37					Oil					582
N$^\epsilon$-Z-Lys-Ala-OMe HCl															345
N$^\epsilon$-Z-N$^\alpha$-Boc-Lys-Ala					98						−14	Aq–1% NaOH			582
N$^\epsilon$-Z-N$^\alpha$-Boc-Lys-Ala-OMe	DCCI	Chf	91											UV	345
N$^\alpha$-Boc-Lys-Ala					37				Me-Eth	106–108	−14	Me			582
N$^\epsilon$-Z-N$^\alpha$-Nps-Lys-α-OBzl-Glu-ONp	MCA								An-Eth	134–136	−49	EtAc			348
N$^\epsilon$-Z-Boc-N$^\alpha$-Lys-Gly-OEt	OSU	MeCl	90		29										547
N$^\epsilon$-Z-Lys-Gly-OEt acetate					59	×									547

(continued)

Table 2 Dipeptides (continued)

Compound	Method	Solvent	% Yield	Chromatography Col	Tl	Ion	Paper	CCD	Recrys. Sol.	M.P. °C	$[\alpha]_D$	Solvent Rotation	Elect.	Other	Ref.
N$^{\epsilon}$-Z-Lys-Gly-OEt HCl															547
N$^{\epsilon}$-Z-Lys-Gly-OTMB HCl			93						Alc-EtAc	166–166.5	+12	DMF			600
N$^{\epsilon}$-Z-N$^{\alpha}$-Nps-Lys-Gly-ONp	MCA								An	110–118 144–146	−45	EtAc			348
N$^{\epsilon}$-Boc-N$^{\alpha}$-Z-Lys-cyclopentyl-Gly hydrazide										173–174	−19	Ac			305
N$^{\epsilon}$-Boc-N$^{\alpha}$-Z-Lys-cyclopentyl-Gly-OMe									Pe	130–131	−8.8	DMF			305
N$^{\epsilon}$-Pht-N$^{\alpha}$-Z-Lys-Gly amide									DMF	215–217	−2.2	DMF			219
N$^{\epsilon}$-Tos-N$^{\alpha}$-Z-Lys-Gly-OEt	ONP	Chf	92						Alc	156	−4.8	Chf			513
N$^{\epsilon}$-Tfa-N$^{\alpha}$-Z-Lys-Gly-OBu^t	OSU	THF	81						EtAc-Pe	70–72	−11	DMF			692
α-ONps-N$^{\epsilon}$-Z-Lys-Gly-OTMB	DCCI		91						Alc	160.5–161	−11	DMF			600
Boc-Lys-Leu	NCA	AQ	45	Si											232
N$^{\beta}$-Ac-N$^{\epsilon}$-(Di-Ac-β-Lys)-β-Lys															526
N$^{\beta}$-Ac-N$^{\epsilon}$-(Di-Ac-β-Lys)-β-Lys-OMe						DO-1									526
N$^{\epsilon}$-Ac-N$^{\beta}$-(Di-Ac-β-Lys)-β-Lys-OBu^t					87										526
N$^{\epsilon}$-Ac-N$^{\beta}$-(Di-Ac-β-Lys)-β-Lys-OMe															526
Z-(N$^{\epsilon}$-Boc-Lys)$_2$										Oil					645
Z-(N$^{\epsilon}$-Boc-Lys)$_2$ Dcha										114	+1.0	Me			645
N$^{\epsilon}$-Z-N$^{\beta}$-(N$^{\epsilon}$-Z-β-Lys)-β-Lys-OBu^t															526
N$^{\epsilon}$-(Z)-N$^{\beta}$-(N$^{\epsilon}$-Z-N$^{\beta}$-Pht-β-Lys)-β-Lys-OBu^t				Si	51										526

N^{ϵ}-(N^{ϵ}-Z-β-Lys)-β-Lys-OBu^{t}															526
N^{ϵ}-(N^{ϵ}-Z-N^{β}-Pht-β-Lys)-N^{α}-Pht-β-Lys-OBu^{t}	DCCI	Acn													526
Boc-(N^{ϵ}-Boc-Lys)$_2$										Oil					645
Boc-(N^{ϵ}-Boc-Lys)$_2$ Dcha										88	−1.0	Me			645
Boc-(N^{ϵ}-Boc-Lys)$_2$-OSu										148	−23	Me			645
(N^{ϵ}-Boc-Lys)$_2$										188	+16	Me			645
N^{ϵ},N^{ϵ}-Bis-Boc-Lys-Lys															645
N^{β}-Decanoyl-N^{ϵ}-(Di-decanoyl-β-Lys)-β-Lys-OMe	DCCI	MeCl		Si					Me	173–174	−1.2	Chf			528
N^{ϵ}-Decanoyl-N^{β}-(Di-decanoyl-β-Lys)-β-Lys-OMe									Me	168–170	−4.8	Chf			528
N^{ϵ}-(Di-Ac-β-Lys)-N^{β}-Ac-β-Lys-OMe															525
N^{β}-(Di-Ac-β-Lys)-N^{ϵ}-Ac-β-Lys-OMe															525
Lys-Lys	NCA														481 726
N^{β}-β-Lys-β-Lys															526
N^{β}-β-Lys-β-Lys-OMe															526
N^{ϵ}-β-Lys-β-Lys							×								526
N^{ϵ}-β-Lys-β-Lys-OMe															526
N^{ϵ}-Z-N^{α}-Nps-Lys-N^{ϵ}-Z-Lys-ONp	MCA								EtAc	157–159	−37	An			348
N^{α},N^{ϵ}-N^{ϵ}-Tris-Boc-Lys-Lys	OSU	DMF													645
N^{α},N^{ϵ},N^{ϵ}-Tris-Boc-Lys-Lys Dcha															645
N^{ϵ}-Tfa-N^{α}-Boc-Lys-Met	OSU	Glyme	80												693
N^{ϵ}-Tfa-N^{α}-Boc-Lys-Met Dcha										94–95	−0.5	Me			693
N^{ϵ}-Tfa-Lys-Met Tfa										98–105	−3.2	Aq			693
N^{ϵ}-N^{α}-Bis-Z-Lys-Phe-OMe	SPy		87						MeCl-Pe	90–92	−12	Chf			365
N^{ϵ}-Boc-N^{α}-Z-Lys-Phe hydrazide					29				Me-Aq	173–174	−34	Me			574
N^{ϵ}-Boc-N^{α}-Z-Lys-Thr	ONP	Di	74						EtAc	130–133	+14	DMF			619
N^{ϵ}-Tfa-N^{α}-Z-Lys-Thr Boc-hydrazide hemihydrate	OSU	THF	81		4 3					149–151	−9.1	DMF			691

(*continued*)

Table 2 Dipeptides (continued)

Compound	Method	Solvent	% Yield	Chromatography Col	Tl	Ion	Paper	CCD	Recrys. Sol.	M.P. °C	$[\alpha]_D$	Solvent Rotation	Elect.	Other	Ref.
N^{ϵ}-Tos-N^{α}-Z-Lys-Thr-OMe	ONP	DMF							Alc-Aq	99–101	+1.9	DMF			297
N^{ϵ}-Tfa-Lys-Thr Boc-hydrazide					4				Me-Eth	134–135	−12	DMF			691
acetate hydrate					3										
N^{ϵ}-Boc-N^{α}-Z-Lys-Tyr	DCCI	MeCl	83							194–196	−77	DMF			490
Methionine															
Met-Asn	NCA	Aq	37–53						Alc-Aq	227–228	+13	10% NH_4OH			164
Met-Glu	NCA	Aq	65							215–216	+14	10% NH_4OH			164
Z-Met-Gly-OEt										96–97	−19	Alc			717
															590
Z-DL-Met-Gly-OEt															415
Boc-Met-Gly-T										161–162					195
Dmpoc-Met-Gly-OEt										78–79					590
N-Thio-Bz-DL-Met-Gly									Eth	111					32
Met-Leu															499
N^{α}-Boc-Met-N^{ϵ}-Z-Lys hydrazide	OPCP	EtAc							Alc-Aq	150–151	−30	Alc			196
N-Thio-Bz-DL-Met-Pro										55–60					32
Z-Met-Val-OMe	SPy		85						MeCl-Pe	103	−26	Alc			365
Boc-Met-Val amide	OPy		88						Alc	185–186					134
N-Methylvaline															
N^{α}-Z-N^{β}-(OBu^t)-MeVal-α,β-Dbu-OMe										122–123	−6.3	Me			27

Norleucine															
N-Z-Nle-Asp	OSU	THF	72						EtAc-Pe	142–144	−9.8	Me			238
α,β-Di-Bu^t-Nle-Asp	OSU	THF	90							Oil	+25	Aq			238
Nle-Asp									Alc-Aq	239–240	+25	Aq			238
Z-Nle-Nle hydrazide					27 43					182–183	−8	DMF			78
Z-Nle-Nle-OMe					31 43					97–99	−23	Alc			78
Norvaline															
Z-Nva-Nva hydrazide					27 43					207–208	−5	DMF			78
Z-Nva-Nva-OMe					31 43					124–125	−30	Alc			78
Ornithine															
N^δ-Tos-N^α-Z-Orn-Gly-OEt			55						Alc	133–134	−6.5	Alc			201
N^δ-Tos-DL-N^α-Z-Orn-Gly-OEt									Alc-Eth	103–105					201
Di-Tos-Orn-Gly-OEt									Alc-Aq	133–134					201
Orn-Gly HBr										129–130					201
DL-Orn-Gly HBr										212–213					201
N^δ-Pht-D-Orn-Gly amide									Me-EtAc	217–218	−43	Aq			730
N^δ-Pht-D-N^α-Tos-Orn-Gly amide									Alc-Aq	202.5–203.5	+19	Ac			730
N^δ-Pht-D-N^α-Tos-Orn-Gly-OEt									Me-Aq	208–210	−6.2	DMF			730
N^δ-Boc-Z-Orn-N^δ-Boc-Orn-OMe			93		9 24					125	−15	Ac-Aq		Mol. wt.	613
Phenylalanine															
Ac-DL-Phe-DL-Ala-OEt	$SiCl_4$	Py	60							186–188					89
Ac-Phe-Ala-OMe	$SiCl_4$	Py	43												89
Boc-Phe-Ala-OMe										98–99	−18				448
Boc-Phe-D-Ala-OMe										63–64	−21				448
Z-Phe-N^G-nitro-Arg									Alc-Aq	185–186	+2	Py			640
Z-Phe-N^G-nitro-Arg-OMe									Me-Eth	160–161	−15	Me			640

(continued)

Table 2 Dipeptides (continued)

Compound	Method	Solvent	% Yield	Chromatography Col	Tl	Ion	Paper	CCD	Recrys. Sol.	M.P. °C	$[\alpha]_D$	Solvent Rotation	Elect.	Other	Ref.
Boc-Phe-N^G-nitro-Arg	OSU	DMF	78						EtAc-Pe	151–154	−13	DMF			279
											−3	Me			640
Boc-Phe-N^G-nitro-Arg-OMe			90								−9	Me			640
Boc-Phe-N^G-nitro-Arg OPic	OTCP	DMF	90								−12				540
CHla-Phe-Arg															151
Phe-Arg	NTA	Aq											×		127
Phe-Arg Tfa											+8	Aq			640
Phe-N^G-nitro-Arg-OBzl HCl										187–189	−6.3	Me_2N-CHO			279
Phe-N^G-nitro-Arg-OMe Tfa															640
Phe-N^G-nitro-Arg Tfa											+9	Aq			640
N-Z-Phe-S-chlorocarbonyl-Cys-OMe									Be	154–155	−5.6	Di			250
N-Z-Phe-Cys-OMe									EtAc-Pe	126–127	−14	Me			250
N-Z-Phe-S-Phoc-Cys-OMe										165.3–165.5	−52	DMF			250
Boc-Phe-S-Bu^t-Cys										138–140 (dec)	−102	80% Me			707
Decanoyl-D-Phe-Cys(SO_3Me)-OMe			40						Me	202–204	+25				188
Phe-S-Bu^t-Cys ½ hydrate											−62	1 *N* HCl			707
N,N′-Bis(Z-Phe)-Cyst-Di-OMe	DCCI	Chf	73						Alc	137–139	−89	DMF			250
Phe-Cyst dihydrate											−68	1 *N* HCl			707
Z-Phe-Glu	OSU		80.9						Aq-Alc	173–175	−4.0	EtAc			257
N^α-Z-Phe-γ-*t*-OBu-Glu-OMe	DCCI	MeCl	98							67–68	+10	Chf			10
Phe-γ-*t*-OBu-Glu-OMe					29										10

Ac-Phe-Gly anilide									Ac-Aq	212.5–213	+22	Ac			563
Z-Phe-Gly	ONP		84.3						Aq-Alc	151–152	−9.5	Ac			257
					126				EtAc-Pe	154–155	−9.0	Ac			699
Z-Phe-Gly-Bzl amide				Seph-adex LH-20	8				Me	143–145					677
Z-Phe-Gly 2,2-Diethyleneglycol ester	MCA	THF													677
Z-Phe-Gy-OEt	A	MeCl	76						EtAc-Pe	106–108					677
	TPP-DDD		89							110–110.5	−17	Alc			397
	ONP	Chf	88						EtAc-Pe	108–110	−24	DMF			196
											−18	Alc			
										112	−17	Alc			717
									Me-Aq	106–108					677
															75
Z-DL-Phe-Gly-OEt	OBHAE	THF	85							97–98					371
Z-Phe-Gly-OMe	AC	Py	88		126				EtAc-Pe	119–120					699
Z-Phe-Gly-T										212–214					195
Z-DL-Phe-Gly-T										208–209					195
Boc-Phe-Gly-OEt										Oil					677
Z(OMe)-D-Phe-Gly-OBzl	DCCI		82		124					115–116	+14	Alc			401
Methyl-3-(N-Z-Phe-Gly)-2-oxo-thiazolidine-4-carboxylate									Alc	177.5–178.5	−82	DMF			250
Nps-Phe-Gly-OBzl(OMe)									EtAc-He	118–119	−4.6	Alc			380
Nvoc-Phe-Gly										167–168	−12				478
Phe-Gly					126					262 (dec)	+104	Aq			699
									Aq-Alc	254–258	+99	Aq			380
Phe-Gly anilide									Alc-Aq	169–170	+67	Ac			563
Phe-Gly-OBzl(OMe), HCl, H_2O									Aq	89.5–93	+14				380

(continued)

Table 2 Dipeptides (continued)

Compound	Method	Solvent	% Yield	Chromatography Col	Tl	Ion	Paper	CCD	Recrys. Sol.	M.P. °C	$[\alpha]_D$	Solvent Rotation	Elect.	Other	Ref.
Phe-Gly-OBzl(OMe), Ts									Chf-Eth	127–128.5	+10				380
Phe-Gly monohydrate									Aq-An	259.2–260.4	+42	Ac			563
											+84	Aq			
Pht-Phe-Gly									Alc	183–185	−149	Alc			563
Pht-Phe-Gly anilide	AC	MeCl	75							211–212	−196	Ac			563
Pht-DL-Phe-Gly-T										260–262					195
Pht-Phe-Gly-OEt	EA	Di	60						Alc	160–161	−146	Alc			564
	AC	MeCl	59						Alc	160.6–161.4	−146	Alc			563
N-Thio-Bz-DL-Phe-Gly									Me	217					32
N-Tos-Phe-3-methyl-azylyl-Gly-OEt									EtAc-He	118–121.5	−52	Me		NMR	437
Z-Phe-Leu-OBu^t									Pe	94–95	−27	Me			656
Z-Phe-Leu-OMe										100–101	−25	Me			717
Z-Phe-Leu-ONp	OSU	THF	82						Alc	115–116	−9.7	DMF			377
Boc-Phe-Leu-ONp	OSU		90						Alc-He	125–126	−7.0	DMF			377
Phe-Leu-OBu^t			97												671
N-Tfa-Phe-Leu-OBu^t	DCCI	MeCl	97						EtAc-He	139–140	−12	Me			671
(Z-Phe-O)$_2$ Cu															630
Boc-Phe-Phe amide	OPy		97						EtAc	208–209					134
Boc-Phe-Phe-OBzl										148–149					671

Boc-Phe-Phe-OMe										114–115	−14				448
Boc-Phe-D-Phe-OMe										127–130	+2.4				448
For-Phe-Phe-OBu^t	OBHAE	THF								114.5–117					371
Tfa-Phe-Phe-OBu^t	OBHAE	THF	62							130–131					371 671
N-Ac-dehydro-Phe-Pro										143–145					79
N-Ac-dehydro-Phe-Pro amide										234.5–235					79
N-Ac-dehydro-Phe-Pro-diketo-piperazine										170–173					79
Z-Phe-Pro	OSU		72						EtAc-He	104–106	−42	Me			257
N-Z-Phe-Pro-OMe	MCA	Chf													496
Z-Phe-Pro-OMe	MCA	Chf													494
Z-D-Phe-Pro-OMe															494 496
Dehydro-Phe-Pro									An-Aq	168–170					79
Dehydro-Phe-Pro-diketopiperazine										170–172					79
For-D-Phe-Pro-OBzl	DCCI	Chf-THF	87							Oil					384
For-D-Phe-Pro Dcha										177–178	−51				384
Phe-Pro										123–124 (dec)					574
Boc-Phe-Z-Thr-OMe	OPy		63						Eth-Pe	83–84					134
N-Tos-Phe-Thr hydrazide									Me	193					437
Boc-Phe-Z-Tyr-OBzl	OPy		92						Eth-Pe	102–103					134
N-Ac-dehydro-Phe-D-Val									99% Alc	218	−24 −36	Alc Py			438
N-Ac-dehydro-Phe-D-Val-OEt										154	−34	Py			438
N-Ac-dehydro-Phe-D-Val-OMe										195	−24	Py			438
N-Ac-dehydro-Phe-D-Val-OPr										151	−39	Py			438

(*continued*)

Table 2 Dipeptides (continued)

Compound	Method	Solvent	% Yield	Chromatography Col	Tl	Ion	Paper	CCD	Recrys. Sol.	M.P. °C	$[\alpha]_D$	Solvent Rotation	Elect.	Other	Ref.
Z-Phe-Val															371
Z-Phe-Val-OHq(5-chloro)	MCA	THF	60						EtAc-He	152–152.5	−40	DMF		Mol. wt.	69
Z-Phe-Val-OBu^t										96–98					671
Z-Phe-Va-ONp									THF	152–158					618
Boc-Phe-Val amide	OPy		90						EtAc-Pe	195–196					134
N-Boc-Phe-Val Dcha									An-He	155–156	−6.2	Me		Mol. wt.	69
Proline															
Z-Pro-Gly-OSu									EtAc	149–150	−59	Me			83
Boc-Pro-Gly-ONp					54				Chf-Eth	153–154					627
Imaa-Pro-Gly	OSU	Di-Aq	72						MeCl-Eth-Pe	168					703 700
Nps-Pro-N$^{\beta}$-Z-α-β-diaminopropionyl-Gly amide									Alc	180–181	−46	DMF			730
Nps-Pro-N$^{\beta}$-Z-α,β-diaminopropionyl-Gly-OEt									EtAc-Pe	172–173	−38	DMF			730
Pro-N$^{\beta}$-Z-α,β-diaminopropionyl-Gly amide						Zerolith FF			Alc-Eth	147–148	−25	DMF			730
Pro-Gly-OMe									Eth-Pe	68–70					676
Pro-Gly-ONp HBr					54				Me-Eth	195–196					627
Z-Pro-His amide															538

Aoc-Pro-3Hyp(*t*-Bu) Dcha	DCCI	Chf	58						EtAc	157–158					319
Z-Pro-Leu			60.7						Ac-Aq An-He	134–135	−75	Chf			257
Boc-Pro-D-Leu-OMe										126–127	−24				448
Boc-Pro-Leu-OMe										80–81	−79				448
Imaa-Pro-Leu										88					704
Imaa-Pro-Leu-OMe															700
Imaa-Pro-Leu-Osu ½ hydrate										82–84					703
N-Tfa-Pro-Leu-OBu^t									EtAc-Pe	88–89					671
Z-Pro-Phe	OSU		87						Alc-Pe	127–128	−49	Chf			257
Z-Pro-Phe-OMe										73–75	−37	Me			717
Z-Pro-Phe-ONBzl									EtAc-Pe	85–87					480
Boc-Pro-Phe-nitro-agmatine	OSU	DMF	83							92–96	−46	DMF			279
Pz-Pro-Phe					126				EtAc-Pe	201–202	$[\alpha]_{578}$ −85	DMF			699
Pz-Pro-Phe-OMe	DCCI	Acn	84							Oil					699
Pro-Phe	NTA								Aq						127
Pro-Phe-ONBzl HBr									EtAc	142–144	+27	Alc			480
Aoc-Pro-Pro					101				EtAc-He						417 319
Aoc-Pro-Pro-OBzl	DCCI	MeCl			95										417
Aoc-Pro-Pro-OSu									EtAc-Pe	111.5–112.5	−104	Alc			569
Z-Pro-Pro										185–186					676
Boc-Pro-Pro									EtAc	186–187				IR NMR	121
	ONP		88							191–193	−138	Chf			530
Boc-Pro-Pro-OBzl	MCA	Chf	98							Oil					121
Pro-Pro					105					144–145	−169	Aq			530
Pro-Pro-OBzl HCl														IR	121
Z-Pro-Val-OMe															415

(*continued*)

Table 2 Dipeptides (continued)

Compound	Method	Solvent	% Yield	Chromatography Col	Tl	Ion	Paper	CCD	Recrys. Sol.	M.P. °C	$[\alpha]_D$	Solvent Rotation	Elect.	Other	Ref.
Trans-4-methyl-Pro-D-Val				Seph-adex G-10		A-C.G. 120	×								222
3-Hydroxyproline															
Z-3-Hyp-Ala									EtAc-Pe	153	−28	Me			675
Z-3-Hyp-Ala hydrazide			53						Me-EtAc-Pe	190					675
Z-3-Hyp-Ala-OMe	MCA	THF								Oil					675
N-Nitroso-4-Hyp-Gly amide									DMF-Alc	197–198	−54	DMF			598
4-Hydroxyproline															
N-Nitroso-4-Hyp-Gly-ONp	MCA		86						DMF-Alc	204.5–205.5 (dec)	−73	DMF			598
N-Nitroso-4-Hyp-Gly-OTMB	ONP	DMF	83						Alc	188.5–189.5	−65	DMF			598
	OPCP		75						Alc	186–187	−67				
Serine															
Z-Ser-Ala-OMe	EEDQ	DMF	58						EtAc	112–114	−31	Me			712
OBu^t-N-Z-Ser-Ala					107				Eth-Pe	101–102	+12	Alc			711
OBu^t-N-Z-Ser-Ala-OMe	OSU	THF	88		27				Eth-Pe	78–79	−14	Me			711
OBu^t-Ser-Ala					49					105	−17	Aq			711
OBz-N-Boc-Ser-N^G-nitro-Arg-OMe											−12	Me			640
OBz-Ser-N^G-nitro-Arg-OMe Tfa salt															640

Z(OMe)-Ser-S-Bzl-Cys	OPCP	DMF-Di	64						Me	149–150	−16	DMF			714
Ser-S-Bzl-Cys									Me	140 (dec)	−43				714
D-Ser-Glu Boc-hydrazide							×		EtAc						492
Z-D-Ser-Gln Boc-hydrazide									Me						492
Z-Ser-Gy Boc-hydrazide									EtAc	112–114	−6.3	Me			83
Z-Ser-Gly-OEt										106–107	−5.9	Alc			717
N-Z-Ser-Gly-ONp	DCCI	Acn	50						DMF-Me	170–171					626
OBzl-N-Boc-Ser-Gly-OEt	DCCI	Chf	86		55 126				Eth	87–88	+25	DMF			58
Boc-Ser-Gly-OBzl	MCA	Chf	65						Chf-Eth	79–79.5				Mol. wt.	626
N,O-Di-Tos-Ser-Gly-OEt									Eth-EtAc	107–108	−2.7	Alc			437
N,O-Di-Tos-Ser-Gly-OMe										114–115	−4.8	Chf			437
Ser-Gly-OBzl HCl									Alc	135–136				Mol. wt. AA	626
Ser-Gly Boc-hydrazide															83
Ser-Gly-ONp HBr									DMF-Acn	172.5–173.5					626
N-Tos-Ser-Gly-OEt	A	EtAc	85						EtAc	108–109	−24	Alc			437
N-Tos-Ser-Gly-OMe	A	EtAc	84							107–108	−21	Alc			437
Z-Ser-Leu-OBu^t									EtAc-Eth	91–93	−42	Me			656
OBzl-N-Boc-Ser-N^δ-Z-Orn hydrazide									Alc-Aq	152–154	−9.5	Ac			261
OBzl-N-Boc-Ser-N^δ-Z-Orn-OMe	DCCI	MeCl	90						EtAc-Pe	86–88	+2.8	Ac			261
OBzl-N-Boc-Ser-D-Phe hydrazide									EtAc-He	114–115	−6.3	Ac			261
OBzl-N-Boc-Ser-D-Phe-OMe	MCA	THF	77						EtAc-He	81.5–82	−19	Ac			261
Ser-Pro									Aq-Alc	196–198 (dec)	−109	Ac		AA	83
Ser-Ser	TMP		0												504
DNS-Ser-Tyr															340

(*continued*)

Table 2 Dipeptides (continued)

Compound	Method	Solvent	% Yield	Chromatography Col	Tl	Ion	Paper	CCD	Recrys. Sol.	M.P. °C	[α]D	Solvent Rotation	Elect.	Other	Ref.
Threonine															
Boc-D-Thr-SBzl-Cys-OMe	MCA	THF	57						EtAc-Pe	79					675
D-Thr-SBzl-Cys-OMe Tfa															675
N-Boc-Thr-Z-Dbu									EtAc-Pe	115–116	−27	DMF			22
N-Boc-Thr-Z-Dbu-OMe	DCCI	THF	82						EtAc-Pe	122–124	−22	DMF			22
Thr-Z-Dbu-OMe Tfa										158–160	−6.2	DMF			22
N,O-Di-Tos-Thr-Gly-OEt									EtAc-Pe	122–122.5	+18	Me			437
N-Tos-Thr-Gly-OEt									EtAc-He	116–116.5	−23	Me			437
N-Z-Thr-Leu-OMe			92							48–53	−36	Chf			10
Thr-Leu-OMe															10
Boc-Thr-N$^{\epsilon}$-Z-Lys-OBzl										82–83	−23				448
N-Z-Thr-Phe									EtAc-He	156–157	+18	DMF			209
N-Z-Thr-Phe hydrazide									Me	207–208	−28 −2.2 −24	HCl DMF Ac			209
Z-Thr-Phe-OMe										107–108	+6.1	Ac			717
N-Z-Thr-Phe-OMe	DCCI	MeCl	80							106–107.5	+3	Ac			209
N-Z-Thr-Phe 1-piperidyl ester										141.5–142.5	+26 +1.4	Chf EtAc			209
N-Z-Thr-Ser-OMe	WRK	Acn	59						EtAc-He	136–137.5	+15 +9.6	Py DMF			209

N-[2-Nitro-3-hydroxy-4-methyl-6-deutero-Bz]-Thr-D-Val-OMe	DCCI	MeCl	75						Me-Chf-Cy	197–199	+1.5 ±0.3	Me			344
Tryptophan															
Z-Trp-Gly-OPh															314
Z-Trp-Gly-T										203–204					195
Z-Trp-Met hydrazide					57				Di-Alc	189–190.5					25
Z-Trp-Met-OMe	MCA	Chf	60		57				EtAc-Pe	134–135	−22	Me			25
	DCCI	DMF	91						EtAc-Pe	132–134					
Boc-Trp-dehydro-Nle	OSU	Alc	66		33						+4.0	Alc			466
Boc-Trp-dehydro-Nle Dcha									Alc-Eth	202–203	0	Alc			466
Boc-Trp-4-dehydro-Nle-5-choro-OHq										120–125	−4.4	Chf			466
Z-Trp-Nva-OMe	MCA	THF	71						EtAc-Pe	144	−25	Me			675
Trp-Nva-OMe HCl													×		675
Boc-Trp-Phe amide	OPy		93						EtAc-Pe	190–191					134
Z(OMe)-Trp-D-Trp									EtAc-Eth-Pe	115–116	−22	Me			342
Z(OMe)-Trp-D-Trp-OMe	DCCI	Chf	90							85–87	−7.2	Me			342
Tyrosine															
N-O-Di-Z-Tyr-Ala-OBzl															132
Tyr-Ala	DCCI		83						Alc	145					132
Z-Tyr-Gln Boc-hydrazide	OTCP	THF	86							145–149	−23	DMF			39
Z-Tyr-Gln hydrazide			83							195–198	−16	DMF			39
Boc-OBzl-Tyr-Gly-OBzl	MCA	THF	93		27				Me-He	101–104	+1.2	Me			375
Nps-Tyr-Gly-OTMB	DCCI		83							205–206	+37	DMF			600
	DCCI	Chf	83							205–206	+37	DMF			600
Tyr-Gly-OTMB HCl									Me-Eth	182–183	+12	DMF			600
OBzl-N-Z-Tyr-Hyp-OBzl(OMe)										Oil	−26				596
OBzl-N-Z-Tyr-Hyp-OTMB										Oil	−28				596
N-Z-Tyr-4-aminobutyryl-Phe									Aq-Me	191–193	−15	DMF			584
Di-Z-Tyr-4-aminobutyryl-Phe-OMe									An-Aq	174–177	−19	DMF			584

(continued)

Table 2 Dipeptides (continued)

Compound	Method	Solvent	% Yield	Chromatography				CCD	Recrys. Sol.	M.P. °C	$[\alpha]_D$	Solvent Rotation	Elect.	Other	Ref.
				Col	Tl	Ion	Paper								
Tyr-4-aminobutyryl-Phe acetate									**Me-Eth**	134–141	+35	Me			584
OBzl-N-Boc-Tyr-Ser-OMe										73–77	+7.7	Me			249
Ac-Tyr-Tyr									Aq	239–240	+19	Alc			612
Ac-Tyr-Tyr-OBzl	A	DMF	69						Alc	194–196	−3.4	DMF			612
O,N-Di-Ac-Tyr-OAc-Tyr-OEt									Alc	199–200					612
OBzl-N-Ac-Tyr-Tyr-OBzl										240–241	+19	Alc			612
	A	DMF	71						Alc	177–179	−3.7	DMF			612
OBzl-Tyr-Tyr-OBzl HCl										94–97					612
OBzl-N-Z-Tyr-Tyr-OBzl									Me-Eth	163–165	−15	DMF			612
OBzl-N-Z-Tyr-Tyr-OEt	MCA	THF							EtAc-Eth	137–139	−12	DMF			612
OBzl-N-Z-Tyr-Tyr hydrazide									DMF-Me	236–237	−13	DMF			612
OBzl-N-Boc-Tyr-Tyr-OBzl	MCA	THF	74						EtAc-Pe	142–143	−3.1	DMF			612
Tyr_2							×		Aq-Eth	284–285 (dec)	+29	Ac			612
Tyr-Tyr-OEt HCl									Alc-Eth	214–216 (dec)	−4.8	DMF			612
Valine															
Boc-Val-Ala-OMe										133–134	−48				448
Boc-Val-D-Ala-OMe										91–92	+8.0				448
Z-Val-N^G-nitro-Arg									Alc-Aq	175–176	−10	Alc			640
Z-Val-N^G-nitro-Arg-OMe									EtAc-Pe	160–161	−26	Di			640
Z-Val-Gly amide	SP		81												313
Z-Val-Gly Boc-hydrazide	DCCI	THF-EtAc	97						EtAc-Pe	142–144	−9.8	Me			39

	MCA	THF	76						Eth-Pe	89–90					
Z-Val-Gly-OBu^t									EtAc-Eth-Pe	146	−4.2	DMF			727
Z-DL-Val-Gly-OEt	OBHAE	THF	94							140–142					371
Z-Val-Gly hydrazide	SP		75												313
Z-Val-Gly-OMe	OSU	THF	83						EtAc-Eth	161–162	−27	Me			223
Z-Val-Gly-T										215–217					195
Nps-Val-Gly-OBzl	DCCI	Chf	96						Eth-Pe	97–98	+52	DMF			727
Tfa-Val-Gly-OEt															415
Val-Gly-OBu^t															727
Val-Gly Boc-hydrazide									EtAc	143–145	+6.9	Me			39
Z-Val-Ile-OMe	SPy		81						MeCl-Pe	116–117	+4	DMF			365
Boc-Val-Ile-OMe	SPy		89						Eth-Pe	147.5–148.5	−3	EtAc			365
Boc-Val-Leu-OMe										126–128	−53				448
Boc-Val-D-Leu-OMe										78	+6.0				448
N-Thio-Bz-DL-Val-Leu									Eth	147–150					32
Tfa-Val-Leu-OBu^t															671
N-Z-Val-N$^\epsilon$-Z-Lys									EtAc-He	140–141	−8.4	Me			94
N-Z-Val-N$^\epsilon$-Z-Lys hydrazide										201–202.5	−4.8	DMF			94
N-Z-Val-N$^\epsilon$-Z-Lys OMe	OSU	Di	80						Me-Aq	125–126.5	−18	Me			94
Z(OMe)-Val-N$^\epsilon$-Z-Lys hydrazide									Me	193–195					712
Z(OMe)-Val-N$^\epsilon$-Z-Lys-OMe	MCA	THF	61						EtAc	140–141	−15	Me			712
Z-Val-Phe	OSU		68.6						EtAc-Pe	174–175	+3.9	DMF			257
Z-Val-Phe Boc-hydrazide	OSU	THF	84						Chf-Pe	162–163	−45	Me			83
Boc-Val-Phe-OMe			91							101					676
Val-Phe Boc-hydrazide acetate											+17 +24	Aq 1 *N* HCl			83
Val-Phe-OMe HCl										196–199					676

(*continued*)

Table 2 Dipeptides (continued)

Compound	Method	Solvent	% Yield	Chromatography: Col	Tl	Ion	Paper	CCD	Recrys. Sol.	M.P. °C	$[\alpha]_D$	Solvent Rotation	Elect.	Other	Ref.
For-2-deutero-D-Val-2-deutero-DL-Pro-OBzl	DCCI										+56	Alc		NMR	343
Z-Val-Ser hydrazide									Aq	246–248 (dec)	−43	1 *N* HCl			39
N-Z-Val-Ser-OMe	DCCI	MeCl	50		54 27				Alc	165–166	−19	Me			58
Z-Val-Ser-OMe	MCA	THF	69						EtAc	171–172	−19	Me			39
N-Aceto-Ac-Val-Val-OMe	DCCI	Di			×					55–58	−53	Alc			42
Z-Val-Val-OPn									EtAc-Cy	137.5–138	−5.6				595
Boc-Val-Val-OBu^t															671
Boc-Val-Val-OMe	SPy		86						Eth-Pe	165–166	−9	EtAc			365
Gallium(III) chloride · divalinate															23
Gallium(III) chloride · di-Val complex															23
Gallium(III) dichloride · Val complex valinate															23
Indium(III) chloride · di-Val complex															23
Indium(III) chloride · di-Val complex hydrate HCl															23
N-Tfa-Val-Val-OBu^t														Glc	671
Val-Val-OBu^t										Oil					491

Chapter 4

Tripeptides

The chemistry and biological activity of pyro-Glu-His-Pro amide, the thyrotropin-releasing hormone (TRH), has generated considerable interest in obtaining related tripeptides during the past few years. The hormone was originally isolated from porcine hypothalami by Schally and colleagues in 1966 (see, e.g., Bøler *et al.*[64]). Subsequently the same substance was isolated from bovine and ovine (sheep) hypothalami and presumably is the same in human hypothalami.[49] As recently mentioned by Beyerman[49] there is a possibility that of the seven or more hormones of the anterior pituitary gland, there is a corresponding releasing hormone in the hypothalamus. Thus, TRH is believed to set off the secretion of thyrotropin and the potential medical importance of this observation has encouraged a substantial effort directed at synthesis of the actual hormone and related tripeptides.[49,64,329,518]

The most recent TRH syntheses were developed through the careful investigations of König and Geiger[329] and offer four new approaches based on protecting amides with the 4,4′-dimethoxybenzhydryl group (Mbh). All involve about five steps with reasonable overall yields. Because of the nicely illustrative nature of these tripeptide syntheses one was selected for outline in Chart 4. A particularly important aspect of each was the formation (88% yield) of pyro-Glu upon cleavage of the Mbh amide-protecting group. The remaining details of this work will appear in the tables of a subsequent volume.

While a number of the tripeptides tabulated below were prepared for structure–activity studies as noted above for TRH, the majority were obtained as intermediates for constructing larger peptides or for other purposes.

Chart 4

329 TRH Synthesis

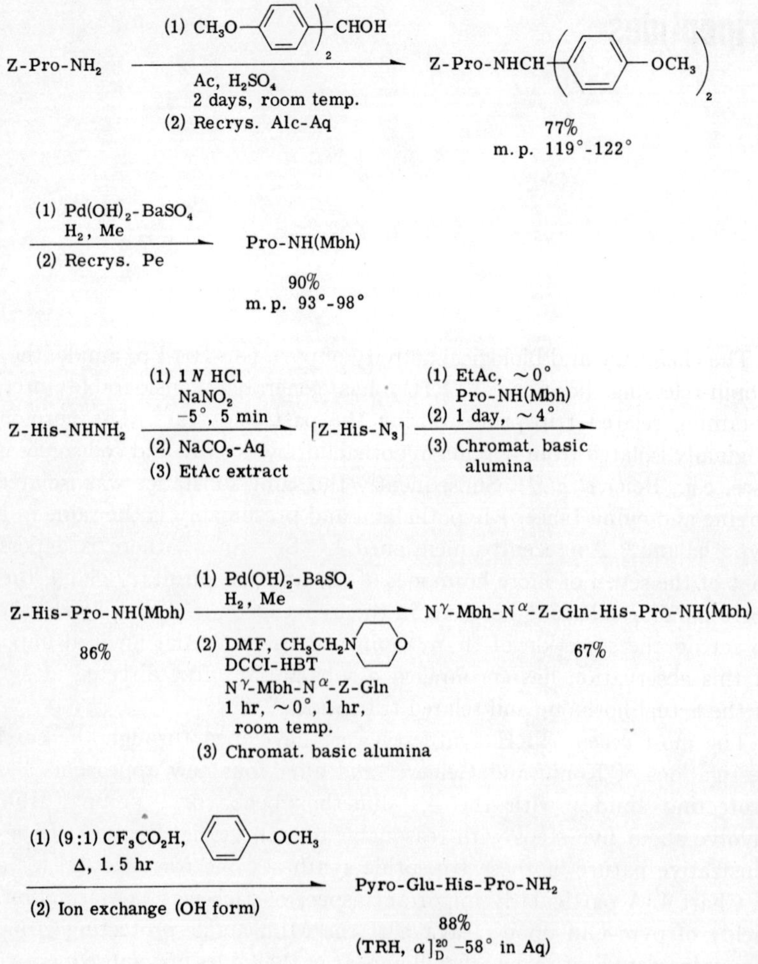

Table 3 Tripeptides

Compound	Method	Solvent	% Yield	Chromatography Col	Tl	Ion	Paper	CCD	Recrys. Sol.	M.P. °C	$[\alpha]_D$	Solvent Rotation	Elect.	Other	Ref.
Alanine															
Ala-Ala-Ala Boc-hydrazide									EtAc	191–193	−19	DMF			197
Ala-Ala-Ala-OEt HCl										144–146 (dec)					186
Ala-Ala-Ala-morpholine amide HCl										135 (dec)					186
Z-Ala-Ala-Ala									EtAc	224–226	−75	TFA			186
Z-Ala-Ala-Ala Boc-hydrazide									Me	248–250	−22	DMF			197
Z-D-Ala-D-Ala-D-Ala Boc-hydrazide															259
Z-Ala-Ala-Ala-OEt									EtAc-He	188–190	−73	TFA			186
Z-Ala-Ala-Ala-morpholine amide									EtAc	198–200	−60	TFA			186
2-OMe-[2-OEt(2-OEt)-Ac-Di-(Ala)-Ala					×					114–116	−80	Aq			186
2-OMe-[2-OEt-(2-OEt)]-Ac-Di-(Ala)-Ala-OEt									Chf-Eth	138.5–140	−71	TFA			186
2-OMe-[2-OEt-(2-OEt)]-Ac-di-(Ala)-Ala-morpholine amide					×					110–111	−50	TFA			186
Ala-Ala-D-Arg HCl									Alc-EtAc		−34	Aq			703
Ala-Nε-Z-Lys-Ala monoformate		Fm	91							209–213	−24	NH_4OH-Ac	×	AA	345
Ala-Lys-Ala Di-HBr										162–165	−26	HCl	×	UV AA	345
Nα-Boc-Ala-Nε-Z-Lys-Ala										94–97	−30	Alc		UV AA	345

(*continued*)

Table 3 Tripeptides (continued)

Compound	Method	Solvent	% Yield	Chromatography Col	Tl	Ion	Paper	CCD	Recrys. Sol.	M.P. °C	$[\alpha]_D$	Solvent Rotation	Elect.	Other	Ref.
N^{α}-Boc-Ala-N^{ϵ}-Z-Lys-Ala OMe	1 + 2 DCCI		56							132–136	−39	Alc		UV	345
N^{α}-Boc-Ala-Lys-Ala										110–114	−35	Aq-NH_4Ac pH 4.9		AA	345
Ala-Phe-Leu	1 + 2 NTA	Aq	63											AA	127
Z-Ala-Phe-Met-OMe										162–163	−53	Me			717
N-Ac-β-(1,3-dioxy-4,4,5,5-tetramethyldihydroimidazol-2-yl)-DL-Ala-Phe-Phe										Oil			×	ESR	664
Ac-Ala-Phe-Pro										213–215	−90	Alc			372
Ac-Ala-Phe-Pro-ONp									EtAc-Eth	174–176	−73	EtAc			372
										148–151	+87	Aq			372
Ala-Phe-Pro									Me-EtAc	159–163	−49	Me			372
Bz-Ala-Phe-Pro									EtAc	136–138	−42	Alc			372
Bz-Ala-Phe-Pro-ONp									EtAc-Eth	112–114	−42	EtAc			372
										149–150	+83	Aq			372
Z-Ala-Phe-Pro										116–119	−51	Alc			372
Boc-Ala-Phe-Pro-OMe	2 + 1 TPP-CTE	Chf	64		×									AA	679
Phenyl-Ac-Ala-Phe-Pro				Si						99–101	−64	Alc			372

Phenyl-Ac-Ala-Phe-Pro-ONp									EtAc	130–132	−65	EtAc			372
										147–149	+86	Aq			372
Ala-Pro-Gly-OMe Tfa					×										367
Ala-Pro-Gly-ONp Tfa															367
Ala-Pro-Gly-OSu Tfa					×										367
Boc-Ala-Pro-Gly									EtAc-He	134–136					367
Boc-Ala-Pro-Gly-OMe	2 + 1 MCA	Glyme	77						EtAc-He	141–142				UV	367
Boc-Ala-Pro-Gly-ONp										191–192.5					367
Boc-Ala-Pro-Gly-OSu									EtAc-He	128.5–130.5					367
N-Ac-β-(1,3,dioxy-4,4,5,5-tetramethyldihydroimidazol-2-yl)-DL-Aa-Pro-Phe										Oil			×	ESR AA	664
CHla-Ala-Pro-Phe															151
β-(1,3-Dioxy-4,4,5,5-tetramethyldihydroimidazole-2-yl)-Ala-Pro-Phe				Si									×	ESR	664
Boc-Ala-D-Thr-Cys									EtAc	127 (dec)					675
Ala-D-Ala-D-Arg HCl									Me-Alc-EtAc		−35	Aq			703
Z-Ala-Ala-D-Arg	2 + 1	Di-Aq	78								−40	Me			703
	OSU														702
Z-Ala-D-Ala-D-Arg															702
	2 + 1 OSU	Di-Aq	90								−1.5	Me			703
Ala-Ala-Gly Boc-hydrazide monoacetate										110–112	−17	Me		AA	39
Ala-Ala-Gly-OPcp HBr										200 (dec)	−17				71

(*continued*)

Table 3 Tripeptides (continued)

Compound	Method	Solvent	% Yield	Chromatography Col	Tl	Ion	Paper	CCD	Recrys. Sol.	M.P. °C	$[\alpha]_D$	Solvent Rotation	Elect.	Other	Ref.
Z-Ala-Ala-Gly	2 + 1 A	Aq-THF	72						Me	210–211	−46	Me			39
Z-Ala_2-Gly										202–205	−63				71
Z-Ala-Ala-Gly Boc-hydrazide									EtAc	151–153	−28	Me			39
Z-D-Ala-D-Ala-Gly-OBu^t															259
Z-Ala-Ala-Gly-OPcp										223–226 (dec)	−42				71
N-Boc-Ala-N^γ-(Dmb)-Gln-Ala-OMe									EtAc-Eth	126–127					491
Ala-N-2-carboxamidoethyl-Gly-Gly						IR-45					+7.0	Aq		IR AA	597
Ala-Gly-Gly							×			217 (dec)	+32 +33	 Aq			71
Ala-Gly-Gly-ONp										131–133					444
Ala-Gly_2-ONp HBr									*i*-Pr-He		+35				71
Ala-Gly_2-OPcp HCl									DMF-EtAc	178–180 (dec)	+26				71
Ala-Gly-Gly-SPh										145–147					444
Ala-Gly_2-OSu HCl									DMF-EtAc		+29				71
Z-Ala-N-2-carboxamidoethyl-Gly-Gly-OTMB	2 + 1 ONP	DMF	94						Alc	172–173	+1.2	DMF		PMR	597
	2 + 1 MCA	DMF	55						Alc	173–173.5	+1.5	DMF		IR PMR	
	2 + 1	DMF	76–95						Alc	172.5–	+1.1	DMF			

	DCCI-NHS								173					
Z-Ala-Gly-Gly									162	−14	DCA			71
Z-Ala-Gly_2-ONp								EtAc-Eth	102–104	−7.5				71
Z-Ala-Gly-Gly-OPn								Alc	155.5–156	−3.7				595
Z-D-Ala-Gly-Gly-OBu^t								Alc-Eth						259
Nps-Ala-Gly-Gly								DMF-Eth	193–194	−39	DMF			71
Nps-Ala-Gly_2-OPcp								DMF-EtAc	162 (dec)	−25	DMF			71
Nps-Ala-Gly_2-OSu								DMF-EtAc	99–100	−29	DMF			71
Tos-Ala-Gly-Gly-OBzl									188–190					444
Z-Ala-Ile-Gly-OMe	1 + 2 ONP	DMF	91		29			EtAc-He	189–190	−59	Me			574
N-Z-Ala-Leu-Ala									194–195	−21	DMF			490
Z-Ala-Leu-Ala-OMe	1 + 2 ONP	EtAc	76					EtAc-Pe		−29 ±0.5	DMF			490
D-Ala-Leu-Arg Tfa salt										−22	Me			640
Z-D-Ala-Leu-N^G-nitro-Arg										−10	Me			640
Z-D-Ala-Leu-N^G-nitro-Arg-OMe										−22	Me			640
Ala-Leu-Met														499
Boc-Ala-D-Thr-SBzl-Cys									171					675
Boc-Ala-D-Thr-SBzl-Cys-OMe	1 + 2 MCA	THF-Aq	75						144					675
N-Z-Ala-Val-Val-OBu^t	1 + 2 ONP	Py	60					EtAc-Pe	150–151					491
β-Alanine														
3-(β-Ala-Gly-amino)-2-Tos-aminopropionyl-Gly								Alc-Aq	198–200 (dec)	−19	10% HCl			223
3-(Z-β-Ala-Gly-amino)-2-Tos-aminopropionyl-Gly-OBzl	2 + 1 DCCI	THF	56					EtAc-Pe	166–168	−21	MeCl			223

(*continued*)

Table 3 Tripeptides (continued)

Compound	Method	Solvent	% Yield	Chromatography Col	Tl	Ion	Paper	CCD	Recrys. Sol.	M.P. °C	$[\alpha]_D$	Solvent Rotation	Elect.	Other	Ref.
Z-β-Ala-Tyr-Ser hydrazide									DMF-Alc	220–224 (dec)	+16	Ac			249
Z-β-Ala-Tyr-Ser OMe										195–196	+2.1	Me			249
α-Aminobutyric acid															
Z-Abu-OBzl-Tyr-Ser hydrazide	1 + 2 DCCI	Acn	72						Alc	151–153	−2.2	Ac			249
Arginine															
Arg-Arg-Leu-OMe Tri-Tfa salt															640
$(Z)_3$-Arg-N^G-nitro-Arg-Leu-OMe									Me	158–169	−11	DMF			640
Arg-γ-OBu^t-Glu-β-OBu^t-Asp Boc-hydrazide diacetate hemihydrate					15				Me-Eth	78–81	−21	DMF			690
Tri-Z-Arg-γ-Bu^t-Glu-β-Bu^t-Asp Boc-hydrazide hemihydrate	1 + 2 OSU	THF	56		35				EtAc	154–156	−12	DMF			690
Arg-Gly-Phe	SP														667
Arg-Gly-Phe amide									Alc-Eth	119–121					667
Decanoyl-Arg-Leu-D-Ala									Me-Eth	263–264	−21	DMF			640
N^G-Nitro-Arg-Leu-D-Ala-OMe HCl															640
N^G-Nitro-Boc-Arg-Leu-D-Ala-OMe									Chf-Eth	106–108	−7.5	DMF			640
N^G-Nitro-decanoyl-Arg-Leu-D-Ala									Alc-Aq	251–253	−17	DMF			640
N^G-Nitro-decanoyl-Arg-Leu-Ala-OMe									EtAc	189–191	−16	DMF			640
Arg-Leu-Arg-OMe Tri-Tfa salt															640
N^G-Nitro-Z-Arg-Leu-N^G-nitro-Arg-OMe									Me	191–192	−14	DMF			640

Tris-Z-Arg-Leu-Gln	1 + 2 ONP	Di-Aq	65							178–179.5 (dec)	−19				710
	OSU	Di-Aq	71		107				DMF-EtAc	185–187	−17				
N^G-Nitro-Boc-Arg-Leu-Phe-OMe									EtAc-Eth	95–96	+4.5	EtAc			640
N^G-Nitro-Arg-Leu-Phe-OMe HCl															640
N^G-Nitro-Z-Arg-Val-Ser-OMe	1 + 2 OPCP	DMF	60		137 54				Alc	185–187	+3	DMF		UV	58
Asparagine															
Z-Asn-Gln-N^{im}-Bzl-His-OBzl	1 + 2 ONP	DMF	63						Me-Aq	184–185	−21	Ac			300
Z-Asn-Gln-His-OMe	1 + 2 ONP	DMF	86						DMF-Me	204–207	−14	Fm			300
Asn-N^ϵ-Tfa-Lys-Gly-OBu^t diacetate									Me-Eth	103–105	−14	DMF			692
Z-Asn-N^ϵ-Tfa-Lys-Gly-OBu^t hemihydrate	MA	Py-Chf	61						EtAc	165	−8.9	DMF		AA	692
Decanoyl-Asn-Phe-N^G-nitro-Arg-OMe									EtAc	205–207	−25	Ac			640
Aspartic acid															
γ-OBzl-Nps-Asp-γ-OBzl-Asp-N^ϵ-Z-Lys-OBzl	1 + 2 MCA		84						EtAc-Eth	137–138	−1.4	DMF			727
α-Bzl-Asp-S-Bzl-Cys-Gly-OPcp HBr									Me-Eth	195–196 (dec)	−15	DMF			336
β-Bzl-Asp-S-Bzl-Cys-Gly-OPcp HBr										196–198 (dec)	−13	DMF			336
β-Bzl-Asp-S-Bzl-Cys-Gly-OPcp Tfa										170 (dec)	−17	DMF			336
α-Bzl-Z-Asp-S-Bzl-Cys-Gly-OPcp	1 + 2 MCA	THF	95		36a				THF-Eth	193–195	−26	DMF			336
β-Bzl-Boc-Asp-S-Bzl-Cys-Gly-OPcp	1 + 2 MCA	THF	90		36a				THF-Me	168–169	−26	DMF			336

(*continued*)

Table 3 Tripeptides (continued)

Compound	Method	Solvent	% Yield	Chromatography Col	Tl	Ion	Paper	CCD	Recrys. Sol.	M.P. °C	$[\alpha]_D$	Solvent Rotation	Elect.	Other	Ref.
β-OBu^t-Boc-Asp-OBu^t-Glu-Gly-OMe										197–198					713
β-OBu^t-Asp-OBu^t-Ser-Ala					108					133 (dec)					711
β-OBu^t-Z-Asp-OBu^t-Ser-Ala Dcha	1 + 2 OSU	DMF-Py	73		86				Eth	154–157	+3.0				711
β-OBu^t-Asp-OBzl-Tyr-OBzl-Ser-Z hydrazide										Oil					671
Cysteine															
S-Bzl-N-Boc-Cys-4-Bzl-Asp-N$^\epsilon$-Z-Lys-OBzl	1 + 2 DCCI-NHS	Di-EtAc	74						EtAc-Pe	99.5–100.5	−20	Me			333
Boc-Cys-Gly-Glu-$(OBu^t)_2$					29						−103	DMF			293
S-Trt-Boc-Cys-Gly-Glu-$(OBu^t)_2$	1 + 2 MCA	THF	61						EtAc-Pe	108–110	+6	Me			293
S-Trt-Cys-Gly-Glu-$(OBu^t)_2$					×										293
S-Trt-N-Trt-Cys-Gly-Glu-$(OBu^t)_2$	1 + 2 DCCI	EtAc	65						EtAc-He	98–110	+8	Me			293
Di-ethyl N,N′-D-Z-Cys-Di-Gly															218
Z-Cys(CH_2CONH_2)-Gly-Leu-OMe									Ac-Eth	95–97	−37	DMF			188
S-Bzl-Cys-Gly-Phe-OMe HCl										166–169					590
S-Bzl-N-Dmpoc-Cys-Gly-Phe-OMe										Oil					590
S-Bzl-N-Z-Cys-Gly-Ser-OMe	1 + 2 ONP	DMF	75						EtAc-Eth	149–151	−29	DMF			300
Z-Cys(CH_2COOMe)-Gly-D-Val-OMe									Me-Aq	103–105	−24	DMF			188

Decanoyl-Cys(CH_2COOMe)-Gly-D-Val-OMe									Me-Aq	93–95	−26	DMF			188
Z-Cys(CH_2COOMe)-Leu-N$^{\epsilon}$-Z-Lys-OMe									EtAc	150–152	−26	DMF			188
Decanoyl-Cys(CH_2COOMe)-Leu-N$^{\epsilon}$-decanoyl-Lys-OMe									Me-Aq	130–131	−29	DMF			188
S-β,β′-Dicarboethoxyethyl-N-for-Cys-N$^{\epsilon}$-Z-Lys-Ala-OBzl	DCCI	MeCl	56	Si	97					102–103	−21	Me			582
Decanoyl-Cys(CH_2COOMe)-Phe-Orn(Pyr)-OMe									EtAc	137–139	−36	DMF			188
S-*p*-Methoxy-Bzl-N-Boc-Cys-Thr-Leu															10
S-*p*-Methoxy-Bzl-N-Boc-Cys-Thr-Leu-OMe	1 + 2 DCCI	MeCl	76	Si						38–42	−41	Chf			10
Z-Cys-OBz-Tyr-Tyr										188–189	−12	Alc			251
Z-Cys-OBz-Tyr-Tyr-OEt										156.5–157	−20	Alc			251
Z-Cys-OBz-Tyr-Tyr amide										204	−34	DMF			251
Cys-OBz-Tyr-Tyr amide										148	+17	DMF			251
Cys-OBz-Tyr-Tyr-OEt hydrate										157	−12	Alc			251
Cys-OBz-Tyr-Tyr HBr										148	+23	Alc			251
Di-Ac-DBu-D-Ala-D-Ala															443
Glutamic acid															
Pyro-Glu-Ala-His amide HCl											−11	Me			578
Pyro-Glu-Ala-His-OMe					×				Me	217–222 (dec)	−23	Ac			578
Pyro-Glu-β-(pyrazolyl-3)-Ala-Pro amide	2 + 1 A	DMF		Sephadex G-50							−62	Aq			239
Pyro-Glu-β-(3-pyrazolyl)-Ala-Pro-amide acetate														AA	177

(continued)

Table 3 Tripeptides (continued)

Compound	Method	Solvent	% Yield	Chromatography				CCD	Recrys. Sol.	M.P. °C	$[\alpha]_D$	Solvent Rotation	Elect.	Other	Ref.
				Col	Tl	Ion	Paper								
Pyro-Glu-Arg-Pro amide					89 91									NMR Mass spec. Mol. wt.	520
Pyro-Glu-Arg-Pro amide acetate	2 + 1 OSU	DMSO-DMF	62			CG-50									177
γ-OBu^t-Glu-Asn-Tyr-OMe															293
γ-OBu^t-Z-Glu-β-OBu^t-Asp-Gly Dcha	1 + 2 OSU	THF-Aq	44						Me-Aq	179–180 (dec)	+62	Aq			39
γ-OBu^t-Glu-β-OBu^t-Asp-Gly Boc-hydrazide Ac											−8.4	Me		AA	39
α-OBzl-Glu-S-Bzl-Cys-Gly-OPcp HBr									Me-Eth	153–155 (dec)	−18	DMF			336
γ-OBzl-Glu-S-Bzl-Cys-Gly-OPcp HBr									Me-Eth	162–163 (dec)	−9.9	DMF			336
γ-OBzl-Glu-S-Bzl-Cys-Gly-OPcp Tfa									Me-Eth	151–153 (dec)	−12	DMF			336
γ-OBzl-Z-Glu-S-Bzl-Cys-Gly-OPcp	1 + 2 MCA		82		36a				THF-Eth	203–205	−22	DMF			336
α-OBzl-Z-Glu-S-Bzl-Cys-Gly-OPcp	1 + 2 MCA				36a				THF-Eth	191–193	−26	DMF			336
γ-OBzl-N-Boc-Glu-S-Bzl-Cys-Gly-OPcp	1 + 2 MCA		73						THE-Me	151–153	−23	DMF			336

α-OEt-Z-Glu-sulfonamide-Cys-Gly-OEt	1 + 2 DCCI	EtAC-DMF	44						EtAc-Alc	118–120					7
γ-Glu-S-Bzl-Cys-Gly-5-OMe-tryptamine									*i*-Pr	125–127	−6.1	DMF			557
γ-Glu-Cys-Gly-5-OMe-tryptamine tartrate									Eth	109–111	+8	Aq			557
α-OMe-Boc-γ-Glu-S-Boc-Cys-Gy-OMe EtAc solvate	MCA	THF	66						EtAc-Pe	92–94	−37	Alc			431
	DCCI	MeCl	34						EtAc-Pe	92–94				IR	
N-Tfa-Glu-S-Bzl-Cys-Gly-5-OMe-tryptamine-OEt									Be	76–79	−39	Me			557
Pyro-Glu-Dbu-Pro amide acetate														AA	177
N-Z-Di(γ-OEt-Glu)-S-Bzh-Cys-OMe	2 + 1 A	Ac	69						EtAc-He	124–125	−42				235
γ-OBzl-Nps-Glu-γ-OBzl-Glu-γ-OBzl-Glu-OBzl	1 + 2 DCCI	Chf		Si						64–65	$[\alpha]_{546}$ −43	Be			628
γ-OBzl-Nps-Glu-γ-OBzl-D-Glu-γ-OBzl-Glu-OBzl									Be	115–117	$[\alpha]_{546}$ −25	Be			628
Z-Glu$_3$(OBu^t)$_4$	OPCP	EtAc	71							79.5–81	−29				410
	MCA	EtAc	75							77.5–79	−30				
Di-Glu-Glu-Tri-γ-OBu^t									Me-Eth-Aq	162–164	+4.7				393
α-OMe-N-Z-Di-Glu-Gu-Tri-γ-OBu^t										119–120	−27				393
α-OMe-N-Z-Di-Glu-D-Glu-Tri-α-OBu^t										96–98	−11				393
Pteroyl-Tri-Glu															184
Di-γ-OBzl-N-Boc-Di-D-Glu-Leu										Oil					216
Di-γ-OBzl-N-Boc-Di-D-Glu-Leu Dcha									Alc-Pe	101–102					216
Di-γ-OBzl-2-methylsulfonyl-ethyl-N-Boc-Di-D-Glu-Leu										83–84	+23	Chf			216

(continued)

Table 3 Tripeptides (continued)

Compound	Method	Solvent	% Yield	Chromatography Col	Tl	Ion	Paper	CCD	Recrys. Sol.	M.P. °C	[α]D	Solvent Rotation	Elect.	Other	Ref.
Di-γ-OBzl-2-methylthio-ethyl-N-Boc-Di-D-Glu-Leu									Eth An-Eth	84–86	+16	Chf			216
Di-γ-OBzl-N-succinimidyl-N-Boc-Di-D-Glu-Leu									An-Eth	94–95 113–114 (dimorphous)	+4.2	Chf			216
Di-γ-OBzl-N-succinimidyl-Di-D-Glu-Leu HCl										70–80	−25	Chf			216
γ-OBzl-Glu-γ-OBzl-Glu-Phe-OTMB HCl									Alc-Eth	169–170	−11	DMF			594
γ-OBzl-Nps-Glu-γ-OBzl-Glu-Phe-OTMB	1 + 2 DCCI		92						Alc	134–135	−14	DMF			594
γ-OBu^t-Glu-Gly-Gly methyl HCl	1 + 2 OPCP	MeCl	85						Me-Eth	76 (dec)	+39	DMF			277
N-Z-Pyro-Glu-Gly-Tyr amide	1 + 2 MCA		67						Me-Eth	132–137 (dec)	−7.2	Me			578
Pyro Glu-Gly-Tyr amide									Me	227–230	+7.9	Me			578
Pyro-Glu-Bzl-His-Ala amide	2 + 1 DCCI	DMF	40						Alc	218–220 (dec)	+7.6	DMF			92
Pyro-Glu-His-Ala amide									Me	240–243 (dec)	−9.9	Me			92
Pyro-Glu-His-Abu amide	2 + 1 DCCI	DMF	35								−29	DMF			92

Pyro-Glu-Bzl-His-Gly amide	2 + 1 DCCI	DMF	34								+6.8	Me			92
Pyro-Glu-His-Gly amide											+0.6	Me			92
Pyro-Glu-His-Ile-OMe	2 + 1 DCCI	DMF	59						Me-Eth	235–238 (dec)	−12	Me			92
Pyro-Glu-His-Ile amide									Me-Eth		−8.5	Me			92
Pyro-Glu-Bzl-His-Leu amide	2 + 1 DCCI	DMF	45						Me	258–260 (dec)	−15	Me-Aq			92
Pyro-Glu-His-Leu amide									Alc-Eth	169–173 (dec)	−19	Me			92
Pyro-Glu-His-Met amide	2 + 1 DCCI	DMF	47						Me-Eth	218–220 (dec)	−37	Aq			92
Pyro-Glu-Bzl-His-Phe amide	2 + 1 DCCI	DMF	60						Me	260–263 (dec)	−27	Ac			92
Pyro-Glu-His-Phe amide									Me	250–255 (dec)	−3.6	Me			92
γ-OBzl-N-Boc-Glu-1-Bzl-His-Pro	1 + 2 OSU		95	Si							−25	DMF			377
N-Z-pyro-Glu-N^{im}-Bzl-His-Pro amide	1 + 2 MCA	THF	90						Chf-EtAc-Eth	149–152	−25	Chf			90
Glu-His-Pro amide															496
Pyro-Glu-Bzl-His-Pro amide	2 + 1 DCCI	DMF	46								−20	Me			92
Pyro-Glu-Bzl-His-Pro-OMe	2 + 1 DCCI	DMF	31								−61				92
Pyro-Glu-N^{im}-Bzl-His-Pro-OMe	SP			Si										Mol. wt. PMR	49
Pyro-Glu-Bzl-His-Pro-$NHCH_3$	2 + 1 DCCI	DMF	44								−16	Me			92

(*continued*)

Table 3 Tripeptides (continued)

Compound	Method	Solvent	% Yield	Chromatography Col	Tl	Ion	Paper	CCD	Recrys. Sol.	M.P. °C	$[\alpha]_D$	Solvent Rotation	Elect.	Other	Ref.
Pyro-Glu-His-Pro amide				Ce	×	A-CG-400			Ac-Eth		−43	95% Ac			248
					×						−50	Aq	×	Mol. wt.	49
											−35	95% Ac			
											−40	Ac		PMR	
															70
															137
	SP		21												90
			68	Si											90
			59		26						−42	Me			64
	SP		73	Sephadex Si							+40	Ac			518
									Me-Eth		−42	Me			92
Pyro-Glu-His-Pro amide HCl									Me-Eth	250 (dec)					64
Pyro-Glu-His-Pro-OMe											−19	Me			92
Pyro-Glu-His-Pro-OMe diacetate			48							105–107	−59	95% Ac	×		49
											−40	Me			
Pyro-Glu-His-Pro-$NHCH_3$									Me-Eth		−45	Me			92
Pyro-Glu-1-methyl-His-Pro amide					23 57									NMR Mass spec.	520

Pyro-Glu-3-methyl-His-Pro amide					23 57									NMR Mass spec.	520
Pyro-Glu-His-Trp					120 99 7				Alc-Eth	188–190 (dec)					91
					120 99 7					188–190 (dec)					579
Pyro-Glu-His-Trp amide	2 + 1 DCCI	DMF	28						Me-Eth	220–230 (dec)	−11	Me			92
Pyro-Glu-His-Trp-OBzl	2 + 1 DCCI-HBT	DMF	72		120 99 7				Alc-EtAc	235–238 (dec)	−6.8	Ac			91
	2 + 1 DCCI-HBT	DMF			120 99 7				Me-EtAc	235–238 (dec)	−6.8	Ac			579
Pyro-Glu-Bzl-His-Val amide	2 + 1 DCCI	DMF	50						Me	285–288 (dec)	−22	Ac			92
Pyro-Glu-His-Val amide									Me-Eth	268–270 (dec)	−17	95% Alc			92
γ-OBzl-N-Boc-Glu-Leu-His-OMe	1 + 2 OSU	Di-EtAc	85						EtAc-Eth	99–101	−27	Me			333
N-Boc-Glu-Leu-His-OMe									Me-Pe	>200 (dec)	−43	Me			333
Pyro-Glu-Lys-Pro amide					57									NMR Mass spec. Mol. wt.	520
Pyro-Glu-Lys-Pro amide acetate	2 + 1 A	DMF	43											AA	177

(*continued*)

Table 3 Tripeptides (continued)

Compound	Method	Solvent	% Yield	Chromatography Col	Tl	Ion	Paper	CCD	Recrys. Sol.	M.P. °C	$[\alpha]_D$	Solvent Rotation	Elect.	Other	Ref.
γ-OBu^t-Z-Glu-Leu-OBu^t-Ser	1 + 2 OSU	DMF-Py							Eth	130–132	−10	Alc			708
γ-OBu^t-Glu-Leu-OBu^t-Ser					49					162–163 (dec)	+19	80% Ac			708
Pyro-Glu-Met-Pro amide					65									NMR Mass spec. Mol. wt.	520
Pyro-Glu-Orn-Pro amide					57									NMR Mass spec. Mol. wt.	520
Pyro-Glu-Phe-His amide									Me	220–223 (dec)	−11	DMF			578
Pyro-Glu-Phe-His-OMe	2 + 1 DCCI	DMF	15						Me	235–240	−0.3	DMF			578
Pyro-Glu-Phe-3-Hyp amide									Me-Eth	169–171 (dec)	−12	DMF			577
Pyro-Glu-Phe-3-Hyp-OMe									Me-Eth	130–132				NMR	577
Pyro-Glu-Phe-Pro amide	2 + 1 DCCI-HBT	DMF		Si	120 99 7				Me-Eth	194–197 (dec)	−44	Me		AA	577

Pyro-Glu-Trp-His amide									Me-Eth		−16	Me			578
Pyro-Glu-Trp-His-OMe															578
Pyro-Glu-Trp-Pro amide	2 + 1 DCCI-HBT	DMF	54		120 99 7				Alc-EtAc	170 (dec)	−23	Me			577
Pyro-Glu-Trp-Tyr amide									Me	250–253 (dec)	−25	DMF			578
Pyro-Glu-O-Bzl-Tyr-Gly amide					×				Me-Eth	197–199 (dec)	−7.8	Ac			578
Pyro-Glu-Tyr-Gly amide											+5.5	Me			578
Pyro-Glu-O-Bzl-Tyr-His-OMe									Me	250 (dec)					578
Pyro-Glu-Tyr-His amide HCl									Me-Eth		+2.7	DMF			578
Pyro-Glu-Tyr-Phe amide									Alc	265 (dec)	−15	DMF			578
Glutamine															
Boc-Gln-Asn-S-Bzl-Cys-OBzl	1 + 2 ONP	DMF	58		×				DMF-An	199–201	−25	DMF			142
Nps-Gln-Asn-S-Bzl-Cys amide	1 + 2 MCA	DMF	60						EtAc-He	214–215					308
Z-Gln-Asp-S-Bzl-Cys-ONBzl	1 + 2 ONP	DMF							DMF-Aq	232–234 (dec)	−27	DMF			243
Z-Gln-β-OBu^t-Asp-Phe-OMe	1 + 2 ONP	DMF	87						Me	186–188	−28	DMF			574
Z-Gln-β-OBu^t-Asp-Phe-OPh										192–194					314
Gln-β-OBu^t-Asp-Phe-OMe HCl					27										574
Boc-Gln-N^{im}-Bzl-His-Leu-OBzl	1 + 2 OPCP	DMF	61							131–134	−28	Me			643
Gln-N^{im}-Bzl-His-Leu-OBzl Tfa															643
Ac-Gln-His-Pro amide					×	A-CG-400			Ac-Eth		−51	Me			248

(*continued*)

Table 3 Tripeptides (continued)

Compound	Method	Solvent	% Yield	Chromatography: Col	Chromatography: Tl	Chromatography: Ion	Chromatography: Paper	CCD	Recrys. Sol.	M.P. °C	$[\alpha]_D$	Solvent Rotation	Elect.	Other	Ref.
Z-Gln-His-Pro amide	1 + 2 ONP	DMF	53	Ce	×						−60 −46	Aq Me			248
Z-Gln-D-Ser-Gln hydrazide															492
Z-Gln-D-Ser-Gln-OMe									Me						492
Pht-Gln-D-Ser-Gln Boc-hydrazide	1 + 2 MCA	DMF													492
Glycine															
Bz-Gly-Ala-Gly-OMe											−44	Me			119
Z-Gly-Ala-Gly-OBzl										161–162					444
Z-Gly-dehydro-Ala-Gly-OEt									EtAc-Eth	130–132				NMR	437
Gly-Ala-Gly, ammonium															444
Z-Gly-Ala-Leu-OBzl	2 + 1 A	DMF	70						Eth-Pe	102	−32	EtAc			263
Z-Gly-Ala-D-Leu-OBzl										125	−1.0	EtAc			263
Gly-Ala-Leu					124				Aq-Me	242–245 (dec)	−84	Aq			263
Gly-Ala-D-Leu					124					243–245 (dec)	−30	Aq			263
Z-Gly-Ala-Pro amide	1 + 2 ONP	DMF							An-Me-Eth	144.5–145.5	−93	Me			574
Gly-DL-Ala-Ser					124										263
Boc-Gly-S-Bzh-Cys-Gly-OBzl	1 + 2 OSU	Chf	90						EtAc-He	123–125	−12	Chf			234

Boc-Gly-S-Trt-Cys-Gly	1 + 2 OSU	Chf	51						Chf-He	190–191	+2.8	Chf			234
Gly-S-Bzh-Cys-Gly										206–207	−13	HMPA			234
Gly-S-Bzh-Cys-Gly-OBzl toluene-sulfonate salt			61							79–80					234
Gly-S-Trt-Cys-Gly										154–155					234
Z-Gly-S-Bzl-Cys-Leu hydrazide									EtAc-Pe	136–138	−19 −23	Me DMF			54
Z-Gly-S-Bzl-Cys-D-Leu hydrazide									Me	142–144	−18 +2.8	DMF Me			54
Z-Gly-S-Bzl-Cys-Leu-OMe	1 + 2 DCCI	EtAc	76						EtAc-Pe	100–102	−42 −36	Me DMF	×		54
Z-Gly-S-Bzl-Cys-D-Leu-OMe	1 + 2 DCCI	EtAc	78							130–131	−8.9 −10	Me DMF			54
Gly-S-Bzl-Cys-Leu-Boc-hydrazide HCl										129–131	−52	Me			54
Gly-S-Bzl-Cys-D-Leu Boc-hydrazide HCl										129–131	−27	Me			54
Gly-S-Bzl-Cys-D-Leu hydrazide 2 HBr													×		54
Gly-S-Bzl-Cys-D-Leu hydrazide 2 HCl											+22	Me			54
Nps-Gly-S-Bzl-Cys-Leu Boc-hydrazide	1 + 2 DCCI	Chf	66							119–121	−53	Me			54
Nps-Gly-S-Bzl-Cys-D-Leu Boc-hydrazide	1 + 2 DCCI	Chf	70						NMe	91–93	+30	Me			54
Z-Di-Gly-Ala	SPh		60							184–185					444
Di-Gly-Ala, ammonium															444
Gly-Gly-β-Ala-SePh HBr															264
Gly-Gly-DL-Ala-SePh HBr															264
Z-Gly-Gly-D-Arg	2 + 1 OSU	Di	69												703

(continued)

Table 3 Tripeptides (continued)

Compound	Method	Solvent	% Yield	Chromatography Col	Tl	Ion	Paper	CCD	Recrys. Sol.	M.P. °C	$[\alpha]_D$	Solvent Rotation	Elect.	Other	Ref.
Dnp-Gly-Gy-Arg															364
Gly-Gly-D-Arg															703
N-(δ-Bromovaleryl)-Gly-Gly-Gly-OEt									Aq-Alc	221–223					88
	2 + 1 CDI	Chf													88
N-($C_2H_4CONH_2$)Gly-Gly-Gly									Aq-Alc	218–219 (dec)					599
[14C]Ac-Gly-Gly-Gly-OEt									Aq	231–232				IR	440
Z-Gly-N-2-cyano-ethyl-Gly-Gly-OTMB	1 + 2 ONP		17						EtAc	131–135				PMR	599
Z-Gly_3-OEt										164–166				IR	718
Z-Gly_3-OMe									Me-Chf	153–155				IR	718
Z-Di-Gly-Gly	1 + 2 ONP	DMF-Alc	87							198					214
Z-Di-Gly-Gly-OEt	1 + 2 ONP		80						Aq-Alc	167					214
Z-Di-Gly-Gly-ONp										220					214
Fe $(Gly)_3$ $(ClO_4)_3 \cdot 6H_2O$															126
Tri-Gly						A-IR4B			Aq-Alc						214
Tri-Gly-OEt HBr															395
Tri-Gly-OEt HCl									Alc-Eth	193					214
									Alc	214–217 (dec)					566

Tri-Gly hydrazide						IRA-400			Alc						566
Di-Gly-Gly-ONp										178–180					444
Di-Gly-Gly-ONp HBr									Alc-Eth	171–172					214
Di-Gly-Gly-SPh										129–131					444
Tri-Gly-SePh HBr									EtAc-Pe	135–136					264
Nps-N-2-cyano-ethyl-Gly-Gly-Gly-OTMB	2 + 1 ONP	DMF	94						Alc	157.5–158.5					599
Pht-Tri-Gly-OEt									Aq-DMF	232.5–233					566
Tos-Di-Gly-Gly-OBzl										174–175					444
Gly-Gly-Glu-γ-OMe-α-SePh HBr															264
Dnp-Gly-Gly-Lys															364
Z-Gly$_2$-Phe hydrazide										134–136	+13	Ac			331
Z-Gly$_2$-Phe-OBzl										81–82	+7.0	Ac			331
Z-Gly-Gly-Phe-OPh-(OBzl)-Chf solvate	2 + 1 SPy	Chf							Chf-Pe	56–88	−18	Chf			110
	MA	Thf-Di-Chf	80												
Decanoyl-Gly-Gly-Pro-OMe															321
Z-Gly-O-(N-Z-Gly)-Ser-OMe									EtAc-Eth	85–87	+15	Chf			540
Gly$_2$-Tyr hydrazide 2 HCl															331
Z-Gly-γ-OBzl-Glu-Gly-OTMB									Alc	151–152	−4.8	DMF		IR	594
Z-Gly-γ-OTMB-Glu-Gly-OTMB									DMF-Alc	192–193	−6.0	DMF			594
Gly-γ-OBzl-Glu-Gly									Aq					IR	594
Z-Gly-γ-OBzl-Glu-Phe-OTMB	1 + 2 ONP	DMF	76						EtAc	160–161	−5.0	DMF		IR	594
Pyro-Glu-His-Thr amide	2 + 1 DCCI	DMF							Alc-Eth		−4.1				92
Gly-Leu-Ala	NCA		86								−59	Aq			296
Z-Gly-Leu-Gly-OMe															415
Z-Gly-D-Leu-N$^{\delta}$-Tos-DL-Orn															495

(*continued*)

Table 3 Tripeptides (continued)

Compound	Method	Solvent	% Yield	Chromatography: Col	Tl	Ion	Paper	CCD	Recrys. Sol.	M.P. °C	$[\alpha]_D$	Solvent Rotation	Elect.	Other	Ref.
Z-Gly-Leu-N^{δ}-Tos-Orn hydrazide	A														496
Z-Gly-D-Leu-N^{δ}-Tos-Orn hydrazide															496
Z-Gly-Leu-N^{δ}-Tos-Orn-OMe															495
Z-Gly-D-Leu-N^{δ}-Tos-Orn-OMe															496
Z-Gly-Leu-N^{δ}-Tos-DL-Orn-OMe															495
Z-Gly-D-Leu-N^{δ}-Tos-DL-Orn-OMe															495
Z-Gly-DL-Leu-N^{δ}-Tos-Orn hydrazide										Oil					494
Z-Gly-DL-Leu-N^{δ}-Tos-Orn-OMe	2 + 1 A	Chf													494
Decanoyl-Gly-Leu-Val-OMe															321
															460
N-Ac-Gly-N^{ϵ}-Z-Lys-Gly-OTMB	1 + 2 OCH_2-CN	DMF	96						Alc-Eth	171–173	−12	DMF			600
N-Ac-Gly-Lys-Gly									Alc-Aq	235–236.5 (dec)	−30	Aq			600
N^{α}-(Z-Gly)-N^{δ}-Tos-DL-Orn-Gly-OEt	A	DMF	47						An-Aq	143–145					201
									An-Aq	143–146					201
	Δ	NMe	83						An-Aq	142–144					201
Gly-Phe-Ala															554
Z-Gly-Phe-Gly-OEt															415
	2 + 1 MCA	EtAc	94						Alc	117–118.5					15

	2 + 1 MCA	THF	85						Alc	120–121					16
	2 + 1 EA								EtAc-Pe	114–116	−12				564
Gly-Phe-Leu	1 + 2 NTA	Aq	64		82						−13	Ac			127
Gly-Phe-[14C]Leu	1 + 2 NTA	Aq			82										127
Z-Gly-Phe-Phe-OBu^t	OBHAE	THF	72							Oil					371
Gly-Phe-Phe amide									Alc-Aq Alc-Eth	117–118 163–164					667
Z-Gly-Phe-Val															371
Z-Gly-Pro-Ala-OPh(OBzl)	MA	Acn	76								−77	EtAc			110
Z-Gly-Pro-Ala-OPh(2,4-dichloro-OBzl)									EtAc-Pe	100–103	−67	Chf			110
Z-Gly-Pro-Ala-OPh(4-chloro-OBzl)									EtAc-Pe	129.5–130.5	−70	EtAc			110
Boc-Gly-4-Hyp-Gly-OTMB	2 + 1 ONP	DMF	78						EtAc	160–160.5	−55	DMF			598
Boc-Gly-4-Hyp-4-Hyp-OTMB									EtAc-Alc	170–171				NMR	598
Decanoyl-Gly-OBz-Ser-N^G-nitro-Arg-OMe											−15	DMF			640
Z-Gly-Thr-Gly-OBzl	2 + 1 A		72						Me	156–157	−16	Di			437
Z-Gly-O-Tos-Thr-Gly-OBzl									EtAc-Eth	115–116	−12	THF			437
N-Ac-Gly-Tyr-Gly OTMB	1 + 2 OCH_2-CN	DMF	67							225.5–226.5	−12	DMF			600
Z-Gly-Tyr-Gly-OBzl	2 + 1 A	Chf	70						EtAc-Eth-Pe	89–91	−5.3	DMF			331
Gly-Tyr-Gly hydrazide 2 HCl															331
Z-Gly-OBzl-Tyr-Tyr-OBzl									EtAc-Pe	151–153	−12	DMF			612

(*continued*)

Table 3 Tripeptides (continued)

Compound	Method	Solvent	% Yield	Chromatography Col	Tl	Ion	Paper	CCD	Recrys. Sol.	M.P. °C	$[\alpha]_D$	Solvent Rotation	Elect.	Other	Ref.
Gly-Tyr_2							×		Aq-Alc	195–197 (dec)	+17	Aq			612
N^{ϵ}-Z-N^{β}-(Z-Gly-Val)-β-Lys-OBu^t															528
N^{ϵ}-Decanoyl-N^{β}-(decanoyl-Gly-Val)-β-Lys-OMe				Si											528
N^{β}-Decanoyl-N^{ϵ}-(decanoyl-Gly-Val)-β-Lys-OMe				Si	×										528
N^{ϵ}-(Decanoyl-Gly-Val)-N^{β}-decanoyl-β-Lys-OMe															525
N^{β}-(Decanoyl-Gly-Val)-N^{ϵ}-decanoyl-β-Lys-OMe															525
N^{β}-(Gly-Val)-β-Lys															528
N^{β}-(Gly-Val)-β-Lys-OMe															528
Histidine															
His-Ala-Gly	NTA														127
N^{α}-Z-His-Nle-Asp-α,β-Di-OBu^t	1 + 2 A	DMF	80						EtAc	155–157 (dec)	−30	Me			238
His-Nle-Asp-α,β-Di-OBu^t-Ac															238
His-Nle-Asp trihydrate									Aq-Me						238
N^{im}-Dnp-His-Pro amide pyro-Glu	SP														518
N^{im}-Bzl-His-Pro-Phe-OPic 3 HBr									Alc						168
Isoleucine															
Ile-Ile-Leu-OBu^t										122–124	−75	Me			671

N-Tfa-Ile-Ile-Leu-OBu^t	1 + 2 OSU	MeCl	83						EtAc-Pe	208	−118	Me			671
Boc-Ile-Pro-Gly	1 + 2 OSU	Glyme-Aq	46							144–145	−53	Me			693
Ile-Pro-Gly Tfa										190–191	+32	0.2 *N* $NaHCO_3$			693
N-Ac-Ile-Val-γ-OBu^t-Glu hydrazide										288 (dec)	−34	DMSO			302
N-Ac-Ile-Val-Glu-α-OEt-γ-OBu^t									Alc	229–231	−29	DMF			302
Z-Ile-Val-Glu-α-OEt-γ-OBu^t															302
Z-Ile-Val-Gly-OBu^t	1 + 2 MCA		67							164–166	−7.9	DMF			727
Ile-Val-Gly						DO-50			Aq-Ac	297–300 (dec)	−16	Aq			727
Ile-Val-Gly-OBu^t										Oil					727
Nps-Ile-Val-Gly-OBzl	1 + 2 DCCI	Chf	86						DMF-Di-EtAc	185–186	+20	DMF			727
Leucine															
Z-Leu-D-Ala-Pro															349
Z-Leu-D-Ala-Pro-OEt	1 + 2 DCCI		77							Oil					349
	2 + 1 A	DMF	56												
Leu-D-Ala-Pro															349
Leu-D-Ala-Pro-OEt HCl										Oil					349
Boc-Leu-S-Bzh-Cys-Gly-OMe	1 + 2 OSU	Glyme	74						Alc-Aq	171–173	−19	Alc		Mol. wt.	69
	1 + 2 DCCI-HBT	Glyme	87							170–172	−19	Alc			
Boc-Leu-Gln-Gly	1 + 2 OSU	Py-Di	36		130				Aq	167–169 (dec)	−22				710

(*continued*)

Table 3 Tripeptides (continued)

Compound	Method	Solvent	% Yield	Chromatography Col	Tl	Ion	Paper	CCD	Recrys. Sol.	M.P. °C	$[\alpha]_D$	Solvent Rotation	Elect.	Other	Ref.
Leu-Gln-Gly HCl					130										710
Z-Leu-Gly-Pro					54				Me-Eth	142–146	−35				705
Boc-Leu-Gly-Pro					54				Me-Eth	143–145	−36	Aq			705
	1 + 2 OSU	Di-Aq	74		54					165	−84				705
Leu-Gly-Val									Aq-Eth		+17	Aq			595
Z-Leu-Gly-Val-OPn									Alc	120–121	−14				595
Z-Leu-Leu Val-OMe										127–129	−54	Me			717
Boc-Leu-N$^\epsilon$-Z-Lys-Gly-OEt	1 + 2 OSU	MeCl-DMF	86–96		29										547
Leu-N$^\epsilon$-Z-Lys-Gly-OEt acetate					59										547
Leu-N$^\epsilon$-Z-Lys-Gly-OEt HCl					59										547
Z-Leu-D-Phe-Gly					124				EtAc-Pe	116–117	+9.3	Me			401
Z-Leu-D-Phe-Gly-OBzl	1 + 2 DCCI	Chf	69		124					106–107	+14	Me			401
Leu-D-Phe-Gly					124					231–232 (dec)	+33	Ac			401
Bpoc-Leu-Sar-Phe-OMe															290
Leu-Sar-Phe hydrate					55 75					102–106	+32	Aq		AA	290
Lysine															
N$^\epsilon$-TrOC-Z-Lys-Ala-Gly									EtAc-Pe	125–126	−3.3	Me			716
N$^\epsilon$-TrOC-Z-Lys-Ala-Gly hydrazide									Me	112–115					716
N$^\epsilon$-TrOC-Z-Lys-Ala-Gly-OMe	1 + 2	DMF	64						EtAc-Eth	105–107	−23	Me			716

	DCCI														
Lys-Ala-Gly				Ce						217–220	−7.7	Ac			716
N$^{\epsilon}$-TrOC-Lys-Ala-Gly						A-CG-4B			Aq-Alc	202–203	+7.0	Ac			716
N$^{\epsilon}$-Boc-Z-Lys-Ala-His-OMe	1 + 2 DCCI-NHS	Di-MeCl	81						Me-EtAc-He	165.5–167	−20	Me			94
Di-decanoyl-β-Lys-Gly-Ala-OMe															525
N$^{\alpha}$-N$^{\epsilon}$-Di-Trt-Lys-Gly-Gly	1 + 2 DCCI	MeCl	64						Be-Hep	122–125					521
N$^{\epsilon}$-N$^{\alpha}$-Z-Lys-Gly-Val-OBu^{t}									Be-Eth	116–117					527
N$^{\epsilon}$-(Z)-N$^{\beta}$-Pht-β-Lys-Gly-Val-OBu^{t}				Si							+13	Di			527
Di-decanoyl-Lys-Gly-Val															527
Di-decanoyl-Lys-Gly-Val-OBu^{t}										Oil					527
Di-decanoyl-Lys-Gly-Val-OMe											−6	Chf			527
Di-decanoyl-β-Lys-Gly-Val-OMe				Si							−12	Di			527
Lys-Gly-Val-OBu^{t}															527
β-Lys-Gly-Val-OBu^{t}															527
N$^{\beta}$-Pht-β-Lys-Gly-Val-OBu^{t}															527
Z-(N$^{\epsilon}$-Boc-Lys)$_3$										107	−12	Me			645
Z-(N$^{\epsilon}$-Boc-Lys)$_3$ Dcha										133	−9.2	Me			645
(N$^{\epsilon}$-Boc-Lys)$_3$										185–187	+7.4	Me			645
Boc-(N$^{\epsilon}$-Boc-Lys)$_3$										Oil					645
Boc-(N$^{\epsilon}$-Boc-Lys)$_3$ Dcha										120–122	−11	Me			645
Boc-(N$^{\epsilon}$-Boc-Lys)$_3$-OSu										96	−18	Me			645
(Lys)$_3$															726
	NCA														481
N$^{\epsilon}$-Boc-Z-Lys-N$^{\epsilon}$-Boc-Lys-Pro	2 + 1 A	DMF-Aq			14 46						−37	Ac			77
N$^{\epsilon}$-Boc-Z-Lys-N$^{\epsilon}$-Boc-Lys-Pro amide															77

(*continued*)

Table 3 Tripeptides (continued)

Compound	Method	Solvent	% Yield	Chromatography Col	Tl	Ion	Paper	CCD	Recrys. Sol.	M.P. °C	$[\alpha]_D$	Solvent Rotation	Elect.	Other	Ref.
N^ϵ-Boc-Z-Lys-N^ϵ-Boc-Lys-Pro-NH$(CH_2)_{15}CH_3$	2 + 1 A	Chf	61						He						77
N^ϵ-Boc-Lys-N^ϵ-Boc-Lys-Pro-NH$(CH_2)_{15}CH_3$										Oil					77
N^ϵ-Boc-Z-Lys-N^ϵ-Boc-Lys-Pro-NHCH$[(CH_2)_{16}CH_3]_2$															77
N^ϵ-Tfa-Z-Lys-N^ϵ-Tfa-Lys-Thr Boc-hydrazide hydrate	1 + 2 OSU	THF	87		4 3 5				EtAc-Pe	180–182	−13	DMF			691
N^ϵ-Tfa-Lys-N^ϵ-Tfa-Lys-Thr Boc-hydrazide acetate					4 3				Me-Eth	115–117	−14	DMF			691
N^ϵ-Boc-Z-Lys-Phe-γ-OBu^t-Glu-OMe	1 + 2 DCCI	MeCl	86							126–127	−26	Chf			10
N^ϵ-Boc-Lys-Phe-γ-OBu^t-Glu-OMe					55										10
N^ϵ-Boc-Z-Lys-Phe-His-OMe	2 + 1 A	DMF-Di		Si	29 27				EtAc	136–137	−15	DMF			574
N^ϵ-Tos-Boc-Lys-Phe-Leu									EtAc-He	100–103	−13	DMF			377
N^ϵ-Tos-Boc-Lys-Phe-Leu-ONp	1 + 2 OSU	DMF	71						Alc	141–141.5	−13	DMF			377
	1 + 2 ONP	THF							Alc	140–141.5					
N^ϵ-Boc-Z-Lys-Tyr-N^{im}-(Boc-TF)-His			38							130–131	−69 ±0.5	DMF			490

N^{ϵ}-Boc-Z-Lys-Tyr-N^{im}-(Boc-TF)-His-OBu^t									Pe		−17	Alc			96
Methionine															
Boc-Met-Cys(CH_2CONH_2)-Leu-OMe			81							Oil					188
Decanoyl-Met-Cys(CH_2CONH_2)-Leu-OMe			71						Alc	185–187	−38				188
Boc-Met-Leu-Gly	SP														313
Boc-Met-Leu-Gly hydrazide	SP														313
Boc-D-Met-Phe-Arg			75								−3.5	Ac			640
Decanoyl-D-Met-Phe-Arg			70								−10	DMF			640
D-Met-Phe-Arg Di-Tfa salt											−9	Ac			640
Norleucine															
Z-Nle-Nle-Pro amide										157–159	−35	DMF			78
Nle-Nle-Pro amide										104–105	−47	DMF			78
Norvaline															
Z-Nva-Nva-Pro amide										156–159	−38	DMF			78
Nva-Nva-Pro amide Tos salt										105–112 (dec)	−50	Me			78
Ornithine															
N^{δ}-Boc-N^{α}-Z-Orn-Gln-N^{δ}-Boc-Orn-OMe	1 + 2 ONP	Py	71						Alc-Eth	189–190	−11	DMF			68
Tri-Orn	NCA														482
Phenylanine															
Boc-Phe-Ala-Gly monohydrate	1 + 2 OSU	Glyme	70						EtAc-Pe	88–90	−17	Me			694
Phe-Ala-Gly										195–200	+1.6	Me			694
Z-Phe-N^G-nitro-Arg-Leu									EtAc-He	171–172	−15	Me			640
Pyro-Glu-Tyr-Pro amide	2 + 1 DCCI		25		120										577
					7										
					120						−32	Me			
					99										
					7										

(*continued*)

Table 3 Tripeptides (continued)

Compound	Method	Solvent	% Yield	Chromatography Col	Tl	Ion	Paper	CCD	Recrys. Sol.	M.P. °C	$[\alpha]_D$	Solvent Rotation	Elect.	Other	Ref.
Pyro-Glu-O-Bzl-Tyr-Trp amide									Me	240–242	−8.4	DMF			578
Pyro-Glu-Tyr-Trp amide									Me-Eth						578
Pyro-Glu-O-Bzl-Tyr-Tyr amide									Me	262–266	−14	DMF			578
γ-OBu^t-N-Z-Glu-Val-Gly-OMe	1 + 2 OPCP	MeCl	78						EtAc-He	138–140	−7.8	DMF			276
γ-OBu^t-Glu-Val-Gly-OMe HCl									Me-Eth	175	+11	DMF			276
Z-Phe-N^G-nitro-Arg-Leu-OMe											−17	Di			640
Boc-Phe-N^G-nitro-Arg-Leu-OMe											−11	Me			640
Decanoyl-Phe-Arg-Leu									Me-Eth	140–143	−15	DMF			640
Decanoyl-Phe-N^G-nitro-Arg-Leu									EtAc	206–208	−14	DMF			640
Decanoyl-Phe-N^G-nitro-Arg-Leu-OMe									EtAc	93–95	−18	DMF			640
Phe-Arg-Leu Tfa salt											+3	Aq			640
Phe-N^G-nitro-Arg-Leu-OMe-HCl															640
Boc-Phe-N^G-nitro-Arg-N$^\epsilon$-hexanoyl-Lys-OMe											−22	Alc			640
Decanoyl-Phe-N^G-nitro-Arg-N$^\epsilon$-hexanoyl-Lys-OMe									Alc	139–141	−10	DMF			640
Phe-N^G-nitro-Arg-N$^\epsilon$-hexanoyl-Lys-OMe Tfa salt															640
Decanoyl-D-Phe-β-OMe-Asp-Cys-(CH_2COOMe)-OMe									Me-An	149–150	−42				188
Z-Phe-Z(OMe)-Cys-Gly-OEt	1 + 2 MCA	THF	61						EtAc	164–165	−47	Me_2N-CHO			488

Z-Phe-S-chlorocarbonyl-Cys-Gly-OMe									Be	153–154 (dec)	−33	Di			250
Z-Phe-Cys-Gly-OMe									EtAc	171–172	−16	DMF			250
Z-Phe-S-Trt-Cys-Gly-OMe	1 + 2 DCCI	MeCl	80						Alc	176–176.5	−13	DMF			250
Propionyl-D-Phe-Cys(CH_2COOMe)-Gly-OMe									Me	152–153.5	−34				188
Propionyl-D-Phe-Cys(SO_3Me)-Gly-OMe									Me	142–144	−20				188
Z-Phe-Gly-S-chlorocarbonyl-Cys-OMe										144–145 (dec)					250
Z-Phe-Gly-Cys-OMe	2 + 1 DCCI	Chf	83						EtAc	135–136	−25	DMF			250
Boc-Phe-Gly-Cys(CH_2COOMe)-OMe									EtAc	75–77	−29				188
Z-Phe-Gly-Gly ½ hydrate	ONP		83						Aq-Alc	99–101	−6.8	Alc			257
Z-DL-Phe-Gly-Gly-T										201–203					195
Z-N-[2-(N-Z-Phe-amino)ethyl]Gly-Phe									Alc-Pe	130–132	−4.4	DMF			26
Z-N-[2-(N-Z-Phe-amino)ethyl]Gly-Phe-OMe									EtAc-Pe	115–120	−10	DMF			26
N-[2-(Phe-amino)ethyl]Gly-Phe									Alc-Eth						26
Boc-Phe-His-Leu hydrazide									Alc-Aq	227–228	−18	Ac			640
Boc-Phe-His-Leu-OMe									EtAc	192–193	−28	DMF			640
Z-D-Phe-Leu-Gly-OMe									EtAc-Cy	157–158	−12				188
Phe-N^{ϵ}-Z-Lys-Ala	1 + 2 NCA	Aq			37						−14	1% NH_4OH			582
Dmpoc-Phe-Met-Gly-OEt										120–122					590
Phe-Pro-N^{δ}-Tos-Orn-OMe HCl									Eth-EtAc						494
Decanoyl-D-Phe-Phe-Arg											−4	DMF			640
Decanoyl-D-Phe-Phe-N^G-nitro-Arg									EtAc	105–107	−13	DMF			640

(*continued*)

Table 3 Tripeptides (continued)

Compound	Method	Solvent	% Yield	Chromatography Col	Tl	Ion	Paper	CCD	Recrys. Sol.	M.P. °C	$[\alpha]_D$	Solvent Rotation	Elect.	Other	Ref.
Decanoyl-D-Phe-Phe-N^G-nitro-Arg-OMe									EtAc	141–143	−21	DMF			640
N-Tfa-Phe-Phe-Leu										215–218					671
N-Tfa-Phe-Phe-Leu-OBu^t	1 + 2 DCCI-NHS	MeCl	79							172–176	−3.3	Ac			671
Z-Phe-Pro-N^δ-Tos-Orn-OMe															496
Z-Phe-Pro-N^δ-Tos-DL-Orn-OMe															495
Z-D-Phe-Pro-N^δ-Tos-Orn-OMe															495
															496
Z-D-Phe-Pro-N^δ-Tos-DL-Orn-OMe															495
D-Phe-Pro-N^δ-Tos-Orn-OMe															496
D-Phe-Pro-N^δ-Tos-DL-Orn-OMe															495
D-Phe-Pro-N^δ-Tos-Orn-OMe HCl															496
D-Phe-Pro-N^δ-Tos-DL-Orn-OMe HCl															495
N-Tos-Phe-Thr-Gly-OEt	2 + 1 A		63						EtAc-He	162–163	−65	Alc			437
N-Tos-Phe-O-Tos-Thr-Gly-OEt									Alc-Eth	121.3–122.5	+7.9	THF			437
Z-Phe-Val-Phe-OBu^t															611
Proline															
Z-Pro-Ala-Phe-OMe										148–151					676
Pro-Ala-Phe-OMe										112–116					676

Z-ω-Bu^t-α-aminosuberyl-Pro-N^ϵ-Tos-Arg-Gly amide											−32	DMF			219
Z-Pro-Gly-Gly	OSU		86						Alc-EtAc-Pe	140–142	−34	DMF			257
Z-Pro-Gly-Phe-OMe															676
Pro-Gly-Phe-OMe															676
Pro-N(im)-Dnp-His-N^α-Boc-pyro-Glu	SP														518
Pro-N(im)-Dnp-His-N^α-Boc-Gln	SP														518
Aoc-Pro-OBu^t-Hyp-Gly Dcha	2 + 1 DCCI	Chf	81						EtAc-Pe	150–151					319
Z-Pro-Leu-Gly	OSU		89						Acn-EtAc-Pe	159–161	−85	Alc			257
Z-Pro-Leu-Gly amide										162–164					62
Z-Pro-Leu-Gly-OEt									EtAc	151–152	−83	Alc			62
Boc-Pro-Leu-Gly amide										150–151	−51	DMF			134
Boc-Pro-Leu-Gly-OMe	1 + 2 OPy	Chf	90						EtAc-Pe	97–98	−49	DMF			134
Pro-Leu-Gly amide															463
															392
Imaa-Pro-Leu-Pro	2 + 1	Di-Aq								137–139					703
	OSU														700
Imaa-Pro-Leu-Pro-OMe															700
Imaa-Pro-Leu-Pro-OSu															700
										91 (dec)					703
Z-ω-Bu^t-α-aminosuberyl-Pro-N^ϵ-Pht-Lys-Gly amide									Alc	112–115	−38	DMF			219
Z-Pro-homo-N^ϵ-Tos-Lys-Gly-OEt	2 + 1 ONP									110–113	−51	Ac			61
Z-Pro-N^ϵ-Pht-Lys-Gly amide	1 + 2 OSU	DMF	77						DMF	173–175	−36	DMF			219

(continued)

Table 3 Tripeptides (continued)

Compound	Method	Solvent	% Yield	Chromatography Col	Tl	Ion	Paper	CCD	Recrys. Sol.	M.P. °C	$[\alpha]_D$	Solvent Rotation	Elect.	Other	Ref.
Z-Pro-N^ϵ-Tos-Lys-Gly-OEt										153–154	−41	Chf			513
Z-Pro-N^ϵ-Boc-Lys-Thr	1 + 2 ONP	Di	83						EtAc	119–123	−34	DMF			619
Z-Pro-N^ϵ-Tos-Lys-Thr-OMe	1 + 2 ONP	DMF	75						Alc-Aq	125–128	−25	DMF			297
Z-Pro-N^δ-Pht-D-Orn-Gly amide	1 + 2 MCA	Chf	60							185–186	−12	DMF			730
Pro-N^δ-Pht-D-Orn-Gly amide						Zero-lith FF									730
Boc-Pro-Phe-N^G-nitro-Arg-OBzl										88–91	−35				279
Boc-Pro-Phe-N^G-nitro-Arg-OPic											−34				540
CHla-Pro-Phe-Arg															151
Pro-Phe-N^G-nitro-Arg-OBzl HCl										122–126	−16				279
Pro-Phe-N^G-nitro-Arg-OPic bis-Tfa											−15	Aq			540
Pz-Pro-Phe-Gly	1 + 2 ONP	Py-Aq	81		126				*i*-Pr-Eth	187 (dec)					699
Pz-Pro-Phe-Gly-OSu					107					153.1 (dec)					699
Aoc-Pro-Pro-Gly															319
Aoc-Pro-Pro-Pro									EtAc-Me						417
Aoc-Pro-Pro-Pro-OBzl	DCCI	MeCl	78						EtAc-He						417
Boc-Pro-Pro-Pro									EtAc	214–215				IR NMR	121

	1 + 2 ONP		93							219	−173	Chf			530
Boc-Pro-Pro-Pro-OBzl	2 + 1 MCA		86							Oil			IR NMR		121
Pro-Pro-hydroxy-Pro-ONp HCl															122
Pro-Pro-Pro					105					122					530
Boc-Pro-Pro-4-Hyp														IR NMR	122
Boc-Pro-Pro-4-Hyp-OBzl	2 + 1 MCA	Chf	88		×					Oil				IR NMR	122
Boc-Pro-OBzl-Ser-Gly-OEt	1 + 2 ONP	Acn	60		55 126					109–110	−9	DMF		AA	58
Boc-Pro-OBzl-Ser-Gly-OBzl	1 + 2 DCCI	EtAc			54 27			×			−27	EtAc			58
Boc-Pro-Ser-Gly														AA	626
Boc-Pro-Ser-Gly-OBzl	1 + 2 MCA		82							48–51				Mol. wt. AA	626
Boc-Pro-Ser-Gly-ONp	1 + 2 MCA	Glyme	75							Oil					626
Pro-OBzl-Ser-Gly-OBzl					54				Alc-Eth						58
Pro-Ser-Gly-ONp Tfa									Acn	143–144					626
Boc-Pro-Val-Phe-OMe	1 + 2 OSU	THF	92		119										676
Pro-Val-Phe-OMe HCl			92						Me-Eth	222–225				Mol. wt.	676
N-Nitroso-4-Hyp-Gly-Gly-OTMB	2 + 1 ONP	DMF	75						Alc	201– 201.5	−37	DMF		AA	598
Sarcosine															
N^{α}-Z-N^{β}-(OBu^{t}-Sar-MeVal)-α,β-Dbu-OMe										109	−26	Me			27

(continued)

Table 3 Tripeptides (continued)

Compound	Method	Solvent	% Yield	Chromatography Col	Tl	Ion	Paper	CCD	Recrys. Sol.	M.P. °C	$[\alpha]_D$	Solvent Rotation	Elect.	Other	Ref.
Serine															
OAc-Ser-Gln-Gly-OMe															208
OAc-Ser-Gln-Gly-OMe HBr															208
OAc-N$^{\alpha}$-Z-Ser-Gln-Gly-OMe	1 + 2 ONP	DMF	52						Aq	210–212	−6.8	N,N-DMF			208
N$^{\alpha}$-Z-Ser-Gln-Gly-OMe	1 + 2 DCCI	DMF	60						Me-Aq	200.5–202	−2.0	DMF			208
Ser-Gln-Gly-OMe															208
OBu^t-Z-D-Ser-Gly-Leu-OMe									EtAc-Hep	96.5–97	−8				188
N-Z-Ser-Met-Asp-D-sulfoxide	1 + 2 A	Eth	45			DO-50			Me-EtAc	156–157	+16	Me			238
Ser-Met-Asp-D-sulfoxide						DO-50			Aq-Ac	242 (dec)	+43	Aq			238
Z-Ser-Phe-Leu-OBu^t									EtAc	160–161	−39	Me			656
Ser-Phe-Leu	1 + 2 NCA	Aq	30		82				Aq-Me						232
OBzl-N-Boc-Ser-D-Phe-Pro-OMe	2 + 1 A	EtAc	~100						Eth-He	62–63	−47	Ac			261
OBzl-Ser-D-Phe-Pro-OMe HCl										189.5–190					261
Boc-Ser-Pro-Gly-ONp	1 + 2 MCA				54					Oil					627
Ser-Pro-Gly-ONp HBr									DMF-Acn	211–212					627
Boc-Ser-Pro-Phe-nitro-agmatine			66						Me-EtAc		−38	DMF			279

Z-Ser-Ser-Gln Boc-hydrazide															492
Z-D-Ser-D-Ser-Gln Boc-hydrazide	1 + 2 A	EtAc							Me-Aq						492
Z-D-Ser-D-Ser-Gln-OMe	1 + 2 A	EtAc- DMF							DMF-Me						492
D-Ser-D-Ser-Gln-OMe							×			Oil					492
Z-Ser-Ser-Gly	1 + 2 A	Aq- THF	64						Me-Eth	169–171 (dec)	−12	Me			83
Z-Ser-Ser-Gly Boc-hydrazide									Acn	111–112	−11	Me			83
Ser-Ser-Gly Boc-hydrazide acetate										Oil	−6.5	Me			83
OBzl-Ser-OBzl-Tyr-Gly-OBzl Tfa					27				Me-Eth	143–145 (dec)	+2.9	Me			91
Boc-OBzl-Ser-OBzl-Tyr-Gly-OBzl					27				EtAc-Pe	92–95	−3.2	Chf			91
Threonine															
Z-Thr-Asn-S-Bz-Cys-OMe	DCCI	DMF	75						DMF-Aq	173–175	−24	DMF			558
Thr-Asn-S-Bz-Cys-OMe HBr															558
α-Bzl-Boc-Thr-Asn-γ-OBzl-Glu	1 + 2 OSU	THF	86		4 3				EtAc-Pe	37–38	−6.3	DMF			689
α-Bzl-Thr-Asn-γ-Bzl-Glu-For							×			103–104	−12	DMF			689
Boc-Thr-N^{ϵ}-Tfa-Lys-Met Dcha	1 + 2 OSU	Glyme- Aq	74							147	−75	Me			693
Thr-N^{ϵ}-Tfa-Lys-Met Tfa										123–128	−11	Aq			693
Thr-Phe-Leu	1 + 2 NCA	Aq	49						Aq		−12	Ac		AA	232
OBu^t-Z-Thr-Phe-Pro															574
OBu^t-Z-Thr-Phe-Pro TEA salt	1 + 2 OSU	DMF								90–93					574
Boc-Thr-Pro-N^{ϵ}-Tos-Lys-Ala													×		205
Z-Thr-Ser-Leu-OBu^t									Chf-Eth	120–121	−37	Me			656
Z-Thr-Val-Gly Boc-hydrazide			50							114–115	−24	Me			39
Thr-Val-Gly Boc-hydrazide acetate			83							140–142	−17	Me			39

(*continued*)

Table 3 Tripeptides (continued)

Compound	Method	Solvent	% Yield	Chromatography Col	Tl	Ion	Paper	CCD	Recrys. Sol.	M.P. °C	$[\alpha]_D$	Solvent Rotation	Elect.	Other	Ref.
Tryptophan															
CHla-Trp-Ala-Pro															151
Boc-Trp-N^G-nitro-Arg-Leu											−27	DMF			640
Boc-Trp-N^G-nitro-Arg-Leu-OMe											−31	DMF			640
Decanoyl-Trp-Arg-Leu											−15	DMF			640
Trp-Arg-Leu Tfa salt											+9	Me			640
Trp-N^G-nitro-Arg-Leu Tfa salt											+11	Me			640
Boc-Trp-Cys(CH_2CONH_2)-Leu-OMe															188
Decanoyl-Trp-Cys(CH_2CONH_2)-Leu-OMe									Alc-Aq	162.5–164	−25				188
Boc-Trp-Gly-Leu-OMe									EtAc-Pe	183.5–184	−31				188
Boc-Trp-Phe-N^G-nitro-Arg			80								−21	DMF			640
Boc-Trp-Phe-N^G-nitro-Arg-OMe											−24	DMF			640
Trp-Phe-Arg Tfa salt											+5	Me			640
Trp-Phe-N^G-nitro-Arg Tfa salt											+34	Me			640
Tyrosine															
OBu^t-Z-Tyr-Ala-Gly-OMe	1 + 3 OPCP	MeCl	70	Si					EtAc-He	122	−7.6	DMF			268
S-Trt-Boc-Tyr-Cys-Asn-OBu^t	1 + 2 MCA	THF							EtAc-Pe	176–178	−8	Me			293
Z-Nps-Tyr-γ-OTMB-Glu-γ-OTMB-Glu-OPcp										166–167		DMF			628

Z-Tyr-γ-OTMB-Glu-γ-OTMB-Glu-OPcp-HCl										154–156					628
Z-Tyr-γ-OBu^t-Gln-Gly Boc-hydrazide	1 + 2 OTCP	DMF-THF	84						EtAc	171–172	−7.4	Me			39
Tyr-γ-OBu^t-Gln-Gly Boc-hydrazide acetate											+1.1	Me			39
OBzl-Z-Tyr-Gly-Gly									Di-Eth-Pe	145–147	−21	DMF			331
OBzl-Z-Tyr-Gly-Gly-OBzl	1 + 2 ONP	Chf	88						EtAc-Eth-Pe	116–118	+4.2	Ac			331
OBzl-Z-Tyr-Gly-Gly hydrazide										209–212	−22	DMF			331
OBzl-Z-Tyr-Gly-Gly-ONp															331
Tyr-Gly-Gly hydrazide 2 HCl															331
OBzl-Z-Tyr-Tyr-Tyr-OBzl	2 + 1 A		71						Me	169–170	−14	DMF			612
Tyr$_3$									Alc	178–181 (dec)	+16	Aq			612
[Tyr^2]-TRF					65									NMR Mass spec.	237
Valine															
Z-Val-N^G-nitro-Arg-N^G-nitro-Arg-OMe									Me	166–167	−14	DMF			640
Val-Arg-Arg-OMe Tri-Tfa salt															640
Val-N^G-nitro-Arg-Leu-OMe HBr															640
Z-Val-γ-OBu^t-Glu-Ala-OMe															299
Boc-Val-γ-OBu^t-Glu-Ala hydrazide									Me	226–227	−49	Me			293
Boc-Val-γ-OBu^t-Glu-Ala-OMe	1 + 2 MCA	THF	72						EtAc-He	134–135	−49	Me			293
Z-Val-γ-OMe-α-Gln-N$^\epsilon$-Tos-Lys-OMe										170.8–171.5	+18	DMF			378
Boc-Val-γ-OBzl-α-Gln-N$^\epsilon$-Tos-Lys-OBzl	1 + 2 OSU	DMF	73						Alc-He	143–143.5	−16	DMF			378

(*continued*)

Table 3 Tripeptides (continued)

Compound	Method	Solvent	% Yield	Chromatography Col	Tl	Ion	Paper	CCD	Recrys. Sol.	M.P. °C	$[\alpha]_D$	Solvent Rotation	Elect.	Other	Ref.
Boc-Val-Gly-Leu-OMe									EtAc-Cy	168–170	+73				188
Z(OMe)-Val-N$^{\delta}$-Z-Orn-Leu-δ-aminovaleryl azide															433
Boc-Val-N$^{\delta}$-Z-Orn-Leu-NHCH$_2$-CH$_2$OH	2 + 1 A	DMF-Di	76						Me-EtAc-Eth	197–198					384
Val-N$^{\delta}$-Z-Orn-Leu-NHCH$_2$CH$_2$OH			94			DO-1			Me-Eth	180–182					384
Z-Val-Phe-Gly									Alc-Aq	207–210	−20	DMF			196
For-2-deutero-D-Val-Pro-N-trideuteromethyl-Gly-OBzl											−19	Alc		NMR	343
For-D-Val-Pro-N-trideuteromethyl-Gly-OBzl	2 + 1 DCCI										−18	Alc		NMR	343
For-2-deutero-D-Val-2-deutero-DL-Pro-Sar-OBzl											+48	Alc			343
For-D-Val-Pro-2,2-dideutero-Sar									Me-EtAc	158–161	−2.8	Alc		NMR	343
Z(OMe)-Val-Ser-S-Bzl-Cys	1 + 2 ONP	Di	67						Me	182–185	−5.0	DMF			714
Val-Ser-S-Bzl-Cys									Me	216–217	−20	Ac			714
Z-Val-Ser-Ser	1 + 2 MCA	THF	47						Me-EtAc	176–178	−26	Me			714
Val-Ser-Ser									Aq-Alc	215–216	+9.9	Aq			714
Val-Tyr-Pro										176–178	−29	Aq			712
Boc-D-Val-Lac-Val-D-Hyv					39						−13	Alc			368
Boc-D-Val-Lac-Val-D-Hyv-OBzl										Oil	−10	Alc			368
D-Val-Lac-Val-D-Hyv-OBzl HCl										Oil	−29	Alc			368

Z-Val-Val Phe Boc-hydrazide	1 + 2 OSU	THF-Aq	83						THF-Aq	215–216	−5.5	DMF			83
Val-Val-Phe Boc-hydrazide acetate									Eth	133–134	−40	Me			83
Gallium(III) dichloride • di-Val complex valinate															23
Gallium(III) chloride • Val complex divalinate															23
Gallium(III) chloride • tri-Val complex															23
Gallium(III) chloride • tri-Val complex • HCl															23

Chapter 5

Tetrapeptides

After the closing date for preparation of the tetrapeptide tables, three papers concerned with tetrapeptide syntheses were viewed. One concerned SP synthesis of pyro-Glu-Tyr-Arg-Trp amide which was found to exhibit detectable luteinizing hormone-releasing factor (LH-RF) activity.[198] Two other papers concerned syntheses of tetrapeptide amides related to the C-terminal Trp-Met-Asp-Phe amide of the gastrins.[113,280] This gastrin tetrapeptide displays marked biological activity and in a new approach to analog synthesis, Jones[280] set out to prepare tetrapeptides whose structures might be predicted by considering the codons (nucleotide triplets) of the messenger RNA for this 4-unit gastrin sequence and the complementary nucleotide sequences predicted by base paring in an antiparallel (4 peptides) or parallel (1 peptide) direction. Unfortunately upon synthesis none exhibited a stimulation or inhibition of gastric acid secretion in anesthetized rat or cat preparations. However, one did show some inhibition of gastric juice volume and acid and pepsin output in a conscious dog with innervated gastric pouches. Since calculations show that for 24-amino acid combinations there will be 331,776 possible tetrapeptides, further refinements in this approach to structure–activity relationships might well be more fruitful in the future.

Synthesis of the cystine derivative of the insulin tetrapeptide A_{6-9} has been described by Zahn and colleagues and its reactivity toward boron trifluoride–glacial acetic acid, hydrazine, and triethylamine has been evaluated.[45]

Table 4 Tetrapeptides

Compound	Method	Solvent	% Yield	Chromatography Col	Tl	Ion	Paper	CCD	Recrys. Sol.	M.P. °C	$[\alpha]_D$	Solvent Rotation	Elect.	Other	Ref.
Alanine															
Tri-Ala-Ala-OEt HCl										240 (dec)					186
Tri-(Ala)-Ala-morpholine amide HCl										245 (dec)					186
Z-Tri-(Ala)-Ala-OEt	2 + 2 MCA	THF	62						Alc	251–253	−95	TFA			186
Z-Tri-(Ala)-Ala-morpholine amide	2 + 2 MCA	THF	40						Alc	270–271	−78	TFA			186
2-OMe-[2-OEt-(2-OEt)]-Ac-Tri-(Ala)-Ala-OEt									Alc	240–241	−89	TFA			186
Z-D-Ala-D-Ala-D-Ala-Gly-OBu^t															259
Ala_2-Gly_2-OPcp HBr										146–149 (dec)	−5.1	DCA		Mol. wt.	71
Z-D-Ala-D-Ala-Gly-Gly-OBu^t	2 + 2 A	Aq	85–88						Alc						259
Z-Ala_2-Gly_2-OPcp	1 + 3 MCA	THF	47						DMF-*i*-Pr	213 (dec)	−24	DCA		Mol. wt.	71
Ala-γ-OBu^t-Glu-Ala-Gly-OPcp HCl									Me-Eth	225	+1	DMF			274
Z-Ala-γ-OBu^t-Glu-Ala-Gly									EtAc-He	183	−5.8	DMF			274
Z-Ala-γ-OBu^t-Glu-Ala-Gly-OMe	1 + 3 OPCP	MeCl	62		22				EtAc-He	191	−8.1	DMF			274
Z-Ala-γ-OBu^t-Glu-Ala-Gly-OPcp									Me	205	−8.7	DMF			274

(*continued*)

Table 4 Tetrapeptides (continued)

Compound	Method	Solvent	% Yield	Chromatography Col	Tl	Ion	Paper	CCD	Recrys. Sol.	M.P. °C	$[\alpha]_D$	Solvent Rotation	Elect.	Other	Ref.
Z-D-Ala-Gly-D-Ala-Gly-OBu^t	2 + 2 ONP	Chf	86						Alc						259
Z-Ala-Gly$_3$	1 + 3 MCA	THF	30						Alc-Eth	166–167	−12	DCA		Mol. wt.	71
Z-Ala-Gly$_3$-OPcp									DMF-Alc	198–200 (dec)	−7.7	DCA		Mol. wt.	71
Ala-Gly$_3$-OPcp HBr										143–145	+20	DCA		Mol. wt.	71
Z-Ala-Leu-Ala-Gly	SP		51												313
Z-Ala-Leu-Ala-N^{im}-Boc-TF-His	3 + 1 DCCI-NHS	DMF	40						Eth-Pe	150–153	−49	Alc			490
Ala-Leu-N$^\epsilon$-Z-Lys-Gly-OEt acetate					59	×									547
Boc-Ala-Leu-N$^\epsilon$-Z-Lys-Gly-OEt					29										547
CHla-Ala-Pro-Phe-Arg															151
N-Boc-Ala-Pro-OBzl-Ser-Gly-OBzl	1 + 3 ONP	Chf	87		54 27										58
N-Boc-Ala-Pro-OBzl-Ser-Gly-OEt	1 + 3 ONP	Acn	75		55 126		×							AA	58
					55 54					132–133	−24	DMF			
N-Boc-Ala-Pro-OBzl-Ser-Gly hydrazide									Me-Eth	83–84	−36	Me			58
Z-Ala-OBu^t-D-Ser-Gly-Leu-OMe									EtAc	172–174	−10	DMF			188

α-Bzl-Ala-Thr-Asn-γ-Bzl-Glu Tfa							×			101–103	−1.2	DMF			689
α-Bzl-Boc-Ala-Thr-Asn-γ-OBzl-Glu	1 + 3 OSU	THF			4 3				Alc-Aq	119–120	−8.3	DMF			689
Z-Ala-Val-Gly-Leu-OMe										Oil					188
β-Alanine															
Z-β-Ala-Gly-(α-methyl)-β-D-Asp-Gly-OBzl	3 + 1 OTCP	DMF	60						Eth-Pe	138–140	−14	Me-EtAc			223
Arginine															
N^{α},N^{δ},N^{ω}-Tris-Z-Arg-β-OBu^t-Asp-OBu^t-Ser-Ala	1 + 3 ONP	DMF	86		34					172–173 (dec)	−6.6				711
Arg-Gly-Phe-Phe	SP													AA	667
Boc-N^G-nitro-Arg-Gln-N^{im}-Bzl-His-Leu-OBzl	1 + 3 MCA	DMF	77		×				DMF-EtAc		−28	Me		UV	643
N^G-nitro-Arg-Gln-N^{im}-Bzl-His-Leu-OBzl Tfa															643
Arg-Leu-Val-Arg Di-Tfa salt											−14	Alc			640
N^G-Nitro-Z-Arg-Leu-Val-N^G-nitro-Arg									Alc-Aq	132–134	−11	DMF			640
N^G-Nitro-Z-Arg-Leu-Val-N^G-nitro-Arg-OMe									Me	197–198	−8	DMF			640
Decanoyl-Arg-Leu-Val-Arg															640
Arg-Val-Val-Phe Boc-hydrazide Diacetate											−33	Aq			83
N^{α}-Z-N^{ω}-nitro-Arg-Val-Val-Phe Boc-hydrazide	1 + 3 MCA	DMF-THF	96						DMF-Aq	238 (dec)	−8.4	DMF			83
Asparagine															
N-Z-Asn [β-bis-(Dmb)-amide]-Ala-Leu-Ala hydrazide									Me-Aq	212–214	+14	HMPA			96
N-Z-Asn-[β-bis-(Dmb)-amide]-Ala-Leu-Ala-OMe									Me	207–209	−12	DMF			96
N-Z-N^{β}-(Dmb)-Asn-Ala-Leu-Ala-OMe									Me	235–236	−9.4	DMF			491

(*continued*)

Table 4 Tetrapeptides (continued)

Compound	Method	Solvent	% Yield	Chromatography Col	Tl	Ion	Paper	CCD	Recrys. Sol.	M.P. °C	$[\alpha]_D$	Solvent Rotation	Elect.	Other	Ref.
Z-Asn-N$^\epsilon$-Boc-Lys-Phe-His hydrazide									Me	201–202	−27	Ac			574
Z-Asn-N$^\epsilon$-Boc-Lys-Phe-His-OMe	1 + 3 ONP	DMF	98						Me	183–184	−28	DMF			574
Z-Asn-ω-Bu^t-α-aminosuberyl-Pro-N$^\alpha$-Tos-Arg-Gly amide										148–150	−31	DMF			219
β-OMe-Nps-Asn-α-aminopimelyl-Pro-Leu-Gly amide									DMF-Eth	175–177	−68	DMF			287
β-OMe-Nps-Asn-α-aminosuberyl-Pro-Leu-Gly amide	1 + 3 DCCI-NHS	DMF	66			DO-50			DMF-Eth	175–177	−70	DMF			289
Aspartic acid															
Nps-[Asp(β-OBzl)]$_3$-N$^\epsilon$-Z-Lys-OBzl										123–125	−25	DMF			727
β-OBu^t-Z-Asp-α-aminosuberyl-Pro-N$^\epsilon$-Pht-Lys-Gly amide									Alc-Aq	188–190	−38	DMF			219
Asp-Tyr-γ-OBu^t-Gln-Gly Boc hydrazide acetate											−4.1	Me			39
Z-Asp-Tyr-γ-Bu^t-Gln-Gly Boc-hydrazide	1 + 3 OTCP	DMF-THF	74							174–176	−21	DMF			39
Cysteine															
Trt-Cys-β-OBu^t-Asn-Trt-Cys-Asn-OBu^t									EtAc-Pe		+11	Me			293
Z-Cys(CH_2COOME)-Cys(CH_2-COOME)-Gly-D-Val-OMe									Me-Aq	150–152	−26	DMF			188

Decanoyl-Cys(CH_2COOME)-Cys-(CH_2COOME)Gly-D-Val-OMe									Me-Aq	104–106	−39	DMF			188
S,S′,N-Z-hemi-Cys-Di(γ-OEt-Glu)-hemi-Cys OMe									Chf-Me-He		−20	DMF			235
N-Z-S-Trt-Cys-Di(γ-OEt-Glu)-S-Bzh-Cys-OMe	1 + 3 DCCI	MeCl	19						Chf-Eth-He	113–116	−26	TFE			235
S,S-Bis[N-Z-hemi-Cys-Di(γ-OEt-Glu)-S-Bzh-Cys OMe]									Chf-Me-Eth						235
S-Bz-N-Z-Cys-ε-Tos-Gly-Asn-Gly									Ac-Aq	191–192	−32	DMF			559
(S-Cys-Gly)$_2$									Aq-Alc		−95	HCl			595
S-Bzl-N-Z-Cys-Di-Gly-Gly															214
S-Bzl-N-Z-Cys-Di-Gly-Gly-OEt	1 + 3 ONP	Aq-THF													214
S-Bzl-N-Z-Cys-Di-Gly-Gly-OEt hydrazide										195	−32	50% Aq-Py			214
S-Bzl-N-Z-Cys-Di-Gly-Gly-SPh										123	−22	DMF			214
Sulfonamido-N-Z-Cys-His-sulfon-amido-Cys-Tyr-OMe									EtAc-Alc	177–179	−16	DMF			7
Decanoyl-Cys(CH_2COOCH_3)-Phe-Leu-Gly-OMe															321
Decanoyl-Cys(CH_2COOME)-D-Phe-Leu-Gly-OMe									Me	180.5–182	−28	DMF			188
Decanoyl-Cys(CH_2CH_2NH-decanoyl)-D-Phe-Leu-Gly-OMe									Me	187–189	−18	DMF			188
S-Bzl-N-Z-Cys-Pro-Leu-Gly amide															62
S-Bzl-N-Boc-Cys-Pro-Leu-Gly amide	1 + 3 OPy	Chf	90							172–173	−52	DMF			134
S-Bzl-Cys-Pro-Leu-Gly amide						×				140–141	−52	Abs. Alc.			463
S-Bzl-Cys-Pro-Leu-Gly amide HBr															463
S-(Boc-propyl)-N-Nps-Cys-Pro-Leu-Gly amide	1 + 3 DCCI	DMF	21						EtAc-Pe	88–91	−60	DMF			285

(*continued*)

Table 4 Tetrapeptides (continued)

Compound	Method	Solvent	% Yield	Chromatography Col	Tl	Ion	Paper	CCD	Recrys. Sol.	M.P. °C	$[\alpha]_D$	Solvent Rotation	Elect.	Other	Ref.
S-β-OMe-carbonylethyl-N-Z-Cys-Pro-Leu-Gly amide									*i*-Pr-Pe	146–148	−56	DMF			287
S-β-OMe-carbonyl-ethyl-homo-N-Z-Cys-Pro-Leu-Gly									EtAc-Pe	135–136	−49	DMF			289
S-(γ-OMe-carbonylpropyl)-N-Z-Cys-Pro-Leu-Gly amide	1 + 3 MCA	Chf	72			DO-50			Aq-DMF	115–117	−56	DMF			285
	1 + 3 OQ(Cl)	DMF	91						Aq-DMF	115–117	−56	DMF			
S-Bzl-N-Z-Cys-Pro-N$^{\epsilon}$-Tos-Lys-Gly										135–136 (dec)	−29	Chf			513
S-Bzl-N-Z-Cys-Pro-N$^{\epsilon}$-Tos-Lys-Gly-OEt	1 + 3 ONP		84							116–117	−25	Chf			513
Decanoyl-Cys(CH_2COOME)-OBu^t-D-Ser-Gly-Leu-OMe									EtAc-Eth	153–154	−20	DMF			188
Decanoyl-Cys(CH_2COOME)-Trp-Gly-Leu-OMe									Ac-Eth	199–200	−15	DMF			188
Sulfonamido-N-Z-Cys-Val-sulfonamido-Cys-Gly-OEt									EtAc-Alc	194–195	−9.0	DMF			7
S-Bzl-Cys-Val-Ser-S-Bzl-Cys									Me	210 (dec)	−20	Ac			714
S-Bzl-N-Z(OMe)-Cys-Val-Ser-S-Bzl-Cys									Me	179–181	−4.0	DMF			714
S,S,S′,S′-Bis[N-Z-hem-Cyst-Di(γ-OEt-Glu)-hemi-Cyst-OMe]									Ac-Aq		−43	DMF			235

Glutamine															
Z-Gln-Ala-Ala-Gly Boc-hydrazide	1 + 3 OTCP	DMF	88						Me	227–228	−37 −11	Me DMF			39
Gln-Ala-Ala-Gly Boc-hydrazide											−36	Me		AA	39
N-Boc-N^{γ}-(Dmb)-Gln-Ala-Val-Val-OBu^t	1 + 3 ONP		68						EtAc	151–152					491
Gln-Ala-Val-Val						A- IRA- 400			Alc		−34	Ac			491
Z-Gln-Gly-Leu-Val amide	2 + 2 DCCI- NHS	DMF	87		47					239.5– 240.5 (dec)	−25				710
Z-Gln-His-Nle-Asp-α,β-Di-OBu^t	1 + 3 OTCP	DMF	76							205 (dec)	−24	DMF			238
Gln-Gly-Leu-Val amide HCl					54										710
N-Z-Gln-N^{ϵ}-Boc-Lys-Phe-γ-OBu^t-Glu-OMe	1 + 3 ONP	DMF	76						Me	219–221	−26	N,N′- DMF			10
Gln-N^{ϵ}-Boc-Lys-Phe-γ-OBu^t-Glu-OMe					55										10
Z-Gln-Thr-Asn-S-Bz-Cys-OMe									DMF- EtAc	229–231 (dec)	−25	DMFA			558
Gln-Thr-Asn-S-Bz-Cys-OMe HBr									DMF-Eth						558
γ-OBu^t-Z-Gln-Thr-Val-Gly Boc-hydrazide	1 + 3 OTCP 1 + 3 OSU	DMF DMF	82 66								−29 −2.9 −28	Me DMF Me			39
γ-OBu^t-Gln-Thr-Val-Gly Boc-hydrazide acetate			92						Me-EtAc	156–158	−21 −26	Me Aq			39
Glutamic acid															
Z-Glu-Arg-Glu-Asp hydrazide Tfa hydrate					15					115–118	−25	DMF			690

(*continued*)

Table 4 Tetrapeptides (continued)

Compound	Method	Solvent	% Yield	Chromatography Col	Tl	Ion	Paper	CCD	Recrys. Sol.	M.P. °C	$[\alpha]_D$	Solvent Rotation	Elect.	Other	Ref.
γ-OBu^t-Z-Glu-Arg-γ-OBu^t-Glu-β-OBu^t-Asp hydrazide Tfa hydrate	1 + 3 OSU	THF	58		15					109–111	−29	Me			690
γ-OBu^t-Glu-Arg-γ-OBu^t-Glu-β-OBu^t-Asp Boc hydrazide diacetate hydrate					15				Me-Eth	80–90	−18	Me			690
Z-Glu$_4$(OBu^t)$_5$	OSU	EtAc	59							84.5–85	−31				410
	MCA	EtAc	75							87–89	−33				
Tetra-γ-OBu^t-Tri-Glu-D-Glu									Aq	146–149	+1.4				393
Tetra-γ-OBu^t α-methyl-N-Z-Tri-Glu-D-Glu									EtAc-Pe	109–112	−11				393
Tri-γ-OBzl-N-Boc-Tri-D-Glu-Leu									Aq-Alc	77–78	+12	Chf			216
Tri-γ-OBzl-2-methylthioethyl-N-Boc-Tri-D-Glu-Leu									Alc-Eth	114–115	+9.5	Chf			216
Tri-γ-OBzl-N-succinimidyl-Tri-D-Glu-Leu HCl									An-Eth	74–76	−29	Chf			216
γ-OBzl-N-Boc-Glu-γ-OBzl-Glu-N^{im}-Bzl-His-Pro-OBzl											−16	DMF			377
γ-OBu^t-Z-Glu-N^δ-Boc-Orn-Gln-N^δ-Boc-Orn-OMe	1 + 3 ONP	Py	73						Alc-Eth	167–168	−15	DMF			68
α-OBzl-Boc-Glu-Thr-Ser-Leu-OBu^t									EtAc	117–119	−32	Me			656
γ-OBu^t-Z-Glu-OBu^t-Tyr-Ala-Gly-OMe	1 + 3 OPCP	MeCl	65	Si					EtAc-He	146	−16	DMF			268
γ-OBu^t-Z-Glu-OBu^t-Tyr-Ala-Gly-OPcp			47						Me	187–188	−13	DMF			268

γ-OBu^t-Glu-OBu^t-Tyr-Ala-Gly-OPcp HCl			95						Me-Eth	172	+2.5	DMF			268
Glycine															
Gly_4-OEt HCl									Aq-Alc	212–213 (dec)				IR Mol. wt.	718
Z-Gly-γ-TMB Glu-γ-TMB-Glu-Phe-OTMB	1 + 3 ONP	DMF								168–170	−2.8	DMF		IR PMR	594
Z-Gly-γ-OBzl-Glu-γ-OBzl-Glu-Phe-OTMB	1 + 3 ONP								Alc-Eth	141–142	−2.4	DMF		IR	594
Gly-Glu-Glu-Phe					×									PMR	594
Z-Tri-Gly-Ala	SPh		54							193–195					444
Tri-Gly-Ala, ammonium										214–215 (dec)					444
N-Z-Tri-Gly-S-Bzl-Cys-OBzl									Aq-DMF	149–150	−19	Di			214
Tri-Gly-S-Bzl-Cys-OBzl HBr										168–169	−34	Alc			214
Tri-Gly-Cys															214
Tetra-Gly Nickel(II) complex															469
Dnp-Gly-Gly-Ile-Arg															364
Decanoyl-Gly-Gly-Leu-Gly-OMe															321
N-Ac-Gly-Gly-Lys-Gly									Aq-Alc	241–242.5 (dec)	−26	Aq			600
N-Ac-Gy-Gly-N$^\epsilon$-Z-Lys-Gly-OTMB	2 + 2 OCH_2-CN	DMF	81												600
Dnp-Gly-Gly-Phe-Arg															364
Z-Gly-Gly-OBzl-Tyr-Tyr-OBzl									Di-Eth	182–183	−4.9	DMF			612
Gly_2-Tyr_2							×			180–183 (dec)	+11	Aq			612
Dnp-Gly-Gly-Val-Arg															364

(*continued*)

Table 4 Tetrapeptides (continued)

Compound	Method	Solvent	% Yield	Chromatography				CCD	Recrys. Sol.	M.P. °C	$[\alpha]_D$	Solvent Rotation	Elect.	Other	Ref.
				Col	Tl	Ion	Paper								
Gly-N$^{\epsilon}$-Z-Lys-Gly-S-Bz-Cys-OMe HCl	2 + 2 DCCI	MeCl	71						Me-EtAc	198–199 (dec)	−26	DMF			235
Decanoyl-Gly-Phe-Gly-Cys(CH_2COOME)-OMe									Me-Aq	86–89	−25	DMF			188
Decanoyl-Gly-Trp-Ala-Ala-OMe															460
Histidine															
Decanoyl-His-N^{G}-nitro-Arg-Leu-Phe-OMe									Alc-Eth	149–151	−13	DMF			640
His-Asp-Gly-Ser															294
N$^{\alpha}$,N^{im}-Bis-Trt-His-Leu-S-Bzh-Cys-Gly-OMe	1 + 3 DCCI-HBT	MeCl	80	Sephadex LH-20	×				Eth-He	95–105	−22	Me			69
N^{im}-Bzl-N$^{\alpha}$-Z-His-Leu-Val-γ-OBu^{t}-Glu										178–182	−17	DMF			299
N^{im}-Bzl-N$^{\alpha}$-Z-His-Leu-Val-Glu-γ-OBu^{t}-α-ONp									EtAc-Pe	113–116	−15	DMF			299
N^{im}-Bzl-N$^{\alpha}$-Z-His-Leu-Val-Glu-α-OEt-γ-OBu^{t}	2 + 2 DCCI	THF	73						Alc	168–172	−20	DMF			299
N^{im}-(Boc-TF)-N$^{\alpha}$-Z-His-N$^{\epsilon}$-Boc-Lys-Tyr-N^{im}-(Boc-TF)-His-OBu^{t}	DCCI-NHS	THF	40							125–132	−16	Alc			96
N^{im}-Bzl-N$^{\alpha}$-Z-His-OAc-Ser-Gln-Gly-OMe	1 + 3 OTCP	DMF	21							197–199					208

N^{im}-Bzl-N^{α}-Z-His-Ser-Gln-Gly-OMe	2 + 2 A	DMF	15						DMF-Eth	247–250 (dec)					208
	1 + 3 OTCP	DMF	30						DMF-Aq	246–248 (dec)					
N^{im}-Bzl-N^{α}-Boc-His-Ser-Gln-Gly-OMe	1 + 3 OTCP	DMF	51						Me-EtAc	188–190 (dec)	−14	Me			208
Isoleucine															
Ile-Ala-Pro-Val															38
Ile-Ala-Val-Gly															434
	SP					DO-50									36
Z-Ile-Gln-Asp-S-Bzl-Cys-ONBzl	1 + 3 ONP								DMF-Aq	240–241.5 (dec)	−23	DMF			243
Boc-Ile-Phe-Ala-Gly hydrate	1 + 3 OSU	Glyme	73						EtAc-Pe	134–136	−38	Me			694
Ile-Phe-Ala-Gly-Tfa hydrate										242–245	+26	DMF			694
Leucine															
Z-Leu-Ala-Ala-Ala Boc-hydrazide	1 + 3 ONP	DMF	89						THF	247–249	−28	DMF			197
Leu-Ala-Ala-Ala Boc-hydrazide									Alc	262–264	−33	DMF			197
Leu-Ala-Gly-Val	SP														551
Z-Leu-N^{δ},N^{ω}-$Adoc_2$-Arg-Pro-Gly amide	2 + 2 DCCI	DMF			27					125–130 (dec)	−31	Me			91
Z-Leu-S-Bzl-Cys-Gly-Ser-OMe	1 + 3 ONP	DMF	85						DMF-Eth	139–141	−26	DMF			300
Boc-Leu-OBu^t-Glu-Asn-Tyr hydrazide									Me-Aq	218–220	−42	Me			293
Boc-Leu-OBu^t-Glu-Asn-Tyr-OMe	DCCI-NHS		88						Me	170–172	−23	Me			293
$(Leu\text{-}Gly)_2$ hydrate									Aq-Alc	243–245	+78	Aq			380

(*continued*)

Table 4 Tetrapeptides (continued)

Compound	Method	Solvent	% Yield	Chromatography				CCD	Recrys. Sol.	M.P. °C	$[\alpha]_D$	Solvent Rotation	Elect.	Other	Ref.
				Col	Tl	Ion	Paper								
Leu-Leu-Glu-Gly	SP														473
Boc-Leu-Leu-Gln-Gly	1 + 3 OSU	DMF-Aq			54					189–190 (dec)	−33				710
Leu-Leu-Gln-Gly															37
Leu-N$^\epsilon$-Z-Lys-Pro-Gly amide Tfa	2 + 2 DCCI-HBT	DMF	88		27						−24	Me			91
Z-Leu-Ser-Ser-Gly Boc-hydrazide	1 + 3 OSU	DMF-THF	72						Me-Acn	189–191 (dec)	−22	Me			83
Leu-Ser-Ser-Gly Boc-hydrazide acetate									Alc		−6.9	Me			83
Lysine															
N$^\epsilon$-Tfa-N$^\alpha$-Z-Lys-Asn-N$^\epsilon$-Tfa-Lys-Gly-OBu^t	1 + 3 OSU	THF	86						EtAc	180–182	−7.4	DMF			692
N$^\epsilon$-Tfa-Lys-Asn-N$^\epsilon$-Tfa-Lys-Gly-OBu^t										145–147	−1.6	DMF			692
N$^\epsilon$-Z-Nps-Lys-Ile-Val-Gly-OBzl	1 + 3 MCA	THF	90						DMF-EtAc	223–225	+23	DMF			727
N$^\epsilon$-Boc-Z-Lys-Ile-Val-Gly-OBu^t	1 + 3 MCA		91							202–203	−17	DMF			727
N$^\alpha$-Boc-(N$^\epsilon$-Boc-Lys)$_4$										92–95	−15	Me			645
N$^\alpha$-Boc-(N$^\epsilon$-Boc-Lys)$_4$ Dcha										128–131	−16	Me			645
N$^\alpha$-Boc-(N$^\epsilon$-Boc-Lys)$_4$-OSu										118–120	−25	Me			645
(Lys)$_4$															726

N^{ϵ}-Tfa-N^{α}-Z-Lys-N^{ϵ}-Tfa-Lys-N^{ϵ}-Tfa-Lys-Thr Boc-hydrazide hydrate	1 + 3 OSU	THF	85		4 5 18				EtAc	206–207	−14	DMF			185
N^{ϵ}-Tfa-Lys-N^{ϵ}-Tfa-Lys-N^{ϵ}-Tfa-Lys-Thr Boc-hydrazide acetate hydrate					4 3				Me-Eth	157–158	−15	DMF			185
CHla-Lys-Trp-Ala-Pro															151
Z-DNS-Lys-Val-Tyr-Pro-OBu^t										127–129					550
DNS-Lys-Val-Tyr-Pro-OBu^t										120–123					550
Methionine															
N-Thio-Bz-DL-Met-Gly-Gly-Gly									Aq-An	170–171					32
Met-N^{ϵ}-Z-Lys-N^G-nitro-Arg-His-OMe	2 + 2 A	DMF-THF		Si						155–160	+9.4	DMF		AA	196
Boc-D-Met-Phe-Gly-Cys(CH_2-COOME)-OMe									EtAc	85–87	−21	DMF			188
Decanoyl-D-Met-Phe-Gly-Cys-(CH_2-COOME)-OMe									Me-Aq	138–140	−25	DMF			188
Ornithine															
Di-decanoyl-Orn-Ala-Ala-Phe-OMe															460
N^{δ}-Boc-Z-Orn-N^{δ}-Boc-Orn-N^{δ}-Boc-Orn-N^{δ}-Boc-Orn-OMe	2 + 2 DCCI-NHS	DMF	74						Me-Aq	152	−18	Me			613
N^{δ}-Boc-Z-Orn-Pro-Val-Gly			100		24						−67	Me			613
N^{δ}-Boc-Z-Orn-Pro-Val-Gly-OEt	1 + 3 ONP	DMF	84		24				Alc-DiPr-Eth	125	−79	Me		Mol. wt.	613
Phenylalanine															
Z-Phe-N^G-nitro-Arg-Leu-D-Ala										215–217	−18	Ac			640
Z-Phe-N^G-nitro-Arg-Leu-D-Ala-OMe									Me-Eth	160–163	−27	Alc			640
Phe-Arg-Leu-D-Ala Tfa salt											−37	Alc			640
Z-Phe-γ-OBu^t-Glu-Ala-Gly-OMe	1 + 3 OPCP	MeCl	78	Si	22				EtAc-He	183–185	−13	DMF			273

(*continued*)

Table 4 Tetrapeptides (continued)

Compound	Method	Solvent	% Yield	Chromatography Col	Tl	Ion	Paper	CCD	Recrys. Sol.	M.P. °C	$[\alpha]_D$	Solvent Rotation	Elect.	Other	Ref.
Z-Phe-γ-OBu^t-Glu-Ala-Gly-OPcp									Me	202–203	−15	DMF			273
Phe-γ-OBu^t-Glu-Ala-Gly-OPcp-HCl										220	+4.2	DMF			273
Boc-Phe-Gln-Asn-S-Bzl-Cys-OBzl									DMF-An	213–216	−23	DMF			142
Propionyl-Phe-Gly-Leu-Ala-OMe															321
Z-D-Phe-Leu-Gly-Cys(CH_2-COOME)-OMe									Me-Eth	157–158	−17	DMF			188
Z-D-Phe-Leu-Gly-Cys(CH_2CH_2-NHZ)-OMe									EtAc	151–156	−10	DMF			188
Decanoyl-D-Phe-Leu-Gly-Cys-(CH_2CH_2NH-decanoyl)-OMe									EtAc	155 163–165	−14	DMF			188
Propionyl-Phe-Leu-Gly-S-Ac-OMe-Cys-OMe															460
Propionyl-D-Phe-Leu-Gly-Cys(CH_2-COOME)-OMe									Alc	177–178	−29	DMF			188
Propionyl-D-Phe-Leu-Gly-Cys(CH_2-CH_2NH-propionyl)-OMe									Me-Eth	128–130	−45	DMF			188
Dmpoc-Phe-Phe-Met-Gly-OEt										169–171					590
Propionyl-Phe-Val-Leu-Ala-OMe															321
Proline															
Pro-Ala-Leu-Ile	SP			Seph-adex		DO-50					−106				130
Imaa-Pro-Gly-Gly-Pro										155 (dec)					703
Imaa-Pro-Gly-Gly-Pro-OSu									Eth-Pe	142–146					703

Z-Pro-Gly-Ser-Gly Boc-hydrazide	2 + 2 OSU	THF	87						Acn	112–115	−29	Me			83
Pro-Gly-Ser-Gly Boc-hydrazide acetate											−30	Me			83
Imaa-Pro-Leu-Gly-Pro	1 + 3 OSU	Di-Aq	46						Eth-Pe	122–125					704
Imaa-Pro-Leu-Gly-Pro-OMe															700
Imaa-Pro-Leu-Gly-Pro-OSu										148–150					704
Pz-Pro-Leu-Gly-Pro	1 + 3 ONP	DMF	91		130				Di-Eth	201–203	−94				705
Pz-Pro-Leu-Gly-Pro-OSu					54 107						−97	Me			705
Imaa-Pro-Leu-Pro-Leu	3 + 1 OSU	Di-Aq	67						EtAc-Eth-Pe	128–130					703
Imaa-Pro-Leu-Pro-Leu-OMe															700
Pz-Pro-Phe-Gly-D-Arg	3 + 1 OSU	DMF	57		54				EtAc-Eth	152–154	−61	Me			699
Z-Pro-Pro-Ala-Phe-OMe			88							63–67					676
Pro-Pro-Ala-Phe-OMe			87							76–79					676
Aoc-Pro-Pro-Gln-N$^{\epsilon}$-Z-Lys hydrazide			70.5						Alc-Eth	104	−69	Alc			569
Aoc-Pro-Pro-Gln-N$^{\epsilon}$-Z-Lys-OMe	2 + 2 OSU	DMF													569
Aoc-Pro-Pro-Pro-Pro			90						EtAc-He						417
Aoc-Pro-Pro-Pro-Pro-OBzl									EtAc-He						417
Boc-Pro-Pro-Pro-Pro									Eth	122–126				IR NMR	121
	ONP		76							178	−186	Chf			530
Boc-Pro-Pro-Pro-Pro-OBzl	2 + 2 MCA		83											IR NMR	121
Pro-(Pro)$_2$-Pro										169–170	−288	Aq			530

(*continued*)

Table 4 Tetrapeptides (continued)

Compound	Method	Solvent	% Yield	Chromatography: Col	Tl	Ion	Paper	CCD	Recrys. Sol.	M.P. °C	$[\alpha]_D$	Solvent Rotation	Elect.	Other	Ref.
3-Hydroxyproline															
Z-α-3-Hyp-Ala-Trp-Nva-OMe				Seph-adex LH-20					EtAc-Pe	180–182	−51	Me			675
α-3-Hyp-Ala-Trp-Nva-OMe Tfa															675
Decanoyl-3-Hyp-N^G-nitro-Arg-Leu-Phe-OMe									EtAc-He	94–95	−32	DMF			640
Serine															
OBzl-N-Boc-Ser-γ-OBzl-Glu-Leu-His-OMe									EtAc-Pe	111–112	−20	Me			333
OBzl-N-Boc-Ser-Glu-Leu-His-OMe	1 + 3 OSU	DMF	79						Me-Aq	>197 (dec)	−18	Me			333
OBu^t-N-Z-Ser-γ-OBu^t-Glu-Leu-OBu^t-Ser					130				Eth	191–193	−11				708
OBu^t-Ser-γ-OBu^t-Glu-Leu-OBu^t-Ser					60						+6.2	Ac			708
OBzl-N-Boc-Ser-Pro-Phe-N^G-nitro-Arg-OPic	1 + 3 DCCI	DMF	89								−38				540
Boc-Ser-Pro-Phe-nitro-Arg OBzl	1 + 3 A	DMF							EtAc-Pe	97–100	−39	DMF			279
CHla-Ser-Pro-Phe-Arg															151
Ser-Pro-Phe-N^G-nitro-Arg-OBzl HCl			99							128–131	−34				279
Threonine															
OBu^t-Z-Thr-Ala-Ile-Gly-OMe	1 + 3 OSU	DMF	80						Alc	229–230	−43	Me			574

Decanoyl-Thr-N^G-nitro-Arg-Leu-Phe-OMe									Alc	163–165	−14	DMF			640
Boc-Thr-Z-Dbu-Thr-Z-Dbu-OMe	DCCI	THF	81						DMF-Aq	167–169	−27	DMF			22
Thr-Z-Dbu-Thr-Z-Dbu-OMe										161–163	−29	DMF	×		22
OBu^t-Z-Thr-Gln-β-OBu^t-Asp-Phe-OMe	1 + 3 OSU	DMF							Alc	155–161	−17	Alc			574
OBu^t-Thr-Gln-β-OBu^t-Asp-Phe-OMe															574
Thr-Lys-Pro-Arg	SP														447
N-Z-Thr-Phe-Thr-Ser hydrazide									Alc	201–202	−20	Ac			209
N-Z-Thr-Phe-Thr-Ser-OMe	2 + 2	DMF	68						Me-Aq	200–	−19	Ac			209
	A									200.5	−2.9	DMF			
N-Z-Thr-Pro-N$^\epsilon$-Boc-Lys-Thr Dcha	1 + 3 A	DMF-THF	60						EtAc-Eth	90–110	−26	DMF			619
N-Z-Thr-Pro-N$^\epsilon$-Tos-Lys-Thr-OMe	1 + 3 DCCI	MeCl	80						Alc Eth-Pe	84–92	−33	DMF			297
Z-Thr-Val-γ-OMe-Gln-N$^\epsilon$-Tos-Lys	1 + 3 DCCI	THF-MeCl	29	Si					EtAc-Me	178–180	+9.5	DMF			378
N-Boc-Thr-Val-γ-O-Bzl-Gln-N$^\epsilon$-Tos-Lys-OBzl	1 + 3 DCCI	MeCl	73							136–137	−6.9	DMF			378
Tryptophan															
CHla-Trp-Ala-Pro-Phe															151
Boc-Trp-Cys(CH_2CONH_2)-Gly-Leu-OMe									EtAc-Eth	130–132	−32	DMF			188
Decanoyl-Trp-Cys(CH_2CONH_2)-Gly-Leu-OMe									EtAc	174–176	−33	DMF			188
Z-Trp-γ-OBu^t-Glu-Ala-Gly									EtAc-He	93	−17	DMF			278
Z-Trp-γ-OBu^t-Glu-Ala-Gly-OMe	1 + 3 OPCP	MeCl	67	Si	22				EtAc-He	150	−23	DMF			278
Z-Trp-γ-OBu^t-Glu-Ala-Gly-OPcp									Me	170	−22	DMF			278
Trp-γ-OBu^t-Glu-Ala-Gly-OPcp HCl										202	−18	DMF			278
Propionyl-Trp-Gly-Leu-Ala-OMe															321
Ac-Trp-Met-Asp-Phe-OMe															391

(*continued*)

Table 4 Tetrapeptides (continued)

Compound	Method	Solvent	% Yield	Chromatography Col	Tl	Ion	Paper	CCD	Recrys. Sol.	M.P. °C	$[\alpha]_D$	Solvent Rotation	Elect.	Other	Ref.
Z-Trp-Met-Asp-Phe amide					57				2-OEt-Alc-Aq	215 (dec)	−36 −41	DMF DMF			25
Z-Trp-Met-β-OBu^t-Asp-Phe amide	2 + 2 A	DMF	61						Alc	191–192					25
Z-Trp-Met-Asp-Phe-OMe TEA salt	2 + 2 A	DMF	67												25
Boc-Trp-Met-Asp-Phe amide	SP														561
Trp-Met-Asp-Phe amide															318
Boc-Trp-4-dehydro-Nle-Asp-Phe amide			59		135 32 6				An	206–208 (dec)	−27	Alc-Aq			466
Boc-Trp-Nle-Asp-Phe amide														AA	466
Boc-Trp-Nle-4,5-3 H-Asp-Phe amide															466
Tyrosine															
Decanoyl-Tyr-N^G-nitro-Arg-Leu-Phe-OMe									EtAc-He	108–109	−19	DMF			640
O-Bu^t-Z-Tyr-OBu^t-γ-Glu-Ala-Gly-OMe	1 + 3 OPCP	MeCl	66	Si					EtAc-He	146	−19	DMF			267
O-Bu^t-Z-Tyr-OBu^t-γ-Glu-Ala-Gly-OPcp									EtAc-He	176	−29	DMF			267
O-Bu^t-Z-Tyr-OBu^t-γ-Glu-Ala-Gly-OPcp									Me	172–173	−21	DMF			267
OMe-Z-Tyr-γ-OBu^t-Glu-Ala-Gly-OMe	1 + 3 OPCP	MeCl	63						EtAc-He	152–154	−24	DMF			272

OMe-Z-Tyr-γ-OBu^t-Glu-Ala-Gly-OPcp HCl									Me	202–205	−21	DMF			272
O-Bu^t-Tyr-OBu^t-γ-Glu-Ala-Gly-OPcp HCl									Me-Eth	230	−22	DMF			267
OMe-Tyr-γ-OBu^t-Glu-Ala-Gly-OPcp HCl									Me-Eth	185	−0.5	DMF			272
OBu^t-Z-Tyr-γ-OBu^t-Glu-Gly-Gly			61	Si					EtAc	132–133	−6.3	DMF			277
OBu^t-Z-Tyr-γ-OBu^t-Glu-Gly-Gly-OMe	1 + 3 OPCP	MeCl	66	Si					EtAc-He	117–118	−4.7	DMF			277
OBu^t-Z-Tyr-γ-OBu^t-Glu-Gly-Gly-OPcp									Me	185–186	−3.8	DMF			277
OBu^t-Tyr-γ-OBu^t-Glu-Gly-Gly-OPcp HCl									Me-Eth	135 (dec)	+4.9	DMF			277
OBu^t-N-Z-Tyr-γ-OBu^t-Glu-Val-Gly									EtAc-He	154–155	−12	DMF			276
OBu^t-N-Z-Tyr-γ-OBu^t-Glu-Val-Gly-OMe	1 + 3 OPCP	MeCl	55	Si					EtAc-He	171	−14	DMF			276
OBu^t-N-Z-Tyr-γ-OBu^t-Glu-Val-Gly-OPcp									Me	188–190	−17	DMF			276
OBu^t-Tyr-γ-OBu^t-Glu-Val-OPcp HCl									Me-Eth	170 (dec)	−9.7				276
Boc-Tyr-Ile-Pro-Gly	1 + 3 OSU	Glyme	70							99–102	−46	DMF		AA	693
Tyr-Ile-Pro-Gly Tfa										135	−67	0.2 *N* $NaHCO_3$			693
OBzl-Boc-Tyr-Ser-Phe-Leu-OBu^t									EtAc-Pe	99–100	−25	Me			656
Valine															
Z-Val-N^G-nitro-Arg-N^G-nitro-Arg-Leu											−12	DMF			640
Z-Val-N^G-nitro-Arg-N^G-nitro-Arg-Leu-OMe									EtAc	140–142	−20	DMF			643
Decanoyl-Val-Arg-Arg-Leu															640
Val-Arg-Arg-Leu Di-Tfa salt											−4	Alc			640

(*continued*)

Table 4 Tetrapeptides (continued)

Compound	Method	Solvent	% Yield	Chromatography: Col	Chromatography: Tl	Chromatography: Ion	Chromatography: Paper	CCD	Recrys. Sol.	M.P. °C	$[\alpha]_D$	Solvent Rotation	Elect.	Other	Ref.
Z-Val-Asn-Gln-N^{im}-Bzl-His-OBzl	1 + 3 ONP	DMF	48						Me-Aq	221–223	−31	Ac			300
Z-Val-Asn-Gln-His-OMe	1 + 3 ONP	DMF	90						DSMO-Me	236	−32	Fm		AA	300
S-Trt-Boc-Val-Cys-Gly-Glu(OBu^t)$_2$	1 + 3 MCA	EtAc	70						Me	213–215	+19	Chf			293
Z-Val-Gly-Ala-Pro amide									Eth	209–211	−31	DMF			574
Val-Gly-Ala-Pro amide															574
Z-Val-His-Pro-Phe-ONBzl	2 + 2 A	EtAc	62	Al_2O_3						86–94	−60	Me			480
Val-His-Pro-Phe									Alc-Eth	180–186				AA	480
Val-His-Pro-Phe-ONBzl 2 HBr										170–180 (dec)	−32	Me			480
Z-Val-N$^\epsilon$-Z-Lys-α-Ala-His-OMe	2 + 2 DCCI	Di-EtAc	54	Al_2O_3						191–192	−23	Me-HCl			89
	2 + 2 A	DMF-THF	34	Al_2O_3						190–192	−25	Me-HCl			
	1 + 3 OSU	Di	74						Me-EtAc-He	191–193	−23	Me-HCl			
N-Boc-Val-N$^\epsilon$-Z-Lys-Ala-His-OMe	1 + 3 OSU	DMF	51						EtAc-Me-Pe	184.5–185.5	−27 −6.6	Me DMF			431
For-2-deutero-D-Val-Pro-N-trideuteromethyl-Gly-N-trideuteromethyl-Val-OMe											−86	Me			343

For-D-Val-Pro-N-trideuteromethyl-Gly-N-methyl-2-deutero-DL-Val-OMe											−16	Me		NMR	343
Boc-Val-Pro-Pro-Gly-OMe					66										676
Val-Pro-Pro-Gly-OMe										90–95			×		676
For-2-deutero-D-Val-2-deutero-DL-Pro-Sar-N-trideutero-Val-OMe											−43	Me		NMR	343
For-D-Val-Pro-2,2-dideutero-Sar-N-MeVal-OMe											−91	Me		NMR	343
Z-Val-Ser-Ser-Pro	2 + 2 A	THF-Aq							Acn	111–116	−75	Me			83
Z-Val-Ser-Ser-Pro Boc-hydrazide hydrate									Me	128–131	−95	Me			83
Val-Ser-Ser-Pro Boc-hydrazide acetate											−90	Me		AA	83

Chapter 6

Pentapeptides

Application of a formaldehyde thioacetal group for protection and introduction of cystine has been demonstrated in a new synthesis of oxidized glutathione.[390] A summary of this interesting approach has been entered in Chart 5 and additional details will be included in the pentapeptide table of a subsequent volume. Another pentapeptide synthesis published in the fall of 1972 corresponded to positions 1–5 of the fibrinogen protein C-chain.[539] The latter substance is of interest from the standpoint of the clotting process which is complex and involves in part release of two of the three protein chains of fibrinogen by the enzyme thrombin. The fibrinogen pentapeptide synthesis was also used to further explore the 4-picolyl ester (OPic) method of peptide bond formation.

Chart 5

390 Oxidized glutathione

S—CH_2CHCO_2H (NHZ)
|
CH_2
|
S—CH_2CHCO_2H (NHZ)

(Z-Djenkolic acid)

(1) Chf, CH_3N (morpholine), $(CH_3)_2CHCH_2OCOCl$, -5°
(2) Gly-OMe, 4 hr, room temp.
(3) Recrys. NMe

→

S—$CH_2CHCONHCH_2CO_2CH_3$ (NHZ)
|
CH_2
|
S—$CH_2CHCONHCH_2CO_2CH_3$ (NHZ)

90%
m.p. 139°-140°

(1) 4 *M* HBr-Ac, 45 min, room temp.
(2) N-Z-Glu-OMe, MCA, DMF
(3) Recrys. Alc

→

NH—(N-Z-γ-Glu-OMe)
|
S—$CH_2CHCONHCH_2CO_2CH_3$
|
CH_2
|
S—$CH_2CHCONHCH_2CO_2CH_3$
|
NH—(N-Z-γ-Glu-OMe)

84%
m.p. 180°-182°

(1) 0.5 *M* Br_2 in Chf, Cool, 4 min
(2) 5% $NaHCO_3$
(3) Recrys. Alc

→

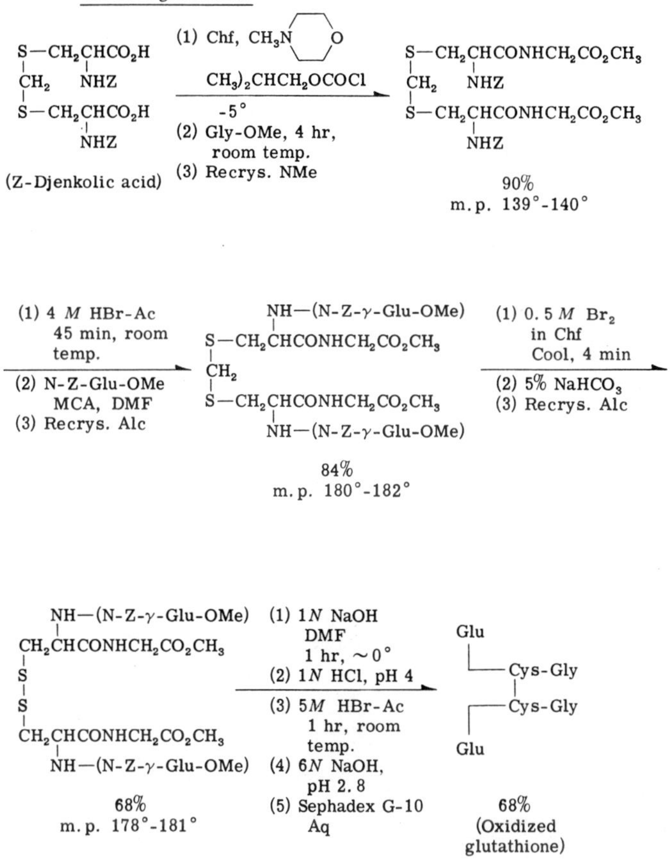

68%
(Oxidized glutathione)

Table 5 Pentapeptides

Compound	Method	Solvent	% Yield	Chromatography Col	Tl	Ion	Paper	CCD	Recrys. Sol.	M.P. °C	$[\alpha]_D$	Solvent Rotation	Elect.	Other	Ref.
Alanine															
$(Ala)_4$-Ala-morpholine amide HCl										>250 (dec)					186
Z-$(Ala)_4$-Ala-OEt									TFE-Alc	250	−107	Tfa			186
Z-$(Ala)_4$-Ala-morpholine amide	2 + 3 MCA	THF	65							>275 (dec)	−92	Tfa			186
2-OMe-[2-OEt-(2-OEt)]-ac-$(Ala)_4$-Ala-OEt									Alc	>250 (dec)	−99	Tfa			186
2-OMe-[2-OEt-(2-OEt)]-Ac-$(Ala)_4$-Ala-morpholine amide									TFE-Alc	250 (dec)	−88	Tfa			186
Ala_2-Gly_3-OPcp-HBr										200 (dec)	−5.6	DCA		Mol. wt.	71
Z-Ala_2-Gly_3-OPcp										202–204	−27	DCA		Mol. wt.	71
Ala-Ala-Leu-N^{ϵ}-Z-Lys-Gly-OEt					59										547
Boc-Ala-Ala-Leu-N^{ϵ}-Z-Lys-Gly-OEt					30										547
Z-Ala-Ala-Val-Gly-Leu-OMe									Eth	231–233	−25	DMF			188
Ala-Gly-Val-S-Bzh-Cys-Ser-OMe HCl															489
Ala-Gly-Val-S-bis-*p*-OMe-phenyl-methyl-Cys-Ser-OMe HCl															489
Ala-Gly-Val-S-chloromercuri-Cys-Ser-OMe HCl										190					489
Ala-Gly-Val-Cys-Ser-OMe HCl										175–178					489

Ala-Gly-Val-S-Trt-Cys-Ser-OMe-HCl										197–199	−31	Me			489
Ala-Pro-Pro-Ala-Phe-OMe															676
Z-Ala-Pro-Pro-Ala-Phe-OMe				Sephadex LH-20											676
Ala-Pro-Pro-Gly-Phe-OMe															676
Z-Ala-Pro-Pro-Gly-Phe-OMe				Sephadex LH-20											676
Ala-Pro-Pro-Val-Phe-OMe HCl															676
Boc-Ala-Pro-Pro-Val-Phe-OMe					119										676
β-Alanine															
Boc-β-Ala-Trp-Met-Asp-Phe amide															47
Arginine															
Arg-Gln-His-Nle-Asp-α,β-di-OBu^t Di-acetate										150 (dec)	−23	DMF		AA	238
N^G-Nitro-Z-Arg-Gln-His-Nle-Asp-α,β-di-OBu^t acetate	1 + 4 MCA	THF	75								−22	DMF			238
Arg-Leu-Gly-Leu-D-Ala acetate									Me-Eth	168–169	−17	Ac			640
N^G-Nitro-Z-Arg-Leu-Gly-Leu-D-Ala											−30	Me			640
N^G-Nitro-Z-Arg-Leu-Gly-Leu-D-Ala-OMe											−32	Di			640
Decanoy-Arg-Leu-Gly-Leu-D-Ala									Alc-Aq	206–207	−19	Me			640
Asparagine															
Asn-S-Bzl-Cys-Pro-Leu-Gly amide HBr															463
Z-Asn-S-Bzl-Cys-Pro-Leu-Gly amide						A-IRA-100				213–214 (dec)	−60	DMF			62
Nps-Asn-S-β-OMe-carbonylethyl-homo-Cys-Pro-Leu-Gly amide	1 + 4 OTCP	DMF	93						DMF-Aq	171–173	−71	DMF			289

(*continued*)

Table 5 Pentapeptides (continued)

Compound	Method	Solvent	% Yield	Chromatography Col	Tl	Ion	Paper	CCD	Recrys. Sol.	M.P. °C	$[\alpha]_D$	Solvent Rotation	Elect.	Other	Ref.
Nps-Asn-S-(γ-OMe-carbonylpropyl)-Cys-Pro-Leu-Gly amide	1 + 4 OTCP	DMF	84						DMF-Aq	191–193	−74	DMF			285
Nps-Asn-S-γ-OMe-carbonylpropyl-homo-Cys-Pro-Leu-Gly amide	2 + 3 DCCI-HBT	DMF	61		64 76 125	DO-50			DMF-Eth	187–191	−67	DMF		Mol. wt.	502
N-Nps-Asn-S-β-OMe-carbonylethyl-Cys-Pro-Leu-Gly amide	1 + 4 OTCP	DMF	98						DMF-Eth	177–180	−79	DMF			287
Z-Asn-S-Bzl-Cys-Pro-homo-N^ϵ-Tos-Lys-Gly amide	1 + 4 ONP	DMF								171–173	−39	DMF			61
N-Z-Asn-S-Bzl-Cys-Pro-N^ϵ-Tos-Lys-Gly	1 + 4 ONP	DMF	82							195–196 (dec)	−42	DMF			513
Asn-N^ϵ-Tfa-Lys-Asn-N^ϵ-Tfa-Lys-Gly-OBu^t acetate									Me-Eth	127–130	−9.6	DMF			692
Z-Asn-N^ϵ-Tfa-Lys-Asn-N^ϵ-Tfa-Lys-Gly-OBu^t trihydrate	1 + 4 MA	Chf	82							152	−12	DMF		AA	692
Aspartic acid															
β-OBu^t-Asp-Arg-Val-Val-Phe Boc-hydrazide diacetate											−11	DMF			83
β-OBu^t-Z-Asp-Arg-Val-Val-Phe Boc-hydrazide acetate hydrate	1 + 4 OSU	DMF	82						Bu-Eth	173 (dec)	−39	Me			83
Nps-[Asp(β-OBzl)]$_4$-N^ϵ-Z-Lys-OBzl									Di-Eth	143–145	−32	DMF			727
Asp-Asp-Tyr-γ-Bu^t-Gln-Gly Boc-hydrazide acetate			98								−15	Ac-Aq		AA	39

Z-Asp-Asp-Tyr-γ-OBu^t-Gln-Gly Boc-hydrazide	1 + 4 OTCP	DMF	76							202–203 (dec)	−25	DMF			39
β-OBzl-Nps-Asp-N$^\epsilon$-Z-Lys-Ile-Val-Gly-OBzl									DMF-EtAc-Eth	231–232	−36	HMPA			727
β-OBu^t-Z-Asp-N$^\epsilon$-Boc-Lys-Ile-Val-Gly-OBu^t hydrate										211–212 (dec)	−21	DMF			727
Cysteine															
Decanoyl-Cys(CH_2COOMe)-Ala-D-Ser(OBu^t)-Gly-Leu-OMe									EtAc	207–208	−13	DMF			188
Decanoyl-Cys(CH_2COOMe)-Ala-D-Ser-Gly-Leu-OMe									EtAc-Eth	168–169	−20	DMF			188
Cys-Asn-OBu^t | Boc-Cys-Gly-Glu(OBu^t)$_2$					×										293
S-Trt-Boc-Cys-S-Bzh-Cys-Gly-Phe-Gly-OBu^t	1 + 4 EDCI	DMF-MeCl	82							220–221	−22	DMF			234
Di-methyl S,S-Bis-N-Pht-hemi-Cyst-Gly-S′,S′-hemi-Cyst											−121	Chf			235
S-*p*-Methoxy-Bzl-N$^\alpha$-Boc-Cys-Thr-Leu-S-*p*-methoxy-Bzl-Cys-Gly-OMe	3 + 2 MCA	THF	78	Si						112–115	−55	Chf			10
S-*p*-Methoxy-Bzl-Cys-Thr-Leu-S-*p*-methoxy-Bzl-Cys-Gly-OMe Tfa															10
Cys-Tyr-Ile-Gln-Asn															86
S-Bzl-Z-Cys-Tyr-Ile-Gln-Cys azide														IR	308
α,β-Diaminobutyric acid															
Z-Dbu-D-Leu-Ile-Z-Dbu-Z-Dbu-OMe Tfa									Di-Eth	155–158			×		22
Boc-Dbu[(+)-6-methyl-octanoyl]-Thr-Z-Dbu-Thr-Z-Dbu hydrazide									DMF-Aq	228–231					22
Boc-Dbu[(+)-6-methyl-octanoyl]-Thr-Z-Dbu-Thr-Z-Dbu-OMe	1 + 4 ONP	DMF	58						DMF-Aq	213–216					22

(*continued*)

Table 5 Pentapeptides (continued)

Compound	Method	Solvent	% Yield	Chromatography Col	Tl	Ion	Paper	CCD	Recrys. Sol.	M.P. °C	$[\alpha]_D$	Solvent Rotation	Elect.	Other	Ref.
Threo-α,β-diaminobutyric acid															
Nα-Z-α,β-Dbu-D-Val-Pro-Sar-MeVal(Nβ-diaminobutyric acid) lactam										145–149	−25	Me			27
Nα-(2-Nitro-3-OBzl-4-methyl-Bz)-α,β-Dbu-D-Val-Pro-Sar-MeVal-(Nβ-Di-aminobutyric acid) lactam										164–168	−35	Me			27
Glutamic acid															
Glu-Arg-Gln-His-Leu							×				−32	Aq	×	AA	643
Glu-homo-Arg-Gln-His-Leu							×				−31	Aq	×	AA	642
Glu-NG-nitro-Arg-Gln-Nim-Bzl-His-Leu	SP					D-IR-45							×		643
Glu-NG-nitro-homo-Arg-Gln-Nim-Bzl-His-Leu	SP	80				D-IR-45									642
OBut-Nps-Glu-NG-nitro-Arg-Gln-Nim-Bzl-His-Leu-OBzl	1 + 4 OPCP	DMF	50	Si	26		×			182–184	−46	Me	×	UV	643
Pyro-Glu-NG-nitro-Arg-Phe-Leu-Pro	SP		65												581
N-Z-(γ-OEt-Gly)$_4$-S-Bzh-Cys-OMe	4 + 1 A								Chf-He		−31	TFE			235
Z-Glu$_5$(OBu^t)$_6$	OSU MCA	EtAc	70 95							102–103.5 75–100	−36 −18 −38				410
Pteroyl-penta-Glu															184

N-Boc-(γ-OBzl-D-Glu)$_4$-Leu									Aq-Alc	132–133	+3.9	Chf			216
Tetra-γ-OBzl-2-methylthio-ethyl-N-Boc-tetra-D-Glu-Leu									Alc	175–176	+5.4	Chf			216
Tetra-γ-OBzl-N-succinimidyl tetra-D-Glu-Leu									Alc-Eth	180 (dec)	−34	Chf			216
N$^{\alpha}$-Boc-Glu-N^{im}-Trt-His-Leu-S-Bzh-Cys-Gly-OMe	1 + 4 ONP	DMF	65												69
	DCCI-HBT	DMF	85												
γ-OBzl-Glu-Leu-Ala-Ala-Ala Boc-hydrazide										257–260	−27	DMF			197
γ-OBzl-Nps-Glu-Leu-Ala-Ala-Ala Boc-hydrazide	1 + 4 ONP	DMF	90						Eth	238–240	−17	DMF			197
γ-OMe-decanoyl-Glu-Trp-Val-Leu-Ala-OMe															321
Glutamine															
Z-Gln-Asn-ω-Bu^{t}-α-aminosuberyl-Pro-N^{G}-Tos-Arg-Gly amide									Alc	168–170	−29	DMF			219
Nps-Gln-Asn-α-aminosuberyl-Pro-Leu-Gly-OMe amide									DMF-Eth	159–164	−49	DMF			289
Nps-γ-OMe-Gln-Asn-α-amino-pimelyl-Pro-Leu-Gly amide					64 71				DMF-Eth	195–197	−52	DMF			287
Z-Gln-Asn-ω-Bu^{t}-α-aminosuberyl-Pro-N$^{\epsilon}$-Pht-Lys-Gly amide									Alc	189–191 (dec)	−37	DMF			219
Z-Gln-Leu-γ-OBu^{t}-Glu-Asn-OBu^{t}-Tyr hydrazide										232–234	−28	DMF			141
Z-Gln-Leu-γ-OBu^{t}-Glu-Asn-OBu^{t}-Tyr-OMe	2 + 3 A	DMF	75						Aq	228–230	+15	DMF			141
Z-Gln-OBu^{t}-Thr-Ala-Ile-Gly-OMe	1 + 4 ONP	DMF	95						Acn-Aq	260–262 (dec)	−26	Ac			574
Gln-OBu^{t}-Thr-Ala-Ile-Gly-OMe										242–245 (dec)					574

(*continued*)

Table 5 Pentapeptides (continued)

Compound	Method	Solvent	% Yield	Chromatography Col	Tl	Ion	Paper	CCD	Recrys. Sol.	M.P. °C	$[\alpha]_D$	Solvent Rotation	Elect.	Other	Ref.
Glycine															
Z-Gly-γ-Bu^t-Gln-Thr-Val-Gly Boc-hydrazide monohydrate	1 + 4 OSU	DMF	77						Me-Aq	203–205	−30 −6.1	Me DMF			39
Gly-γ-Bu^t-Gln-Thr-Val-Gly Boc-hydrazide acetate										220–225	−29	Me			39
Z-Gly-Gly-D-Ala-Gly-Gly-OBu^t															259
(Gly)$_4$-S-Bzl-Cys-OBzl HCl									Alc	174	−40	Aq			214
N-Trt-(Gly)$_4$-S-Bzl-Cys-OBzl									An	165	−21	DMF			214
(Gly)$_4$-Gly-atocuprate-disodium 4,5-hydrate															56
Z-Gly$_3$-OBzl-Tyr-Tyr-OBzl										167–169	−6.7	DMF			612
Gly$_3$-Tyr$_2$							×			185–188 (dec)	+11	Aq			612
Gly-Ile-Thr-Trp-(DNPS)-Lys															680
Z-Gly-Pro-Pro-Ala-Phe-OMe										80–83					676
Gly-Pro-Pro-Ala-Phe-OMe			95 70							Oil					676
Z-Gly-Pro-Pro-Gly-Phe-OMe			55	Seph-adex LH-20										UV	676
Gly-Pro-Pro-Gly-Phe-OMe			97												676
Histidine															
His-Ala-Asp-Gly-Ser															294
His-Ala-Glu-Gly-Ser															294

His-Gly-Glu-Ser-Phe															294
N^{im}-Bzl-N^{α}-Z-His-Leu-Val-γ-OBu^{t}-Glu-Ala hydrazide										243–244	−17	DMF			299
N^{im}-Bzl-N^{α}-Z-His-Leu-Val-γ-OBu^{t}-Glu-Ala-OMe	2 + 3 A	MeCl	75				×		Me	180–181	−22	DMF		AA	299
N^{im}-Bzl-N^{α}-Boc-His-Thr-Val-γ-OBzl-Gln-N^{ϵ}-Tos-Lys-OBzl	1 + 4 DCCI-NHS	DMF	61						Alc	105–107	−16	DMF			378
Isoleucine															
Z-Ile-Leu-Ser-Ser-Gly Boc-hydrazide	1 + 4 OSU	Di-DMF	78						Di-Aq	232–234 (dec)	−41	Me			83
Ile-Leu-Ser-Ser-Gly Boc-hydrazide acetate											−24	Me		AA	83
N^{ϵ}-Tfa-Z-Ile-Lys-N^{ϵ}-Tfa-Lys-N^{ϵ}-Tfa-Lys-Thr Boc-hydrazide	1 + 4 OSU	THF-Acn	93		4 5				EtAc	236–237	−18	DMF		AA	691
Z-Ile-N^{ϵ}-Tfa-Lys-N^{ϵ}-Tfa-Lys-N^{ϵ}-Tfa-Lys-Thr hydrazide Tfa					5					200–203	−9.1	DMF			691
Ile-N^{ϵ}-Tfa-Lys-N^{ϵ}-Tfa-Lys-N^{ϵ}-Tfa-Lys-Thr Boc-hydrazide acetate					3					147–149	−13	DMF			691
Boc-Ile-Pro-Pro-Ala-Phe-OMe	1 + 4 MCA	THF	62	Sephadex LH-20						89–93				AA	676
Ile-Pro-Pro-Ala-Phe-OMe	1 + 4 MCA	THF	68							94–97				AA	676
Leucine															
Boc-Leu-Gln-Gly-Leu-Val amide	1 + 4 OSU	DMF-Py	86							233–235 (dec)	−20	DMF			710
	3 + 2 DMF-NHS	DMF	78		48					234–235 (dec)	−19	DMF			710
Leu-Gln-Gly-Leu-Val amide HCl					54						−33	Me			710
Boc-Leu-Pro-Pro-Ala-Phe-OMe			68							94–97					676

(*continued*)

Table 5 Pentapeptides (continued)

Compound	Method	Solvent	% Yield	Chromatography Col	Tl	Ion	Paper	CCD	Recrys. Sol.	M.P. °C	$[\alpha]_D$	Solvent Rotation	Elect.	Other	Ref.
Leu-Pro-Pro-Ala-Phe-OMe Tfa salt			95												676
Leu-Ser-Phe-Met-Ala					126 122 55 116						−19	Fm		AA	498
Leu-Ser-Phe-Met-Ala															543
Leu-OBzl-Thr-Ala-Leu-Gly															38
Boc-Leu-Val-S-Trt-Cys-Gly-Glu-(OBu^t)$_2$	2 + 3 DCCI	DMF	43		29 43										293
Trt-Leu-Val-S-Trt-Cys-Gly-Glu-(OBu^t)$_2$	2 + 3 DCCI-NHS	DMF	37	Si					EtAc-He	155–159	0	Me			293
Lysine															
N$^\epsilon$-Z-Boc-Lys-Ala-Leu-N$^\epsilon$-Z-Lys-Gly-OEt	1 + 4 EDCI	MeCl	45–76		59 29				Alc-Eth	184–188	−28	Pn			547
N$^\epsilon$-Z-Boc-Lys-Ala-Leu-N$^\epsilon$-Z-Lys-Gly-OEt	1 + 4 OSU	MeCl			29										547
N$^\epsilon$-Z-Lys-Ala-Leu-N$^\epsilon$-Z-Lys-Gly-OEt acetate					59										547
N$^\alpha$,N$^\epsilon$-Bis-Boc-Lys-Ala-Pro-OBzl-Ser-Gly-OEt	1 + 4 ONP	Acn	83		55 126		×		Me-Eth	131–132	−30	DMF		AA	58
N$^\alpha$,N$^\epsilon$-Bis-Boc-Lys-Ala-Pro-OBzl-Ser-Gly hydrazide					55 137				Me-Eth	100–105	−66	Me			58

N^{α},N^{ϵ}-Bis-Boc-Lys-Ala-Pro-Ser-Gly hydrazide					55 126 137				EtAc-Eth	130	−38	DMF		NMR	58
N^{α},N^{ϵ}-Bis-Boc-Lys-Ala-Pro-Ser-Gly-OPcp					54 27				Me	167–173	−35	DMF			58
N^{ϵ}-Tfa-N^{α}-Boc-Lys-Ala-Thr-Asn-γ-OBzl-Glu-OBzl	1 + 4 OSU	THF-DMF	89							175–176	−8.9	DMF		AA	689
N^{ϵ}-Tfa-Lys-Ala-Thr-Asn-γ-OBzl-Glu-OBzl Tfa hydrate							×			181–182	−1.3	DMF			689
$(Lys)_5$															726
N^{α}-Nonanoyl-Lys-N^{ϵ}-Z-Lys-Thr-N^{ϵ}-Z-Lys-D-Leu															423
CHla-Lys-Trp-Ala-Pro-Phe															151
N^{ϵ}-Tfa-Lys-Tyr-Ile-Pro-Gly Tfa salt										138–139	−46	$NaHCO_3$			693
N^{ϵ}-Tfa-N^{α}-Boc-Lys-Tyr-Ile-Pro-Gly	1 + 4 OSU	Glyme								128	−38	DMF		AA	693
N^{ϵ}-Tos-Boc-Lys-OBzl-Tyr-Ser-Phe-Leu									EtAc	148–150	−34	Me			656
N^{ϵ}-Z-N^{α}-Boc-Lys-Val-N^{ϵ}-Z-Lys-Ala-His-OMe	1 + 4 OSU	DMF-Di	86						EtAc-Me-Pe	189.5–191 (dec)	−12	DMF			93
Methionine															
Boc-Met-Glu-N^{ϵ}-Boc-Lys-Phe-γ-OBu^t-Glu									Me-Aq	212–215	−21	DMF			10
Boc-Met-Glu-N^{ϵ}-Boc-Lys-Phe-γ-OBu^t-Glu-OMe	1 + 4 MCA	THF	84						Me	223–225	−23	DMF			10
Boc-Met-Ile-Phe-Ala-Gly dihydrate	1 + 4 OSU	Aq-Glyme	69							189–190	−21	DMF		AA	694
Met-Ile-Phe-Ala-Gly Tfa hydrate										232–236	+82	DMF			694
Phenylalanine															
Boc-Phe-His-Leu-N^{G}-nitro-Arg-Leu-OBz									Alc-Aq	146–149	−23	DMF			640

(*continued*)

Table 5 Pentapeptides (continued)

Compound	Method	Solvent	% Yield	Chromatography Col	Tl	Ion	Paper	CCD	Recrys. Sol.	M.P. °C	$[\alpha]_D$	Solvent Rotation	Elect.	Other	Ref.
Decanoyl-Phe-His-Leu-Arg-Leu															640
Phe-His-Leu-Arg-Leu											−17	Me			640
Phe-His-Leu-N^G-nitro-Arg-Leu-OBz															640
Boc-Phe-OBzl-Ser-Pro-Phe-N^G-nitro-Arg-OPic	DCCI	MeCl	83								−37				540
Boc-Phe-Ser-Pro-Phe-N^G-nitro-Arg-OBzl										114–116	−37				279
CHla-Phe-Ser-Pro-Phe-Arg															151
Phe-Ser-Pro-Phe-N^G-nitro-Arg-OBzl HCl										136–138	−33				279
Z-Phe-Tyr-Thr-propyl-N^ϵ-Boc-Lys-Thr	1 + 4 ONP	Di	96							144–150	−35	DMF			619
Z-Phe-Val-Asn-Gln-N^{im}-Bzl-His									DMSO-Aq	223–225	−3.4	DMSO			300
Z-Phe-Val-Asn-Gln-N^{im}-Bzl-His-OBzl	1 + 4 ONP	DMF	71						DMSO-Me	232–233	−27	Ac			300
Z-Phe-Val-Asn-Gln-His-OMe	1 + 4 ONP	DMF	82						DMSO-Me	239–240					300
N-Pht-Phe-Val-Gln-Trp-Leu hemihydrate	2 + 3 DCCI-NHS	DMF	75						Alc-Aq	226–227 (dec)	−32				698
Proline															
Imaa-Pro-Leu-Gly-Pro-Ala										130–140 (dec)					703

Imaa-Pro-Leu-Gly-Pro-Ala-OMe															700
Imaa-Pro-Leu-Gly-Pro-Ala-OSu									EtAc-Pe	101 (dec)					703
Boc-Pro-α-OBzl-Glu-Thr-Ser-Leu										95–100	−43	Me			656
Imaa-Pro-Gly-Gly-Pro-Arg										152 (dec)					703
Imaa-Pro-Gly-Gly-Pro-D-Arg	4 + 1 OSU	Di-Aq	25							170 (dec)					703
Imaa-Pro-Leu-Gly-Pro-Arg	4 + 1 OSU	Di	64						EtAc-Eth	170 (dec)					704
															700
															702
Imaa-Pro-Leu-Gly-Pro-D-Arg															702
	4 + 1 OSU	Di	65							168 (dec)				AA	704
Pz-Pro-Leu-Gly-Pro-Arg															702
Pz-Pro-Leu-Gly-Pro-D-Arg	4 + 1 OSU	Di-Aq	92		54				Bu^t-Di- Eth	158–159	−90	Me		AA	705
															702
Pz-Pro-Leu-Gly-Pro-D-Arg-OMe HCl	4 + 1 DCCI- NHS	DMF	88		54										705
Imaa-Pro-Leu-Gly-Pro-Orn														AA	703
Imaa-Pro-Leu-Gly-Pro-D-Orn															703
CHla-Pro-Lys-Trp-Ala-Pro															151
Aoc-Pro-Pro-Pro-Pro-Pro									Me-EtAc						417
Aoc-Pro-Pro-Pro-Pro-Pro OBzl	DCCI	Chf	78						EtAc						417
Boc-Pro-Pro-Pro-Pro-Pro														IR NMR	121
	ONP		81							224	−208	Chf			530
Boc-Pro-Pro-Pro-Pro-Pro OBzl	2 + 3 MCA	EtAc	53						EtAc	196–197				IR NMR	121
Pro-(Pro)$_3$-Pro					68					186	−331	Aq			530

(*continued*)

Table 5 Pentapeptides (continued)

Compound	Method	Solvent	% Yield	Chromatography Col	Tl	Ion	Paper	CCD	Recrys. Sol.	M.P. °C	$[\alpha]_D$	Solvent Rotation	Elect.	Other	Ref.
Sarcosine															
Sar-Lys-Trp-Ala-Pro	SP		56												581
O-(Z-Sar-N-MeVal)-N-(2-nitro-3-OBzl-4-methyl-Bz)-Thr-D-Val-Pro										115–120	−2.3	Me			408
O-(Z-Sar-N-MeVal)-Thr-D-Val-Pro															408
Sar-Nle-Phe-Ala-Pro	SP		52												581
N^{α}-Z-N^{β}-(OBu^t-Sar-MeVal)-α,β-Dbu-D-Val-Pro-ONp										72–76	−31	Me			27
Serine															
Z(OMe)-Ser-S-Bzl-Cys-Val-Ser-S-Bzl-Cys	1 + 4 OPCP	DMF	74						Me	190–193	−20	DMF			714
Ser-S-Bzl-Cys-Val-Ser-S-Bzl-Cys										229–231	+24	DMSO			714
						A-CG-4B								AA	714
Ser-Leu-Phe-Met-Ala					126 131 122 53						−29	Fm		AA	498
															543
OBzl-Boc-Ser-N^{δ}-Z-Orn-OBzl-Ser-D-Phe-Pro									Me-Aq	135.5–136	−35	Ac			261
OBzl-Boc-Ser-N^{δ}-Z-Orn-OBzl-Ser-D-Phe-Pro hydrazide			96								−37	Ac			261

OBzl-Boc-Ser-N$^{\delta}$-Z-Orn-OBzl-Ser-D-Phe-Pro-OMe	A	EtAc	89							138.5–139	−41	Ac			261
OBzl-Boc-Ser-N$^{\delta}$-Z-Orn-OBzl-Ser-D-Phe-Pro-ONp									EtAc-Eth	151–152	−36	Ac		IR	261
OBzl-Ser-N$^{\delta}$-Z-Orn-OBzl-Ser-D-Phe-Pro-OMe HCl									Me-Eth	168					261
Z-Ser-Ser-Val-Ser-Ser	2 + 3 A	DMF-Aq	40						Me	189–191	−3.9	DMF			714
Ser-Ser-Val-Ser-Ser									Aq-Alc	194–196	−58	Aq		AA	714
Threonine															
N-[2-Nitro-3-OBzl-4-methyl-6-deutero-Bz]-Thr-D-allo-Ile-Pro-Sar-N-MeVal-OBzl	1 + 4 WRK	NMe	75	Al_2O_3							−54 ±2	Me			344
OBu^t-Z-Thr-OBu^t-Ser-γ-OBu^t-Glu-Leu-OBu^t-Ser	1 + 4 OSU	DMF	83						EtAc	204–206	+15	DMF			708
OBu^t-Nps-Thr-OBu^t-Ser-γ-OBu^t-Glu-Leu-OBu^t-Ser	1 + 4 OSU	DMF			86				EtAc-Eth	218–220 (dec)	−12				708
N-[2-Nitro-3-OBzl-4-methyl-6-deutero-Bz]-Thr-2-deutero-D-Val-Pro-N-trideuteromethyl-Gly-N-trideutero-MeVal-OMe	1 + 4 WRK			Al_2O_3							−51	Me		NMR	343
N-[2-Nitro-3-OBzl-4-methyl-6-deutero-Bz]-Thr-D-Val-Pro-N-trideuteromethyl-Gly-N-methyl-2-deutero-DL-Val-OMe	1 + 4 WRK										−12	Me		NMR	343
N-(4-Bromo-2-nitro-3-OBzl-Bz)-Thr-D-Val-Pro-Sar-N-MeVal-OMe	1 + 4 WRK	NMe	55		29						−53	Me			552
N-(4-Bromo-2-nitro-3-hydroxy-Bz)-Thr-D-Val-Pro-Sar-N-MeVal											−72	Me			552
N-(4-Bromo-2-nitro-3-hydroxy-Bz)-Thr-D-Val-Pro-Sar-N-MeVal-OMe					27						−28	Me			552

(*continued*)

Table 5 Pentapeptides (continued)

Compound	Method	Solvent	% Yield	Chromatography				CCD	Recrys. Sol.	M.P. °C	$[\alpha]_D$	Solvent Rotation	Elect.	Other	Ref.
				Col	Tl	Ion	Paper								
N-[2-Nitro-3-OBzl-4-methyl-6-deutero-Bz]-2-deutero-Thr-D-Val-Pro-Sar-N-MeVal-OMe											−151	Me		NMR	343
N-[2-Nitro-3-OBzl-4-methyl-6-deutero-Bz]-Thr-2-deutero-D-Val-2-deutero-DL-Pro-Sar-N-trideutero-MeVal-OMe	1 + 4 WRK		60								−25	Me		NMR	343
N-[2-Nitro-3-OBzl-4-ethyl-6-deutero-Bz]-Thr-D-Val-Pro-2,2-dideutero-Sar-N-MeVal-OMe	1 + 4 WRK			Al_2O_3							−48	Me		NMR	343
N-[2-Nitro-3-OBzl-4-methyl-6-deutero-Bz]-Thr-D-Val-Pro-Sar-N-MeVal											−40 ±2	Me			344
N-[2-Nitro-3-OBzl-4-methyl-6-deutero-Bz]-Thr-D-Val-Pro-Sar-N-MeVal-OMe	1 + 4 WRK	NMe	65	Al_2O_3							−53 ±2	Me			344
Tryptophan															
CHla-Trp-Ala-Pro-Phe-Arg															151
Tyrosine															
S-Bzl-β-mercapto-propionyl-Tyr-Ile-Gln-Asp-S-Bzl-Cys										249–251 (dec)	−28	DMF		AA	243
S-Bzl-β-mercapto-propionyl-Tyr-Ile-Gly-Asp-S-Bzl-Cys									DMF-Aq	239.5–241.5 (dec)	−23	DMF			243

OBzl-Z-Tyr-Ile-Gln-Asp-S-Bzl-Cys-ONBzl	1 + 4 ONP								DMF-Aq	247–248	−18	DMF			243
OBzl-Boc-Tyr-Phe-Gln-Asn-S-Bzl-Cys-OBzl										229–231	−19	DMF			2
OBzl-Tyr-Phe-Gln-Asn-S-Bzl-Cys-OBzl															142
OBzl-Boc-Tyr-Pro_2-Phe_2															678
OBzl-Boc-Tyr-Pro_2-Phe_2-OMe	1 + 4 DCCI-NHS	DMF-THF	60			Sephadex LH-20								AA	678
OBu^t-Z-Tyr-OBu^t-Thr-Gln-β-OBu^t-Asp-Phe-OMe	1 + 4 MCA	THF	75						EtAc-Eth	169–173	−4	DMF			574
OBu^t-Tyr-OBu^t-Thr-Gln-β-OBu^t-Asp-Phe-OMe															574
OBzl-Z-Tyr-Thr-Pro-N^ϵ-Tos-Lys-Thr-OMe	1 + 4 ONP	DMF	70							109–113	−32	DMF			297
N,O-Bis-Z-Tyr-Thr-Pro-N^ϵ-Boc-Lys-Thr	1 + 4 ONP	Di	76						EtAc-Eth	120–125	−38	DMF			619
N,O-Bis-Z-Tyr-Thr-Pro-N^ϵ-Boc-Lys-Thr Dcha									EtAc	159–162	−18	DMF			619
Valine															
Boc-Val-α-OBu^t-Glu-Ala-Leu-Tyr					×				EtAc-He						293
Boc-Val-α-OBu^t-Glu-Ala-Leu-Tyr hydrazide									Me	215–216	−46	Me			293
Boc-Val-α-OBu^t-Glu-Ala-Leu-Tyr-OMe	3 + 2 A	Chf	82						EtAc-Pe	191–192	−43	Me			293
Z(OMe)-Val-N^ϵ-Z-Lys-Val-Tyr-Pro	SP														712
Z(OMe)-Val-N^ϵ-Z-Lys-Val-Tyr-Pro	2 + 3 A	DMF	71						EtAc-Eth	127–129	−20	DMF		AA	712
Z-Val-DNS-Lys-Val-Tyr-Pro-OBu^t										120–122					550
Val-DNS-Lys-Val-Tyr-Pro-OBu^t										118–121					550

(*continued*)

Table 5 Pentapeptides (continued)

Compound	Method	Solvent	% Yield	Chromatography Col	Tl	Ion	Paper	CCD	Recrys. Sol.	M.P. °C	$[\alpha]_D$	Solvent Rotation	Elect.	Other	Ref.
Z(OMe)-Val-N$^\delta$-Z-Orn-Leu-D-Ala-Pro	2 + 3 A	DMF	68						Me-Eth	137–139	−36	Me			349
Z-(OMe)-Val-N$^\delta$-Z-Orn-Leu-D-Ala-Pro-OEt										163–165	−21	DMF			349
Z(OMe)-Val-N$^\delta$-Z-Orn-Leu-D-Ala-Pro hydrazide									Di-Eth	158–160	−20	DMF			349
Val-N$^\delta$-Z-Orn-Leu-D-Ala-Pro HCl										147–149 (dec)	−36	DMF			349
Boc-Val-N$^\delta$-Z-Orn-Leu-D-Phe-Gly	2 + 3 A	DMF-DMSO	76		124				Me-Eth	182–183	+3.8	DMF			401
Val-N$^\delta$-Z-Orn-Leu-D-Phe-Gly					124 25					192–193	−4.3	Ac			401
Z(OMe)-Val-N$^\delta$-Z-Orn-Leu-D-Phe-Pro hydrazide					124				DMF-Aq	131–132	−33	DMF			401
Val-N$^\delta$-Z-Orn-Leu-D-Phe-Pro															433
Val-N$^\delta$-Z-Orn-Leu-D-Phe-Pro amide						DO-1			Me-Eth	118–122					385
Val-N$^\delta$-Z-Orn-Leu-D-Phe-Pro-$NHCH_2CH_2OH$						DO-1				102–104	−67				385
Val-N$^\delta$-Z-Orn-Leu-D-Phe-Pro-Fm salt															385
Z-Val-Pro-Pro-Phe-Phe-OMe	3 + 2 MCA	THF	52												676
Val-$(Pro)_2$-$(Phe)_2$-OMe															676

Chapter 7

Hexapeptides

The ONP method has been used by Hruby and colleagues[244] in a synthesis of the 20-membered cyclic disulfide designated tocinoic acid which encompasses the tocin ring of oxytocin. In addition to the oxytocin-(1–6)-hexapeptide amide the vasopressin-(1–6)-hexapeptide (Chart 6) amide has also been recently synthesized.[729] In terms of biological activity the conclusion was reached that the tripeptide side chain of the naturally occurring nonapeptide hormones is necessary for the pressor effects as neither amide showed any influence on the despinalized rat blood pressure. Otherwise, both exhibit some of the usual oxytocin and vasopressin activity.

The syntheses of some eight Orn-containing[493] hexapeptides and two Lys-containing[109] hexapeptides were received too late to be included in the tabular summary of this chapter. The Orn peptides were obtained for configurational inspection and the Lys peptides for sequential polypeptide syntheses.

Chart 6

729

Vasopressin-(1-6)-hexapeptide   Oxytocin-(1-6)-hexapeptide

Table 6 Hexapeptides

Compound	Method	Solvent	% Yield	Chromatography: Col	Chromatography: Tl	Chromatography: Ion	Chromatography: Paper	CCD	Recrys. Sol.	M.P. °C	$[\alpha]_D$	Solvent Rotation	Elect.	Other	Ref.
Alanine															
$(Ala)_5$-Ala morpholine amide HCl										>250 (dec)					186
Z-$(Ala)_5$-Ala-OEt									TFE	>275 (dec)	−119	TFA			186
Z-$(Ala)_5$-Ala morpholine amide										>275 (dec)	−103	TFA			186
D-Ala-Ala-D-Ala-Ala-Ala-Ala															259
D-Ala-Ala-D-Ala-Ala-D-Ala-Ala															259
D-Ala-D-Ala-D-Ala-Ala-Ala-Ala															259
Z-Ala-D-Ala-Ala-D-Ala-D-Ala-D-Ala Boc-hydrazide															259
2-OMe-[2-OEt-(2-OEt)]-Ac-$(Ala)_5$-Ala-OEt										>250 (dec)	−108	TFA			186
2-OMe-[2-OEt-(2-OEt)]-Ac-$(Ala)_5$-Ala morpholine amide									TFA-Alc	>275 (dec)	−95	TFA			186
Ala_3-Gly_3-OPcp HBr										300 (dec)	−25	DCA		Mol. wt.	71
D-Ala-D-Ala-D-Ala-Gly-Gly-Gly															259
Z-Ala_3-Gly_3-OPcp										220 (dec)	−39	DCA		Mol. wt.	71
D-Ala-D-Ala-Gly-Gly-Gly-Gly															259
Ala-Asn-N^{ϵ}-Tfa-Lys-Asn-N^{ϵ}-Tfa-Lys-Gly-OBu^t acetate hydrate									Me-Eth	212	−11	DMF			692

Z-Ala-Asn-N$^{\epsilon}$-Tfa-Lys-Asn-N$^{\epsilon}$-Tfa-Lys-Gly-OBu^{t} hydrate	1 + 5 OSU	DMF	65							220–223	−12	DMF		AA	692
Z-Ala-β-OBu^{t}-Asp-Arg-Val-Val-Phe Boc-hydrazide acetate	1 + 5 OSU	DMF	77						Eth	189 (dec)	−19	DMF		AA	83
Z-D-Ala-Gly-D-Ala-Gly-D-Ala-Gly															259
D-Ala-Gly-D-Ala-Gly-D-Ala-Gly															259
Z-D-Ala-Gly-D-Ala-Gly-D-Ala-Gly-OBu^{t}															259
D-Ala-Gly-D-Ala-Gly-Gly-Gly															259
Ala-Di-Gly-Ala-Gly-Gly, ammonium										200 (dec)					444
Z-Ala-Di-Gly-Ala-Gly-Gly	SPh		50.5							163–164					444
D-Ala-Gly-Gly-Gly-Gly-Gly															259
Ala-(Gly)$_4$-Gly, ammonium										220 (dec)					444
Z-Ala-(Gly)$_4$-Gly	SPh		52							233–235					444
										214–215					444
Boc-(Ala-Pro-Gly)$_2$										165–166					367
Boc-(Ala-Pro-Gly)$_2$-OMe	MCA	Glyme-Chf													367
						D-50W × 12			EtAc	147–149				UV	367
α-Aminobutyric acid															
p-Hydroxyphenylpropionic acid (O-sulfate)-Abu-Gly-Trp-Nle-Asp-Phe amide															44
p-Hydroxysuccinamic acid (O-sulfate)-Abu-Gly-Trp-Nle-Asp-Phe amide															44

(*continued*)

Table 6 Hexapeptides (continued)

Compound	Method	Solvent	% Yield	Chromatography				CCD	Recrys. Sol.	M.P. °C	$[\alpha]_D$	Solvent Rotation	Elect.	Other	Ref.
				Col	Tl	Ion	Paper								
Asparagine															
N^2-Z-N^4-methyl-Asn-Asn-S-Bzl-Cys-Pro-Leu-Gly amide	1 + 5 ONP	DMF				R-RG1				256–258 (dec)	−68	Fm-Aq			224
N^α-Boc-Asn-Glu-N^{im}-Trt-His-Leu-S-Bzh-Cys-Gly-OMe	1 + 5 ONP	DMF	84												69
	DCCI-HBT	DMF	65												
Aspartic acid															
Asp-Arg-Val-Tyr-Ile-His	SP			Ce	57 126 111						−4.5	Ac		AA	317
β-OBu^t-Asp-Gly-γ-OBu^t-Gln-Thr-Val-Gly Boc-hydrazide acetate			91								−24	Me		AA	39
β-OBu^t-Z-Asp-Gly-γ-Bu^t-Gln-Thr-Val-Gly Boc-hydrazide	1 + 5 OSU	DMF	87						Me-Aq	184–186	−24 −11	Me DMF			39
Z-[Asp(β-OBu^t)]$_2$-N^ϵ-Boc-Lys-Ile-Val-Gly-OBu^t			84							208–212	−23	DMF			727
Nps-[Asp(β-OBzl)]$_2$-N^ϵ-Z-Lys-Ile-Val-Gly-OBzl									DMSO-Me-Eth	222	+16	HMPA			727
γ-OBzl-Boc-Asp-Pro-α-OBzl-Glu-Thr-Ser-Leu										100–102	−35	Me			656
Cysteine															
Decanoyl-Cys(CH_2COOMe)-Ala-Ala-Val-Gly-Leu-OMe									Ac-Eth	254–256	−30				188

Boc-Cys-Asn-OBu^t | Boc-Val-Cys-Gly-Glu(OBu^t)$_2$					×										293
Cys-Asn-OBu^t | Boc-Val-Cys-Gly-Glu(OBu^t)$_2$					×										293
(S-Sulfonamide-Z-Cys-Cyst-S-sulfonamide-Cys-OEt)$_2$									EtAc-Alc	183–184	−16	DMF			7
S-Trt-N-Z-Cys-(γ-OEt-Glu)$_4$-S-Bzh-Cys-OMe	1 + 5 DCCI		33						Chf-He		−21	TFE			235
S,S′,N-Z-Hemi-Cys-(γ-ethyl-Glu)$_4$-hemi-Cys-OMe											−46	DMF			235
S,S-Bis[N-Z-hemi-Cys-(γ-OEt-Glu)$_4$-S-Bzh-Cys-OMe]															235
S,S,S′,S′-Bis[N-Z-hemi-Cys-(γ-OEt-Glu)$_4$-hemi-Cys-OMe]											−69	DMF			235
S-Bzl(OMe)-Cys-γ-OBzl-Glu-Leu-Ala-Ala-Ala Boc-hydrazide										286–288	+24	DMF			197
S-Bzl(OMe)-N^α-Nps-Cys-γ-OBzl-Glu-Leu-Ala-Ala-Ala Boc-hydrazide	1 + 5 ONP	DMF	69						DMF-Eth	232–234	−14	DMF			197
Di-methyl N,N′-bis[(N-Pht-S-Bzh-Cys-Gly)-Cyst]	DCCI	MeCl	71	Si							−58	Chf			235
S,S′,N-Z-Hemi-Cys-Gly-N^ε-Z-Lys-Gly-hemi-Cys-Gly										215–217	−51	DMF			235
S,S′,N-Z-Hemi-Cys-Gly-N^ε-Boc-Lys-Gly-hemi-Cys-Gly OBu^t									Chf-Me-Eth-He	194.5–196 (dec)	−43	DMA			235
S′,S′-Bis(Bu^t-N-Z-S-Bzh-Cys-Gly-N^ε-Boc-Lys-Gly-hemi-Cys-Gly)										183–185	−53	DMA			235

(continued)

Table 6 Hexapeptides (continued)

Compound	Method	Solvent	% Yield	Chromatography Col	Tl	Ion	Paper	CCD	Recrys. Sol.	M.P. °C	$[\alpha]_D$	Solvent Rotation	Elect.	Other	Ref.
S,S-Bis-N-Z-hemi-Cys-Gly-N$^\epsilon$-Z-Lys-Gly-S′,S′-hemi-Cys-Gly									DMF-Eth	226–229	−67	DMF		AA	235
S,S,S′,S′-Bis(N-Z-hemi-Cys-Gly-N$^\epsilon$-Boc-Lys-Gly-hemi-Cys-Gly-OBu^t)				Si	×					202–204 (dec)	−77	DMA		AA	235
N-Z-Cys-Gly-N$^\epsilon$-Z-Lys-Gly-Gly-S-Bz-Cys-OMe									Me-Eth	156–158	−17	DMF			235
S,S′,N-Z-hemi-Cys-Gly-N$^\epsilon$-Z-Lys-Gly-Gly-hemi-Cys-OMe										205–208	−45	DMF			235
S,S-Bis(methyl N-Z-hemi-Cys-Gly N$^\epsilon$-Z-Lys-Gly-Gly-S-Bz-Cys)										144–146	−50	DMF			235
S,S,S,S′-Bis(methyl N-Z-hemi-Cys-Gly-N$^\epsilon$-Z-Lys-Gly-Gly-hemi-Cys-di-OMe										197–201	−74	DMF			235
S′,S′,-Bis(methyl N-Z-S-Trt-Cys-Gly-N$^\epsilon$-Z-Lys-Gly-Gly-hemi-Cys)									DMF-Acn	218–219	−18	DMF			235
S-Chloromercuri-N-Z-Cys-Gly-N$^\epsilon$-Z-Lys-Gly-Gly-S-Bz-Cys-OMe										199–200 (dec)					235
S-Trt-N-Z-Cys-Gly-N$^\epsilon$-Z-Lys-Gly-Gly-S-Bz-Cys-OMe	1 + 5 WRK	ACN	90							179–181	−5.2	DMF			235
	1 + 5 EDCI	DMF	26							176–179					
	1 + 5 MCA	DMF	16							178–180					

S-Bzl-Z-Cys-Tyr-Ile-Gln-Asn-S-Bzl-Cys amide										249–250					308
S-Bzl-Z-Cys-Tyr-Ile-Gln-Asn-S-Bzl-Cys-OMe	3 + 3 A	DMF	90							244–245					308
S-Bzl-Z-Cys-Tyr-Ile-Gln-Asn-S-Bzl-Cys-ONBzl	1 + 5 ONP								DMF-Aq	252–253 (dec)	−32	DMF			243
Cys-Tyr-Ile-Gln-Asn-Cys-NH_2 S————————S														AA	308
S-Bzl-Z-Cys-OBzl-Tyr-Phe-Gln-Asn-S-Bzl-Cys-OBzl	1 + 5 ONP	Alc								243–246	−32	DMF			142
S-Bzl-Z-Cys-OBzl-Tyr-Phe-Gln-Asn-S-Bzl-Cys-OBzl															142
S-Bzl-Z-Cys-Tyr-Phe-Gln-Asn-S-Bzl-Cys amide	3 + 3 A	DMF	51		28 125					193–194					307
Cys-Tyr-Phe-Gln-Asn-Cys(pressinoic acid)						Seph-adex G-25			Aq		−11	Ac		AA	142
S————————S Cys-Tyr-Phe-Gln-Asn-Cys-NH_2	3 + 3 A						×	3					×		307
(Z-Cys-Tyr-Tyr)$_2$										183–185	−41	Alc			251
(Z-Cys-Tyr-Tyr amide)$_2$										207–210	−96	DMF			251
(Z-Cys-Tyr-Tyr-OEt)$_2$										173	−84	DMF			251
(Cys-Tyr-Tyr)$_2$ dihydrate										168	−53	DMF			251
(Cys-Tyr-Tyr-OEt)$_2$										206–207	−61	DMF			251
Glutamic acid															
Z-γ-OBu^t-Glu-Arg-Gln-His-Nle-Asp-Di-OBu^t-Di-Ac	1 + 5 OSU	DMF									−36	Bu^t Me-Aq		AA	238
γ-OBu^t-Glu-Arg-Gln-His-Nle-Asp-Di-OBu^t-Di-Ac											−20	DMF		AA	238

(*continued*)

Table 6 Hexapeptides (continued)

Compound	Method	Solvent	% Yield	Chromatography: Col	Tl	Ion	Paper	CCD	Recrys. Sol.	M.P. °C	$[\alpha]_D$	Solvent Rotation	Elect.	Other	Ref.
Z-Glu-Asn-S-Bzl-Cys-Pro-Leu-Gly amide									Alc-Aq	233–234 (dec) 247–248 (dec)	−54	DMF			62
Z-Glu$_6$(OBu^t)$_7$	OPCP MCA	EtAc EtAc	67 71							115–118 123–125	−40 −40				410
γ-OBzl-Boc-Glu-His-Phe-Arg-Trp-Gly	1 + 5 ONP	DMF	89								−24	Ac			249
Pyro-Glu-His-Trp-OBzl-Ser-OBzl-Tyr-Gly					120 99 7				Alc	171–175 (dec)	−0.9	Ac			91
Pyro-Glu-His-Trp-OBzl-Ser-OBzl-Tyr-Gly-OBzl	3 + 3 DCCI-HBT	DMF	69		120 99 7				Alc	197–200 (dec)	−2.8	Ac			91
γ-OBu^t-Z-Glu-Thr-Ala-Ala-Ala-N$^\epsilon$-Boc-Lys hydrazide										229 (dec)	−12	DMF			238
γ-OBu^t-Boc-Glu-Thr-Ala-Ala-Ala-N$^\epsilon$-Boc-Lys hydrazide	1 + 5 OSU	DMF	93							240–241 (dec)	−45	Ac			238
Glutamine															
Gln-Asn-S-Bzl-Cys-Pro-Leu-Gly amide HBr															463
Gln-Asn-S-γ-carboxypropylhomo-Cys-Pro-Leu-Gly amide					64 76	DO-50								AA	502

N^{γ}-Mbh-Boc-Gln-N^{β}-Mbh-Asn-S-Bzl-Cys-Pro-Leu-Gly hydrazide															685
Nps-Gln-Asn-S-β-carbo-OMe-ethyl-Cys-Pro-Leu-Gly amide	1 + 5 OTCP	DMF	62		64 71				DMF-Eth	210–212	−47	DMF			287
Nps-Gln-Asn-S-(γ-OMe carbonyl-propyl)-Cys-Pro-Leu-Gly amide	1 + 5 OTCP	DMF	84						DMF-Eth	180–182	−56	DMF			285
Nps-Gln-Asn-S-β-OMe-carbonyl-ethyl-homo-Cys-Pro-Leu-Gly amide	1 + 5 OTCP	DMF	66						DMF-Eth	157–160	−52	DMF			289
Z-Gln-Asn-S-Bzl-Cys-Pro-homo-N^{ϵ}-Tos-Lys-Gly amide	1 + 5 ONP		88							198–201	−42	DMF			61
Z-Gln-Asn-S-Bzl-Cys-Pro-N^{ϵ}-Tos-Lys-Gly	1 + 5 ONP	DMF	86							202 (dec)	−41	DMF			513
Boc-Gln-N^{γ}-Pht-Dbu-S-Bzl-Cys-Pro-Leu-Gly amide	2 + 4 DCCI-NHS	DMF	55						Alc-Aq	205–208 (dec)	−71	Ac			221
Nps-Gln-Asn-S-γ-OMe-carbonyl-propyl-homo-Cys-Pro-Leu-Gly amide	1 + 5 OTCP	DMF	80		64 76 125				Me-Eth	170–174	−21	DMF		Mol. wt.	502
Boc-Gln-N^{ϵ}-Tos-Lys-OBzl-Tyr-Ser-Phe-Leu										196–198	−23	Me			656
Z-Gln-D-Ser-Gln-D-Ser-D-Ser-Gln-OMe	3 + 3 A	DMF													492
Glycine															
$(Gly)_5$-Ala, ammonium										240 (dec)					444
Z-Gly-S-Bzl-Cys-Leu-Gly-S-Bzl-Cys-Leu Boc-hydrazide	3 + 3 A	DMF	81						Me-Aq	208–211	−43	Me			54
Z-Gly-S-Bzl-Cys-Leu-Gly-S-Bzl-Cys-Leu hydrazide									Alc	196–199	−23	DMF			54
Z-Gly-S-Bzl-Cys-Leu-Gly-S-Bzl-Cys-Leu-OMe	3 + 3 A	Eth	79						Aq-Alc	182–184	−43	DMF	×		54

(*continued*)

Table 6 Hexapeptides (continued)

Compound	Method	Solvent	% Yield	Chromatography Col	Tl	Ion	Paper	CCD	Recrys. Sol.	M.P. °C	$[\alpha]_D$	Solvent Rotation	Elect.	Other	Ref.
Z-Gly-S-Bzl-Cys-D-Leu-Gly-S-Bzl-Cys-D-Leu Boc-hydrazide									NMe	174–176	+8.6	Me			54
Z-Gly-Gly-D-Ala-D-Ala-D-Ala-Gly															259
Z-Gly-Gly-D-Ala-D-Ala-D-Ala-Gly-OBu^t															259
Z-Gly-Gly-D-Ala-D-Ala-Gly-Gly															259
Z-Gly-Gly-D-Ala-Gly-D-Ala-Gly															259
Z-Gly-Gly-D-Ala-Gly-D-Ala-Gly-OBu^t															259
Z-Gly-Gly-D-Ala-D-Ala-Gly-Gly-OBu^t															259
Z-(Gly)$_3$-Ala-Gly-Gly	SPh		53.8							>195					444
Z-Gly-Gly-Gly-D-Ala-Gly-Gly															259
(Gly)$_3$-Ala-Gly-Gly-OBzl										214–215					444
Z-Gly-Gly-Gly-D-Ala-Gly-Gly-OBu^t															259
(Gly)$_3$-Ala-Gly-Gly, ammonium										216–218 (dec)					444
Z-(Gly)$_5$-Ala	SPh		52							225–228					444
Z-Gly-Gly-Val-His-Pro-Phe-ONBzl	2 + 4 OSU	DMF	70						Me-EtAc-Eth	94–102 (dec)	−47	Me			480
Gly-Gly-Val-His-Pro-Phe-ONBzl Di-HBr									Me-EtAc-Eth	150–158 (dec)					480
Z-Gly-Leu-N$^\delta$-Tos-Orn-Phe-Pro-N$^\delta$-Tos-Orn-OMe															496

Z-Gly-Leu-N$^{\delta}$-Tos-Orn-Phe-Pro-N$^{\delta}$-Tos-DL-Orn-OMe															494
Z-Gly-Leu-N$^{\delta}$-Tos-Orn-D-Phe-Pro-N$^{\delta}$-Tos-Orn-OMe															496
Z-Gly-Leu-N$^{\delta}$-Tos-Orn-D-Phe-Pro-N$^{\delta}$-Tos-DL-Orn-OMe															495
Z-Gly-Leu-N$^{\delta}$-Tos-DL-Orn-Phe-Pro-N$^{\delta}$-Tos-Orn-OMe															494
Z-Gly-Leu-N$^{\delta}$-Tos-DL-Orn-Phe-Pro-N$^{\delta}$-Tos-DL-Orn-OMe															495
Z-Gly-Leu-N$^{\delta}$-Tos-DL-Orn-D-Phe-Pro-N$^{\delta}$-Tos-Orn-OMe															495
Z-Gly-D-Leu-N$^{\delta}$-Tos-Orn-Phe-Pro-N$^{\delta}$-Tos-Orn OMe															496
Z-Gly-DL-Leu-N$^{\delta}$-Tos-Orn-Phe-Pro-N$^{\delta}$-Tos-Orn-OMe	3 + 3 A	EtAc													494
Z-Gly-D-Leu-N$^{\delta}$-Tos-Orn-D-Phe-Pro-N$^{\delta}$-Tos-Orn-OMe															496
Z-Gly-D-Leu-N$^{\delta}$-Tos-DL-Orn-D-Phe-Pro-N$^{\delta}$-Tos-DL-Orn	3 + 3 A														495
Z-Gly-DL-Leu-N$^{\delta}$-Tos-Orn-D-Phe-Pro-N$^{\delta}$-Tos-Orn-OMe															494
Z-N-[2-(Boc-amino)-ethyl]-Gly-Phe-Ser-Pro-Phe-N^{G}-nitro-Arg-OBzl	1 + 5 OSU	DMF							An-Eth	100–110					26
Boc-Gly-Phe-OBzl-Ser-Pro-Phe-N^{G}-nitro-Arg-OPic	OTCP	DMF	93												540
Boc-Gly-Phe-Ser-Pro-Phe-N^{G}-nitro-Arg-OBzl										113–116	−33	DMF			279
CHla-Gly-Phe-Ser-Pro-Phe-Arg															151
Gly-Phe-Ser-Pro-Phe-N^{G}-nitro-Arg-OBzl HCl										145–149	−37	DMF			279

(*continued*)

Table 6 Hexapeptides (continued)

Compound	Method	Solvent	% Yield	Chromatography Col	Tl	Ion	Paper	CCD	Recrys. Sol.	M.P. °C	$[\alpha]_D$	Solvent Rotation	Elect.	Other	Ref.
Z-$(Gly_2$-$Tyr)_2$-OBzl	3 + 3 A		64						DMF-EtAc	115–116	−5	DMF			331
Z-$(Gly_2$-$Tyr)_2$-OEt	3 + 3 A		41						Alc	124–126	+19	Ac			331
Z-$(Gly_2$-$Tyr)_2$ hydrazide										151–153	−6.4	DMF			331
$(Gly_2Tyr)_2$									Aq-Alc	238–242 (dec)	−2.8	Aq			331
$(Gly_2$-$Tyr)_2$ hydrazide 2 HCl															331
Gly-Tyr-Val-His-Pro-Phe-OBzl 2 HBr									Me-EtAc-Eth	158–162 (dec)	−35	Me			480
Z-Gly-Tyr-Val-His-Pro-Phe-ONBzl	2 + 4 A	EtAc	64						Me-Eth	121–123 (dec)	−50	Me			480
Histidine															
His-Leu-Ser-Phe-Met-Ala					126 122 55 116						−29	Fm		AA	498
Isoleucine															
Aoc-Ile-Gln-Asn-ω-Bu^t-α-aminosuberyl-Pro-N^G-Tos-Arg-Gly amide									Aq	175–179 (dec)	−31	DMF			219
Ile-Gln-Asn-α-aminopimelyl-Pro-Leu-Gly amide					64 71	D-50			DMF-Eth		−46	DMF		AA	287

Ile-Gln-Asn-α-aminosuberyl-Pro-Leu-Gly amide			73	DO-50		D-50			DMF-Eth		−39	DMF		AA	289
Nps-Ile-Gln-Asn-ω-OMe-α-amino-pimelyl-Pro-Leu-Gly amide					64 71 125				DMF-Eth	242–244	−70	DMF			287
Nps-Ile-Gln-Asn-ω-OMe-α-amino-suberyl-Pro-Leu-Gly amide									DMF-Eth	252–255	−66	DMF			289
Aoc-Ile-Gln-Asn-ω-OBu^t-α-amino-suberyl-Pro-N$^\epsilon$-Pht-Lys-Gly amide									Alc-Aq	192–194 (dec)	−37	DMF			219
Leucine															
Z-Leu-Ala-Ala-Leu-Ala-Gly					59										547
Z-Leu-Ala-Ala-Leu-Ala-Gly-OEt	OSU	MeCl			29				Alc-DMF-Aq	254–256	−26	DMF			547
Leu-Ala-Ala-Leu-Ala-Gly-OEt	1 + 5 OSU	MeCl-DMF			29 59	DO-2									547
Boc-Leu-Ala-Ala-Leu-N$^\epsilon$-Z-Lys-Gly					59				Alc-EtAc-He	170–173 (dec)	−23	DMF			547
Boc-Leu-Ala-Ala-Leu-N$^\epsilon$-Z-Lys-Gly-OEt	1 + 5 OSU	MeCl	89		29				Alc-Eth	213–215	−27	DMF			547
Leu-Ala-Ala-Leu-N$^\epsilon$-Z-Lys-Gly-OEt					59	DO-2									547
(Z)-Leu-Cys-Gly-OMe (dimer)									Me	193–195	−91	DMF			618
Boc-Leu-Leu-Gln-Gly-Leu-Val amide	4 + 2 DCCI-NHS	DMF			108					255 (dec)	−32	DMF			710
Boc-Leu-Leu-Gln-Gly-Leu-Val amide hydrate	1 + 5 OSU	DMF	90		130					254 (dec)	−31				710
Leu-Leu-Gln-Gly-Leu-Val amide	SP														35
Leu-Leu-Gln-Gly-Leu-Val amide HCl					54		×								710
Z-Leu-N$^\epsilon$-Z-Lys-Ala-Leu-Ala-Gly					59				Alc-Eth	237–239					547

(continued)

Table 6 Hexapeptides (continued)

Compound	Method	Solvent	% Yield	Chromatography Col	Tl	Ion	Paper	CCD	Recrys. Sol.	M.P. °C	$[\alpha]_D$	Solvent Rotation	Elect.	Other	Ref.
Z-Leu-N^{ϵ}-Z-Lys-Ala-Leu-Ala-Gly-OEt	1 + 5 OSU	MeCl	80		29				Alc-Eth	219–221	−23	DMF			547
Boc-Leu-N^{ϵ}-Z-Lys-Ala-Leu-Ala-Gly					59				Alc-Aq	180–182 (dec)	−24	DMF			547
Boc-Leu-N^{ϵ}-Z-Lys-Ala-Leu-Ala-Gly-OEt	OSU	MeCl			29				Alc-Aq	217–218	−26	DMF			547
Leu-N^{ϵ}-Z-Lys-Ala-Leu-Ala-Gly-OEt					59	DO-2									547
Boc-Leu-N^{ϵ}-Z-Lys-Ala-Leu-N^{ϵ}-Z-Lys-Gly					59				DMF-Aq	180–183					547
Boc-Leu-N^{ϵ}-Z-Lys-Ala-Leu-N^{ϵ}-Z-Lys-Gly-OEt					29				Alc-Aq	218–221	−25	DMF			547
Leu-N^{ϵ}-Z-Lys-Ala-Leu-N^{ϵ}-Z-Lys-Gly-OEt					59	DO-2									547
Leu-Ser-Phe-Met-Ala-Ile															543
Z-Leu-Tyr-Leu-Val-S-Bzl-Cys-Gly hydrazide									DMF-Aq	285 (dec)	−43	DMF			299
Lysine															
N^{ϵ}-N^{α}-Boc-Lys-Ala-Pro-Ser-Gly-Arg					55					191–193	−20	DMF			58
N^{ϵ},N^{α}-Boc-Lys-Ala-Pro-Ser-Gly-Arg-OBzl	5 + 1 A	DMF	85		55					94–95					58
Lys-Ala-Pro-Ser-Gly-Arg						A-IRC-50	×		Me-EtAc	170–175 (dec)	−21	DMF		AA	58

N^{α},N^{ϵ}-Di-Boc-Lys-γ-OBu^{t}-Glu-Thr-Ala-Ala-Ala azide									DMF-Eth	230 (dec)	−16	DMF		AA	67
N^{α}-Z-N^{ϵ}-Boc-Lys-cyclopentyl-Gly-γ-Bu^{t}-Glu-N^{δ}-Boc-Orn-Gln-His-OMe									DMF-Aq and DMF-Eth	214	−13	DMF			67
N^{ϵ}-Tfa-N^{α}-Boc-Lys-N^{ϵ}-Tfa-Lys-Ala-Thr-Asn-γ-OBzl-Glu-OBzl	1 + 5 OSU	DMF-THF	90							211–212	−12	DMF		AA	689
N^{ϵ}-Tfa-Lys-N^{ϵ}-Tfa-Lys-Ala-Thr-Asn-γ-OBzl-Glu-OBzl-Tfa							×			205–206	−1.5	DMF			689
$(Lys)_6$															726
N^{α}-Boc-N^{ϵ}-Tfa-Lys-N^{ϵ}-Tfa-Lys-Tyr-Ile-Pro-Gly	1 + 5 OSU	Glyme	82							131–133	−31	DMF		AA	693
N^{ϵ}-Tfa-Lys-N^{ϵ}-Tfa-Lys-Tyr-Ile-Pro-Gly Tfa										140–141	−53	0.2 *N* $NaHCO_3$			693
N^{ϵ}-Tfa-N^{α}-Boc-Lys-Met-Ile-Phe-Ala-Gly dihydrate	1 + 5 OSU	Glyme	68							268–272	−23	DMF		AA	694
N^{ϵ}-Tfa-Lys-Met-Ile-Phe-Ala-Gly-Tfa dihydrate										260–263	−2.0	DMF			694
N^{ϵ}-Boc-N^{α}-Z-Lys-Phe-γ-OBu^{t}-Glu-N^{δ}-Boc-Orn-Gln-N^{δ}-Boc-Orn-OMe	2 + 4 A	EtAc	88							214–216 (dec)	−18	DMF			68
N^{ϵ}-Boc-Lys-Phe-γ-OBu^{t}-Glu-N^{δ}-Boc-Orn-Gln-N^{δ}-Boc-Orn-OMe acetate										200–201 (dec)	−7.4	Ac			68
CHla-Lys-Trp-Ala-Pro-Phe-Arg															151
Phenylalanine															
(Propionyl-D-Phe-Cys-Gly-OMe)-(propionyl-D-Phe-Cys-Gly-OMe)									Me	199–200	−99				188
Aoc-Phe-Gln-Asn-ω-OBu^{t}-α-amino-suberyl-Pro-N^{G}-Tos-Arg-Gly amide									Alc-EtAc	142–146	−30	DMF			219

(*continued*)

Table 6 Hexapeptides (continued)

Compound	Method	Solvent	% Yield	Chromatography Col	Tl	Ion	Paper	CCD	Recrys. Sol.	M.P. °C	$[\alpha]_D$	Solvent Rotation	Elect.	Other	Ref.
Aoc-Phe-Gln-Asn-ω-OBu^t-α-amino-suberyl-Pro-N^ε-Pht-Lys-Gly amide									Alc-Aq	188–190 (dec)	−35		DMF		219
Phe-Phe-Leu-Ile-Ile-Leu HCl			69		×					292 (dec)	−52	Ac			671
N-Tfa-Phe-Phe-Leu-Ile-Ile-Leu-OBu^t	3 + 3 DCCI-NHS	DMF	91							278 (dec)	−42	HMPA			671
Boc-(Phe)$_2$-(Pro)$_2$-(Phe)$_2$-OMe	1 + 5 OSU	DMF	98												676
Z-Phe-Ser-Val-Ser-Ser-Pro Boc-hydrazide	2 + 4 A	DMF-THF	68							215–216 (dec)	−32	DMF			83
Phe-Ser-Val-Ser-Ser-Pro Boc-hydrazide acetate											−74	Me			83
Z-Phe-Tyr-Thr-Pro-N^ε-Tos-Lys-Thr-OMe	1 + 5 ONP	DMF	90						Me-Eth	139–141	−41	DMF			297
Proline															
Z-Pro-Ile-Leu-Ser-Ser-Gly Boc-hydrazide	1 + 5 OSU	DMF-THF							Me-Aq	238–239 (dec)	−69	Me			83
Pro-Ile-Leu-Ser-Ser-Gly Boc-hydrazide acetate											−56	Me		AA	83
I maa-Pro-Leu-Gly-Pro-Ala-D-Arg															703 702
Pz-Pro-Leu-Gly-Pro-Ala-D-Arg															702

Imaa-Pro-Leu-Pro-Leu-Gly-Pro	3 + 3 OSU	Aq	88							152					703
Imaa-Pro-Leu-Pro-Leu-Gly-Pro-OSu															703
Boc-Pro-N^{ϵ}-Z-Lys-Val-N^{ϵ}-Z-Lys-Ala-His-OMe	1 + 5 OSU	DMF	84						EtAc-Me-Pe	199–199.5 (dec)	−24	DMF			93
Aoc-Pro-Pro-Pro-Pro-Pro-Pro															417
Aoc-Pro-Pro-Pro-Pro-Pro-Pro-OBzl	3 + 3 DCCI	Chf	69												417
Boc-Pro-$(Pro)_4$-Pro															530
					×					229–231				IR NMR	121
Boc-Pro-Pro-Pro-Pro-Pro-Pro-OBzl	2 + 4 MCA	EtAc	94						EtAc	221–229				IR NMR	121
Pro-$(Pro)_4$-Pro					105					206–208	−378	Aq			530
Serine															
Z-Ser-Gly-Pro-Gly-Ser-Gly Boc-hydrazide	2 + 4 A	DMF	80							168–169 (dec)	−22	DMF			83
Ser-Gly-Pro-Gly-Ser-Gly Boc-hydrazide acetate											−11	Me			83
Threonine															
Z-Thr-Ala-Ala-Ala-N^{ϵ} Boc-Lys hydrazide											−6.5	DMF			238
Z-Thr-Ala-Gln-Ala-Ala-Gly Boc-hydrazide	2 + 4 A	DMF	78						DMF-EtAc	267–270 (dec)	−14	DMF		AA	39
Thr-Ala-Gln-Ala-Ala-Gly Boc-hydrazide acetate			95								−73 −52	Aq DMF-Aq		AA	39
Boc-Thr-Ala-Ile-Gly-Val-Gly	SP		57												313
OBu^t-Z-Thr-OBu^t-Tyr-OBu^t-Thr-Gln-β-OBu^t-Asp-Phe hydrazide			27							226–229 (dec)	+5	DMF			574

(*continued*)

Table 6 Hexapeptides (continued)

Compound	Method	Solvent	% Yield	Chromatography				CCD	Recrys. Sol.	M.P. °C	$[\alpha]_D$	Solvent Rotation	Elect.	Other	Ref.
				Col	Tl	Ion	Paper								
OBu^t-Z-Thr-OBu^t-Tyr-OBu^t-Thr-Gln-β-OBu^t-Asp-Phe-OMe			27 19							180–182	+9	DMF			574
OBu^t-Z-Thr-Tyr-OBu^t-Thr-Gln-β-OBu^t-Asp-Phe-OPh										211–214					314
Tryptophan															
Pentanoyl-Trp-Leu-Val-Pro-Leu-Ala-OMe															321
Propionyl-Trp-Leu-Val-Pro-Leu-Ala-OMe															321
Tyrosine															
Z-Tyr-Asp-Asp-Tyr-γ-OBu^t-Glu-Gly Boc-hydrazide	1 + 5 OTCP	DMF	77						Me-Aq	208–209 (dec)	−28	DMF			39
Tyr-Asp-Asp-Tyr-γ-OBu^t-Glu-Gly Boc-hydrazide acetate														AA	39
Boc-Tyr-Cys-Asn-OBu^t | Boc-Cys-Gly-Glu(OBu^t)$_2$					×										293
Boc-Tyr-Cys-Asn-OBu^t | Boc-Tyr-Cys-Asn-OBu^t									Me	187–189	−24	Me			293
OBu^t-Z-Tyr-Gln-Leu-γ-OBu^t-Glu-Asn-OBu^t-Tyr hydrazide										235	−29	DMF			141

Tyr-(-O-sulfate)-Gly-Trp-Met-Asp-Phe amide															44
Valine															
Boc-Val-Ala-Phe-Ala-Gly-Gly-4-carboxy-phenylhydrazide	SP			Seph-adex LH-20	69										677
Z-Val-[Asp(β-OBzl)]$_4$-N$^\epsilon$-Z-Lys-OBzl					×				Di-Eth	151–158	−37	DMF			727
Val-(Asp)$_4$-Lys	SP														311
Val-[Asp]$_4$-Lys			91						Aq-Alc	265–272 (dec)	−38	Aq			727
ω-Amidopimelyl-Val-Tyr-Ile-His-Pro-Phe															190
2,6-Diamino-4-hexynoyl-Val-Tyr-Ile-His-Pro-Phe															190
D-2,6-Diamino-4-hexynoyl-Val-Tyr-Ile-His-Pro-Phe															190
2-Amino-6-dimethylamino-4-hexynoyl-Val-Tyr-Ile-His-Pro-Phe															190
D-2-Amino-6-dimethylamino-4-hexynoyl-Val-Tyr-Ile-His-Pro-Phe															190
α-Amino-ε-trimethylammoniocaproyl-Val-Tyr-Ile-His-Pro-Phe															190
δ-Guanidinovaleroyl-Val-Tyr-Ile-His-Pro-Phe															190
δ-Nitroguanidinovaleroyl-Val-Tyr-Ile-His-Pro-Phe															190
ε-Trimethylammoniocaproyl-Val-Tyr-Ile-His-Pro-Phe															190
δ-Ureidovaleroyl-Val-Tyr-Ile-His-Pro-Phe															190
Val-Tyr-Ile-His-Pro-Phe															190

(continued)

Table 6 Hexapeptides (continued)

Compound	Method	Solvent	% Yield	Chromatography				CCD	Recrys. Sol.	M.P. °C	$[\alpha]_D$	Solvent Rotation	Elect.	Other	Ref.
				Col	Tl	Ion	Paper								
Z-Val-Tyr-Val-His-Pro-Phe-ONBzl	2 + 4 A	EtAc	70						Me-Eth	154–159 (dec)	−69	Me			480
Val-Tyr-Val-His-Pro-Phe				Sephadex G-10						185–190 (dec)				AA	480
Val-Tyr-Val-His-Pro-Phe-ONBzl										130–140 (dec)	−55	Me			479
Val-Tyr-Val-His-Pro-Phe-ONBzl 2 HBr										170–174 (dec)	−34	Me			480

Chapter 8

Heptapeptides

The structure developed by Li and colleagues for the human pituitary growth hormone has stimulated attempts to synthesize this very important protein. Recent work by the Li group toward this objective has been concerned with the 53–67 segment and, e.g., the heptapeptide Boc-Gln-β-OBzl-Asp-Pro-γ-OBzl-Glu-Thr-Ser-Leu.[656]

Table 7 Heptapeptides

Compound	Method	Solvent	% Yield	Chromatography				CCD	Recrys. Sol.	M.P. °C	$[\alpha]_D$	Solvent Rotation	Elect.	Other	Ref.
				Col	Tl	Ion	Paper								
Alanine															
Z-$(Ala)_6$-Ala-OEt									TFE-Alc	275 (dec)	−125	Tfa			186
Z-$(Ala)_6$-Ala-morpholine amide										>275 (dec)	−108	Tfa			186
2-OMe-[2-OEt-(2-OEt)]-Ac-$(Ala)_6$-Ala-OEt										>250 (dec)	−119	Tfa			186
2-OMe-[2-OEt-(2-OEt)]-Ac-$(Ala)_6$-Ala-morpholine amide									TFE-Alc	>275 (dec)	−102	Tfa			186
Boc-Ala-D-Thr-Ala-α-Hyp-Ala-Trp-Nva-OMe				Sephadex LH-20									×	UV	138
ε-Aminocaproic acid															
ε-Acp-Val-Tyr-Ile-His-Pro-Phe															190
Arginine															
N^G-Nitro-Arg-S-OBzl(OMe)-Cys-γ-OBzl-Glu-Leu-Ala-Ala-Ala Boc-hydrazide									DMF-Alc	237–239	−17	DMF			196
N^G-Nitro-N^α-Nps-Arg-S-OBzl(OMe)-Cys-γ-OBzl-Glu-Leu-Ala-Ala-Ala Boc-hydrazide	1 + 6 ONP	DMF	72						EtAc	226–229	−26			AA	197
Arg-Leu-Leu-Gln-Gly-Leu-Val amide acetate HBr											−34	Ac			710

N^{α},N^{δ},N^{ω}-Tris-Z-Arg-Leu-Leu-Gln-Gly-Leu-Val amide	1 + 6 ONP	DMF-HMPA	82							257 (dec)					710
Arg-Val-Tyr-Ile-His-Pro-Phe															190
D-Arg-Val-Tyr-Ile-His-Pro-Phe															190
Homo-Arg-Val-Tyr-Ile-His-Pro-Phe															190
D-Homo-Arg-Val-Tyr-Ile-His-Pro-Phe															190
N^{G}-Nitro-Arg-Val-Tyr-Ile-His-Pro-Phe															190
N^{G}-Nitro-D-Arg-Val-Tyr-Ile-His-Pro-Phe															190
Asparagine															
Boc-Asn-Pro-N^{ϵ}-Z-Lys-Val-N^{ϵ}-Z-Lys-Ala-His-OMe	1+ 6 ONP	DMF	42						*i*-Pr-Eth	150 (dec)	−61	Me			93
Aspartic acid															
Boc-β-OBzl-Asp-Ala-Asn-N^{ϵ}-Tfa-Lys-Asn-N^{ϵ}-Tfa-Lys-Gly-OBu^{t} hydrate	1 + 6 MA	DMF	75							175–177	−11	DMF		AA	692
Z-[Asp(β-OBu^{t})]$_3$-N^{ϵ}-Boc-Lys-Ile-Val-Gly-OBu^{t}										213–214 (dec)	−23	DMF			727
Nps-[Asp(β-OBzl)]$_3$-N^{ϵ}-Z-Lys-Ile-Val-Gly-OBzl									DMSO-Me	214–216	+22	HMPA			727
β-OBu^{t}-Z-Asp-Ile-O-Bu^{t}-Thr-Ala-O-Bu^{t}-Ser-Val-Gly-OPh										260–262 (dec)					314
Z-β-OBu^{t}-Asp-Phe-Ser-Val-Ser-Ser-Pro Boc-Hydrazide	1 + 6 OSU	DMF-THF	81						Me-Aq	200–201 (dec)	−35	DMF			83
β-OBu^{t}-Asp-Phe-Ser-Val-Ser-Ser-Pro Boc-hydrazide acetate											−57	Me		AA	83
δ-Aminovaleric acid															
δ-Ave-Val-Tyr-Ile-His-Pro-Phe															190

(*continued*)

Table 7 Heptapeptides (continued)

Compound	Method	Solvent	% Yield	Chromatography Col	Tl	Ion	Paper	CCD	Recrys. Sol.	M.P. °C	$[\alpha]_D$	Solvent Rotation	Elect.	Other	Ref.
Citrullene															
Cit-Val Tyr-Ile-His-Pro-Phe															190
D-Cit-Val-Tyr-Ile-His-Pro-Phe															190
Cysteine															
Boc-Cys-Asn-OBu^t | Leu-Val-Cys-Gly-Glu(OBu^t)$_2$					×										293
Boc-Cys-Asn-OBu^t | Boc-Leu-Val-Cys-Gly-Glu(OBu^t)$_2$					×										293
S-Bzl-N-Z-Cys-(Gly)$_5$-Gly	5 + 2 SPh	Ac	73							244–246 (dec)	−19	DMF			214
S-Bzl-N-Z-Cys-(Gly)$_5$-Gly-OEt	3 + 4 A	DMF								246 (dec)	−20	DMF			214
S-Bzl-N-Z-Cys-(Gly)$_5$-Gly-ONp										232–234 (dec)	−18	DMSO			214
Glutamic acid															
Boc-α-OBzl-Glu-Gln-N$^\epsilon$-Tos-Lys-OBzl-Tyr-Ser-Phe-Leu										195–197	−26	Me			656
Z-Glu$_7$(OBu^t)$_8$	OPCP	EtAc	32							95–100	−42				410
	MCA	EtAc	70							130–132	−44				
Glutamine															
γ-Bu^t-Z-Gln-β-OBu^t-Asp-Gly-γ-Bu^t-Gln-Thr-Val-Gly Boc-hydrazide	1 + 6 OSU	DMF	82						Me-Aq	204–206 (dec)	−31	Me			39

γ-Bu^t-Gln-β-OBu^t-Asp-Gly-γ-Bu^t-Gln-Thr-Val-Gly Boc-hydrazide acetate											−23	Me		AA	39
Boc-Gln-γ-OBzl-Asp-Pro-α-OBzl-Glu-Thr-Ser-Leu										115–118	−56	Me			656
DL-N-Hepta-Gln										225					320
Z-Gln-Thr-Ala-Ile-Gly-Val-Gly	SP		62												313
Glycine															
Z-(Gly)$_6$-S-Bzl-Cys-OBzl										277–279 (dec)	−21	DMF			214
(Gly)$_6$-S-Bzl-Cys-OBzl										261–263 (dec)	−24	DMSO			214
(Gly)$_6$-S-Bzl-Cys-OBzl-HBr										216–217 (dec)	−20	DMF			214
Succinamyl-Gly-Val-Tyr-Ile-His-Pro-Phe					54 17 42								×	AA	284
Succinyl-Gly-Val-Tyr-Ile-His-Pro-Phe					54 17 57 42								×	AA	284
Z-Gly-Val-Tyr-Val-His-Pro-Phe-ONBzl	1 + 6 DCCI-NHS	DMF	79		57					174–179 (dec)	−25	DMF			479
Gly-Val-Tyr-Val-His-Pro-Phe						Seph-adex C-25				210–211 (dec)	−57 −21	Alc Ac		AA	479
Gly-Val-Tyr-Val-His-Pro-Phe-ONBzl 2 HBr					57				Eth	141–148	−49	Me			479
Hydantoyl-Gly-Val-Tyr-Val-His-Pro-Phe					57					188–191 (dec)	−17	Ac			479

(*continued*)

Table 7 Heptapeptides (continued)

Compound	Method	Solvent	% Yield	Chromatography Col	Tl	Ion	Paper	CCD	Recrys. Sol.	M.P. °C	$[\alpha]_D$	Solvent Rotation	Elect.	Other	Ref.
Hydantoyl-Gly-Val-Tyr-Val-His-Pro-Phe-ONBzl					57				Me-Aq	169–175 (dec)	−35	DMF			479
Propionyl-Gly-Val-Tyr-Val-His-Pro-Phe					57					194–197 (dec)	−18	Ac			479
Isoleucine															
Z-Ile-N^4-methyl-Asn-Asn-S-Bzl-Cys-Pro-Leu-Gly amide										261–263 (dec)	−73	Fm-Aq			224
Z-Ile-Gln-Asn-S-Bzl-Cys-Pro-Leu-Gly amide									Alc-Aq	241–243 (dec)					62
Ile-Gln-Asn-S-Bzl-Cys-Pro-Leu-Gly amide HBr															463
Ile-Gln-Asn-S-β-carboxyethyl-Cys-Pro-Leu-Gly amide					64 71	DO-50			DMF-Eth		−12	DMF		AA	287
Ile-Gln-Asn-S-β-carboxyethyl-homo-Cys-Pro-Leu-Gly amide				DO-50					DMF-Eth		−49	DMF		AA	289
Ile-Gln-Asn-S-[γ-carboxypropyl]-Cys-Pro-Leu-Gly amide						DO-50			DMF-Eth		−51	DMF		AA	285
Ile-Gln-Asn-S-γ-carboxypropyl-homo-Cys-Pro-Leu-Gly amide					64 76	DO-50			DMF-Eth		−50	DMF		AA	502
Nps-Ile-Gln-Asn-S-β-carbo-OMe-ethyl-Cys-Pro-Leu-Gly amide	1 + 6 OSU	DMF	79		64 71				DMF-Eth	245–247	−74	DMF			287
Nps-Ile-Gln-Asn-S-β-OMe-carbonyl-ethyl-homo-Cys-Pro-Leu-Gly amide	1 + 6 OSU	DMF	85						DMF-Eth	230–233	−62	DMF			289

Nps-Ile-Gln-Asn-S-γ-carboxypropyl-homo-Cys-Pro-Leu-Gly amide					64 76 125				DMF-Eth	186–189	−57	DMF			502
Nps-Ile-Gln-Asn-S-(γ-OMe-carbonylpropyl)Cys-Pro-Leu-Gly amide	1 + 6 OSU	DMF	79						DMF-Eth	247–249	−73	DMF			285
Nps-Ile-Gln-Asn-S-γ-OMe-carbonyl-propyl-homo-Cys-Pro-Leu-Gly amide	1 + 6 OSU	DMF	70		64 76 125				DMF-Eth	252–254	−65	DMF			502
Z-Ile-Gln-Asn-S-Bzl-Cys-Pro-N$^{\epsilon}$-Tos-Lys-Gly	1 + 6 ONP	DMF	85							224.5–225.5 (dec)	−39	DMF			513
Boc-Ile-Gln-N$^{\gamma}$-Pht-Dbu-S-Bzl-Cys-Pro-Leu-Gly amide	1 + 6 ONP	DMF	63						Alc	203–206 (dec)	−74	Ac			221
Leucine															
Boc-Leu-N$^{\epsilon}$-Tfa-Lys-N$^{\epsilon}$-Tfa-Lys-Ala-Thr-Asn-γ-OBzl-Glu-OBzl	1 + 6 OSU	DMF	89							217–218	−14	DMF		AA	689
Leu-N$^{\epsilon}$-Tfa-Lys-N$^{\epsilon}$-Tfa-Lys-Ala-Thr-Asn-γ-OBzl-Glu-OBzl Tfa							×			215	−3.5	DMF			689
Boc-Leu-N$^{\epsilon}$-Z-Lys-N$^{\epsilon}$-Z-Lys-Thr-N^{2}-non-Lys-N$^{\epsilon}$-Z-Lys-D-Leu	SP		76		81 125					146–149	−13	Me		AA	732
D-Leu-Val-Tyr-Ile-His-Pro-Phe															190
Leu-Val-Tyr-Ile-His-Pro-Phe															190
Lysine															
Lys-Ala-Gly-Leu-Gly-Trp-Leu	SP														454
(Lys)$_7$															726
N$^{\epsilon}$-Tos-N$^{\alpha}$-Boc-Lys-Phe-Leu-γ-OBzl-Glu-γ-OBzl-Glu-N^{im}-Bzl-His-Pro-OBzl	3 + 4 DCCI-NHS	MeCl	60						Alc	213–214.5	−22	DMF			377
N$^{\epsilon}$-Boc-Z-Lys-Pro-Val-Gly-N$^{\epsilon}$-Boc-Lys-N$^{\epsilon}$-Boc-Lys-N$^{\epsilon}$-Boc-Lys-NH-$(CH_2)_4$-NH-Boc	6 + 1 DCCI-HBT	DMF	74						Acn	150–155	−24	DMF			171

(*continued*)

Table 7 Heptapeptides (continued)

Compound	Method	Solvent	% Yield	Chromatography Col	Tl	Ion	Paper	CCD	Recrys. Sol.	M.P. °C	$[\alpha]_D$	Solvent Rotation	Elect.	Other	Ref.
N^{ϵ}-Boc-Lys-Pro-Val-Gly-N^{ϵ}-Boc-Lys-N^{ϵ}-Boc-Lys-N^{ϵ}-Boc-Lys-NH-$(CH_2)_4$-NH-Boc-Tos															171
N^{α}-Boc-N^{ϵ}-Z-Lys-Thr-N^{α}-nonanoyl-Lys-N^{ϵ}-Z-Lys-Thr-N^{ϵ}-Z-Lys-D-Leu	SP														422
N^{ϵ}-For-Z-Lys-Tyr-Asp-Asp-Tyr-γ-Bu^t-Gln-Gly Boc-hydrazide	1 + 6 OTCP	DMF	68							210–211 (dec)	−25	DMF			39
N^{ϵ}-For-Lys-Tyr-Asp-Asp-Tyr-γ-Bu^t-Gln-Gly Boc-hydrazide acetate			96								−15	Ac-Aq		AA	39
Homo-Lys-Val-Tyr-Ile-His-Pro-Phe															190
D-Homo-Lys-Val-Tyr-Ile-His-Pro-Phe															190
Lys-Val-Tyr-Ile-His-Pro-Phe															190
D-Lys-Val-Tyr-Ile-His-Pro-Phe															190
Methionine															
Met-Ala-Ile-Pro-Pro-Lys-Lys					126						−60	Fm			498
Boc-Met-γ-Bzl-Glu-His-Phe-Arg-Trp-Gly	1 + 6 ONP	DMF	8								−19	DMF			249
Ornithine															
Orn-Val-Tyr-Ile-His-Pro-Phe															190
D-Orn-Val-Tyr-Ile-His-Pro-Phe															190

Phenylalanine															
Z-Phe-Gln-Asn-S-Bzl-Cys-Pro-N-Tos-homo-Lys-Gly amide										201–204	−38	DMF			61
Z-Phe-γ-OBu^t-Glu-Arg-Gln-His-Nle-Asp-Di-OBu^t triacetate	1 + 6 OSU	DMF	82							206–207 (dec)	−33	Bu^t-Me-Aq		AA	238
Phe-Glu-Arg-Gln-His-Nle-Asp														AA	238
Z-Phe-Phe-Tyr-Thr-Pro-N^ε-Boc-Lys-Thr			58						EtAc-Eth	164–166	−43	DMF			619
Z-Phe-Phe-Tyr-Thr-Pro-N^ε-Tos-Lys-Thr-OMe									Alc-Aq	163–165	−39	DMF			297
Phe-Phe-Tyr-Thr-Pro-N^ε-Tos-Lys-Ala					73					148–151 (dec)	−48	DMF	×		636
p-Sulfonyl-Phe-Thr-Gly-Trp-Met-Asp-Phe amide															44
p-Sulfonamido-Phe-Val-Gly-Trp-Met-Asp-Phe amide															44
Proline															
Boc-Pro-Gly-Phe-OBzl-Ser-Pro-Phe-N^G-nitro-Arg-OPic	OTCP	DMF	90								−44				540
Boc-Pro-Gly-Phe-Ser-Pro-Phe-N^G-nitro-Arg-OBzl										122–124	−40				279
CHla-Pro-Gly-Phe-Ser-Pro-Phe-Arg															151
Pro-Gly-Phe-Ser-Pro-Phe-N^G-nitro-Arg-OBzl HCl										132–139	−43				279
Imaa-Pro-Leu-Gly-Pro-Ala-Ala-D-Arg	4 + 3 OSU	Di-Aq	63							168 (dec)					703
Imaa-Pro-Leu-Gly-Pro-Ala-D-Ala-D-Arg	4 + 3 OSU	Di-Aq	47							171 (dec)					703
Imaa-Pro-Leu-Gly-Pro-Gly-Gly-D-Arg	4 + 3 OSU	Di-Aq	63							198 (dec)					702 703

(*continued*)

Table 7 Heptapeptides (continued)

Compound	Method	Solvent	% Yield	Chromatography Col	Tl	Ion	Paper	CCD	Recrys. Sol.	M.P. °C	$[\alpha]_D$	Solvent Rotation	Elect.	Other	Ref.
Imaa-Pro-Leu-Pro-Leu-Gly-Pro-Arg	6 + 1 OSU	Di-Aq	50							184 (dec)				AA	703
Imaa-Pro-Leu-Pro-Leu-Gly-Pro-Arg															702
Imaa-Pro-Leu-Pro-Leu-Gly-Pro-D-Arg	6 + 1 OSU	Di-Aq	43							187 (dec)				AA	703
															702
Boc-Pro-Pro-Gly-Phe-Ser-Pro-Phe-nitro-agmatine				Si					Me-Eth	135–137	−49	DMF			279
Boc-Pro-(Pro)$_5$-Pro	3 + 4 ONP		75							267	−242	Chf			530
4-(Nitroguanidino)Valeryl-Pro-Pro-Gly-Phe-Ser-Pro-Phe-nitro-agmatine										143–146	−44	DMF			279
4-Guanidino-Valeryl-Pro-Pro-Gly-Phe-Ser-Pro-Phe-agmatine											−99	Aq	×	AA	279
Pro-(Pro)$_5$-Pro										226–227	−393	Aq			530
3-Hydroxyproline															
[*i*-Boc-α-3-Hyp-Ala-Trp-Nva-(ind-2) [Ala-D-Thr-Ala(NH_2)-yl-(β)]-sulfide				Seph-adex LH-20										AA	138
[*i*-Boc-α-3-Hyp-Ala-Trp-Nva-(ind-2)]-[Boc-Ala-D-Thr-Ala-(NH_2)-yl-(β)]-sulfide					×										138

[*i*-Boc-α-3-Hyp-Ala-Trp-Nva-(OMe)-(ind-2)]-[Boc-Ala-D-Thr-Ala(NH_2)-yl-(β)]-sulfide				Seph-adex LH-20									×	AA	138
[α-3-Hyp-Ala-Trp-Nva(OMe)-(ind-2)]-[Boc-Ala-D-Thr-Ala-yl-(β)]-sulfide				Seph-adex LH-20							−33	Ac	×	UV	138
Serine															
OBzl-Boc-Ser-OBzl-Tyr-Gly-Leu-N^G-nitro-Arg-Pro-Gly amide	SP				120 99 7				Me-Eth		−43	Me		AA	579
OBzl-Ser-OBzl-Tyr-Gly-Leu-N^G-nitro-Arg-Pro-Gly amide Tfa salt					120 7										579
Threonine															
Boc-Thr-$N^ε$-Tfa-Lys-Met-Ile-Phe-Ala-Gly	1 + 6 OSU	Glyme-Aq	85						Aq	224–225	−16	DMF		AA	694
Thr-$N^ε$-Tfa-Lys-Met-Ile-Phe-Ala-Gly Tfa trihydrate										210–212	−5	DMF			694
Tryptophan															
Z-Trp-Pro-Ile-Leu-Ser-Ser-Gly Boc-hydrazide	1 + 6 OTCP	DMF	76			DO-50			Alc-Eth	160–162 (dec)	−53	Me			83
Trp-Pro-Ile-Leu-Ser-Ser-Gly Boc-hydrazide acetate											−55	Me		AA	83
Tyrosine															
Boc-Tyr(O-sulfate)-Abu-Gly-Trp-Met-Asp-Phe amide															44
Boc-Tyr-(O-sulfate)-Abu-Gly-Trp-Nle-Asp-Phe amide															44
OBu^t-N-Aoc-Tyr-Ile-Gln-Asn-α-aminosuberyl-Pro-N^G-Tos-Arg-Gly amide									Alc-Aq	202–204 (dec)	−25	DMF			219

(*continued*)

Table 7 Heptapeptides (continued)

Compound	Method	Solvent	% Yield	Chromatography Col	Tl	Ion	Paper	CCD	Recrys. Sol.	M.P. °C	$[\alpha]_D$	Solvent Rotation	Elect.	Other	Ref.
OBu^t-N-Aoc-Tyr-Ile-Gln-Asn-ω-ONp-α-aminosuberyl-Pro-N^G-Tos-Arg-Gly amide										202–204 (dec)	−25	DMF			219
Tyr-Ile-Gln-Asn-α-aminosuberyl-Pro-N^G-Tos-Arg-Gly amide						A-IR-55 Seph-adex C-25			Alc-Eth		−27	DMF			219
OBu^t-Boc-Tyr-Ile-Gln-Asn-α-amino-pimelyl-Pro-Leu-Gly amide	1 + 6 OSU	DMF	78						DMF-Eth	211–212	−33	DMF			287
OBu^t-Boc-Tyr-Ile-Gln-Asn-α-amino-suberyl-Pro-Leu-Gly amide									DMF-Eth	201–203	−31	DMF	×		289
Tyr-Ile-Gln-Asn-α-aminopimelyl-Pro-Leu-Gly amide lactam				Bio-gel P-2 and P-4	64 71 125			×	Me-Eth		−31	Ac	×	AA	287
Tyr-Ile-Gln-Asn-α-aminosuberyl-Pro-Leu-Gly amide				Bio-gel P-2 and P 4				×	Me-Eth		−67	Ac		AA	289

OBu^t-N-Aoc-Tyr-Ile-Gln-Asn-α-aminosuberyl-Pro-N$^\epsilon$-Pht-Gly-Gly amide										234–236 (dec)	−40	Ac			219
O-Bu^t-N-Aoc-Tyr-Ile-Gln-Asn-ω-ONp-α-aminosuberyl-Pro-N$^\epsilon$-Pht-Lys-Gly amide										229–233 (dec)	−26	DMF			219
Boc-Tyr(-O-sulfate)-Nle-Gly-Trp-Met-Asp-Phe amide															44
β-Bzl-mercaptopropionyl-Tyr-Phe-Gln-Asn(Bzl)-Cys-N$^\xi$-Tos-homo-Lys-Gly amide					41 54					210–213	−40	DMF			361
OBu^t-N-Aoc-Tyr-Phe-Gln-Asn-α-aminosuberyl-Pro-N^G-Tos-Arg-Gly amide									Alc-Aq	176–179 (dec)	−29	DMF			219
OBu^t-N-Aoc-Tyr-Phe-Gln-Asn-ω-ONp-α-aminosuberyl-Pro-N^G-Tos-Arg-Gly amide									DMF-Eth	171–176 (dec)	−29	DMF			219
Tyr-Phe-Gln-Asn-α-aminosuberyl-Pro-N^G-Tos-Arg-Gly amide lactam						A-IR-45 Sephadex C-25			Alc		−35	DMF			219
OBu^t-N-Aoc-Tyr-Phe-Gln-Asn-α-aminosuberyl-Pro-N$^\epsilon$-Pht-Lys-Gly amide									Alc	230–232 (dec)	−37	DMF			219
OBu^t-N-Aoc-Tyr-Phe-Gln-Asn-ω-ONp-α-aminosuberyl-Pro-N$^\epsilon$-Pht-Lys-Gly amide									DMF-Eth	217–220 (dec)	−29	DMF			219
Boc-Tyr(-O-sulfate)-Phe-Gly-Trp-Met-Asp-Phe amide															44

(*continued*)

Table 7 Heptapeptides (continued)

Compound	Method	Solvent	% Yield	Chromatography				CCD	Recrys. Sol.	M.P. °C	[α]D	Solvent Rotation	Elect.	Other	Ref.
				Col	Tl	Ion	Paper								
Boc-Tyr(-O-sulfate)-Thr-Gly-Trp-Met-Asp-Phe amide															44
Tyr-Thr-Gly-Trp-Met-Asp-Phe amide															44
Boc-Tyr-(-O-sulfate)-Thr-Gly-Trp-Nle-Asp-Phe amide															44
Boc-Tyr(-O-sulfate)-Trp-Gly-Trp-Met-Asp-Phe amide															44
Boc-Tyr-(-O-sulfate)-Tyr-Gly-Trp-Met-Asp-Phe amide															44
Boc-Tyr-(-O-sulfate)-Tyr-(-O-sulfate)-Gly-Trp-Met-Asp-Phe amide															44
Boc-3-chloro-Tyr (-O-sulfate)-Val-Gly-Trp-Met-Asp-Phe amide															44
Boc-D-Tyr (-O-sulfate)-Val-Gly-Trp-Met-Asp-Phe amide															44
Boc-3,5-dibromo-Tyr (-O-sulfate)-Val-Gly-Trp-Met-Asp-Phe amide															44
Boc-*m*-Tyr (-O-sulfate)-Val-Gly-Trp-Met-Asp-Phe amide															44
Boc-Tyr (-O-sulfate)-Val-Gly-Trp-Met-Asp-Phe amide															44

Valine															
Val-His-Phe-Asp-Ala-OBzl-Ser-Val	1 + 6 NCA	Aq												AA	127
	1 + 6 OSU	DMF	32	Si											
Val-Lys-Val-Tyr-Pro-Ala-Gly	SP										−67	Aq		AA	712

Chapter 9

Octapeptides

The orderly and unambiguous synthesis of disulfide bridges in insulin-type peptides represents a challenge under active investigation, e.g., by Hiskey and colleagues. A nicely developed recent contribution to this problem involved synthesis of the A_{6-13} segment of ovine insulin utilizing the sulfenylthiocyanate technique and OSU coupling procedures.[233] This contribution appeared after the closing date for the tables of this chapter. Because of the chemically sophisticated treatment of Cys and the new selective reactions introduced in this fragment condensation synthesis, a detailed outline has been presented in Chart 7.

For reasons noted above a synthesis of the octapeptide corresponding to the A_{14-21} segment of ovine insulin has been completed.[236] The Bpoc-blocking group in conjunction with OSU and A coupling methods were evaluated. Conversion of a hydrazide to the corresponding azide in the presence of Bpoc-protecting groups was given detailed treatment and it was established that this blocking group was stable to excess hydrogen chloride in THF-DMF mixtures at temperatures ranging from $-40°$ to $-20°$. By this means, azide preparation and the subsequent coupling reaction were performed in good yields. Also in this series of experiments, cleavage of an otherwise refractory NPS blocking group was achieved using *o*-nitrophenylsulfenyl chloride in the presence of 2-mercaptoethanol.

A series of angiotensin II analogs have been synthesized by Fessler using an SP procedure and screened as antagonists of angiotensin II on vascular smooth muscle.[143] These included [Asn^1, Val^5, Abu^8] angiotensin II and all were found to be potent inhibitors of the vascular effects of angiotensin II.

Chart 7

233 Insulin A_{6-13}

Z-Ala-Gly-OBzl —[10% Pd/C, H_2 / Alc-Ac, 4 hr]→ Ala-Gly
90%
m. p. 234°-235°

S-Bzh-Cys —[(1) 4*N* NaOH to pH 10. 2, Di-Aq, Boc-N_3 / (2) DCHA, Eth / (3) Recrys. Chf-He]→ S-Bzh-N-Boc-Cys DCHA salt —[(1) Free acid, glyme, DCCI-NHS / 2 hr ~0°, 10 hr ~27° / (2) Recrys. EtAc-He]→
79%
m. p. 157°-158°

S-Bzh-N-Boc-Cys-OSu —[Ala-Gly, glyme / Aq-$KHCO_3$, 2. 5 hr, room temp.]→ S-Bzh-N-Boc-Cys-Ala-Gly —[(1) Ac, BF_3· etherate, 30 min, room temp. / (2) 6% NaOAc / (3) Recrys. DMA-EtAc]→
76% ; 85%, m. p. 124°-125°

S-Bzh-Cys-Ala-Gly —[(1) S-Trt-N-Boc-Cys, glyme, DCCI-NHS, 2 hr ~0°, 1 day, room temp. / (2) ~4% $KHCO_3$, 6 hr, room temp.]→ S-Trt-N-Boc-Cys-S-Bzh-Cys-Ala-Gly
66%, d. p. 178°-180° ; PART A 94%, d. p. 170°-175°

Z-Ser + Leu-OBzl Tos salt —[DMA-EtAc, EDCI, NMM / 2 hr ~0°, 1 day, room temp.]→ Z-Ser-Leu-OBzl —[(1) ⫫, H_2SO_4, Chf, 27°, 12 hr / (2) Chromat. SiO_2 / (3) Recrys. He]→
73%
m. p. 81°-83°
57% by DCCI-NHS

OBu^t-N-Z-Ser-Leu-OBzl —[10% Pd/C, H_2, Alc-Aq / 2. 3 hr, room temp.]→ OBu^t-Ser-Leu —[(1) N, S-Trt-Cys, glyme, DCCI-NHS, 2 hr ~0°, 11 hr, room temp. / (2) Chf, NMM, 12 hr, room temp. / (3) Chromat. SiO_2 / (4) Recrys. Eth-He]→
84%, m. p. 69°-70° ; 90%, m. p. 96°-98° or 63% and 18. 5% dioxopiperazine with Ac as solvent

Chart 7 (cont.)

N, S-Trt-Cys-OBu^t-Ser-Leu → S-Trt-Cys-OBu^t-Ser-Leu Tos salt

(above arrow) (1) 80% Ac, 10 min, ~95°
(below arrow) (2) $CH_3C_6H_5SO_3H$, Eth

N, S-Trt-Cys-OBu^t-Ser-Leu: ~90%, m.p. 110°-114°

S-Trt-Cys-OBu^t-Ser-Leu Tos salt: 62%, m.p. 146°-147°

(above arrow) (1) NPS-Val-OSu, glyme, NMM, 2 days, room temp.
(below arrow) (2) Chromat. SiO_2
(3) Recrys. Eth-He

→ NPS-Val-S-Trt-Cys-OBu^t-Ser-Leu →

(above arrow) Chf, $HSCH_2CH_2OH$
(below arrow) 6.5 *N* HCl in Di-Eth, 5 min, room temp.

NPS-Val-S-Trt-Cys-OBu^t-Ser-Leu: 72%, m.p. 184°-186°, bright yellow

Val-S-Trt-Cys-OBu^t-Ser-Leu HCl salt →

(above arrow) (1) DCCI-NHS, DMA, 27°, 1 day
(below arrow) (2) Wash *i*-Pr (removes urea)
(3) Part A, DMA-NMM, 2 days, room temp.

Val-S-Trt-Cys-OBu^t-Ser-Leu HCl salt: 98%, m.p. 137°-138°

S-Trt-N-Boc-Cys-S-Bzh-Cys-Ala-Gly-Val-S-Trt-Cys-OBu^t-Ser-Leu

84%, d.p. 222°

(above arrow) (1) $Br_2 + Pb(SCN)_2 \rightarrow (SCN)_2$, EtAc
(2) Chf-Ac, ~0°, 3 days, room temp.
(below arrow) (3) Recrys. Me-Aq (SCN polymer side product sol. in hot Me)

→

Chart 7 (cont.)

S————————————S

N-Boc-Cys-S-Bzl-Cys-Ala-Gly-Val-Cys-OBu^t-Ser-Leu

87%
m.p. 219°-221°

mol. wt. 1075. Found 1050 osometrically

(1) 1:1 Tfa-Chf, −20°, 10 hr; (2) Recrys. DMA-Aq → de-N-Boc derivative (OBu^t group stable for ~10 hr at −20°)

90%
m.p. 218°-220°
$\alpha]_D$ −12° (DMF, 0.9 c)

Tfa, 50 min, room temp. → de-N-Boc and OBu^t

93%
m.p. 219°-221°
$\alpha]_D$ −13° (DMF, 0.5 c)

Table 8 Octapeptides

Compound	Method	Solvent	% Yield	Chromatography Col	Tl	Ion	Paper	CCD	Recrys. Sol.	M.P. °C	$[\alpha]_D$	Solvent Rotation	Elect.	Other	Ref.
Alanine															
Z-(Ala)$_7$-Ala morpholine amide	3 + 5 MA	DMSO-THF	38						Fipr-Aq	>275 (dec)	−117	Tfa			186
2-OMe-[2-OEt-(2-OEt)]-Ac-(Ala)$_7$-Ala-morpholine amide									TFE-Alc	>275 (dec)	−104	Tfa			186
Arginine															
N-[2-(Arg-Pro-amino)-ethyl]-Gly-Phe-Ser-Pro-Phe-Arg							×						×	AA	26
N-Z-N-[2-(N-Z-N^G-Nitro-Arg-Pro-amino)-ethyl]-Gly-Phe-Ser-Pro-Phe-N^G-nitro-Arg-OBzl	1 + 7 MCA	DMF		Si					An-Eth	115–125					26
Arg-Pro-Pro-Gly-Phe-Ser-Pro-Phe-agmatine Tri-HCl										175–185	−104	Aq	×	AA	279
Z-Nitro-Arg-Pro-Pro-Gly-Phe-Ser-Pro-Phe-nitro-agmatine			47.5						Me-Eth	139–143	−56	DMF			279
Asparagine															
Asn-Arg-Val-Tyr-Ala-His-Pro-Phe	SP					×							×	AA	283
Asn-Arg-Val-Tyr-Abu-His-Pro-Phe	SP			Sephadex G-25		×							×	AA	283
Asn-Arg-Val-Tyr-Aib-His-Pro-Phe	SP					×								AA	282
Asn-Arg-Val-Tyr-α-cyclohexyl-Gly-His-Pro-Phe	SP			Sephadex G-25		×							×	AA	283

Asn-Arg-Val-Tyr-α-cyclopentyl-Gly-His-Pro-Phe	SP			Seph-adex G-25		×							×	AA	283
Asn-Arg-Val-Tyr-cyclo-Leu-His-Pro-Phe													×		282
Asn-Arg-Val-Tyr-Nle-His-Pro-Phe	SP			Seph-adex G-25		×							×		283
Asn-Arg-Val-Tyr-Pro-His-Pro-Phe	SP			Seph-adex		×							×	AA	282
Asn-Arg-Val-Tyr-D-Pro-His-Pro-Phe	SP			Seph-adex G-25		×							×	AA	282
Asn-Arg-Val-Tyr-Sar-His-Pro-Phe	SP					×							×	AA	282
Asn-Arg-Val-Tyr-α-amino-β-ethyl-Val-His-Pro-Phe	SP			Seph-adex G-25		×							×	AA	283
Asn-Arg-Val-Tyr-Val-His-Pro-Phe															190
Asn-Gln-Tyr-Val-Orn(δ-Z)-Leu-D-Phe-Pro									DMF-EtAc-Eth	285–287 (dec)	−20	DMF			342
Asn-Gly-Val-Tyr-Ile-His-Pro-Phe	1 + 7 ONP	DMF	80		54 17				EtAc				×	AA	284
Asn-Gly-Val-Tyr-Val-His-Pro-Phe					57					205–208 (dec)				AA	479
Z-Asn-Gly-Val-Tyr-Val-His-Pro-Phe-ONBzl	1 + 7 ONP	DMF	82		57				Eth	180–190					479
Boc-Asn-Pro-N^{ϵ}-Tfa-Lys-N^{ϵ}-Tfa-Lys-Tyr-Ile-Pro-Gly										138–140	−56	DMF		AA	693
Aspartic acid															
N^{α}-Z-Asp-[β-bis(Dmb-amide)]-Ala-Leu-Ala-N^{im}-(Boc-TF)-His-N^{ϵ}-Boc-Lys-Tyr-N^{im}-trifluoro-1-Boc-ethyl-His-OBu^t								5		115–123	−15	Alc		AA	96

(*continued*)

Table 8 Octapeptides (continued)

Compound	Method	Solvent	% Yield	Chromatography				CCD	Recrys. Sol.	M.P. °C	$[\alpha]_D$	Solvent Rotation	Elect.	Other	Ref.
				Col	Tl	Ion	Paper								
β-Asp-Arg-Val-Tyr-Val-His-Pro-Phe				Sephadex G-15						237				AA	524
β-OBzl-Z-Asp-N^G-nitro-Arg-Val-OBzl-Tyr-Val-N^{im}-Bzl-His-Pro-Phe-OPic						Sulphoethyl Sephadex G-25									168
β-OPic-Z-Asp-N^G-nitro-Arg-Val-OBzl-Tyr-Val-N^{im}-Bzl-His-Pro-Phe-OPic	1 + 7 OTCP	DMF	91												168
Z-[Asp(β-OBu^t)]$_4$-N^ϵ-Boc-Lys-Ile-Val-Gly-OBu^t										211–213 (dec)	−26	DMF			727
Nps-[Asp(β-OBzl)]$_4$-N^ϵ-Z-Lys-Ile-Val-Gly-OBzl									DMSO-Me	221–224	+22	HMPA			727
β-OBzl-Z-Asp-Gly-Val-Tyr-Ile-His-Pro-Phe	1 + 7 ONP	DMF			54 17								×	AA	284
Pyro-Asp-Tyr-(-O-sulfate)-Thr-Gly-Trp-Met-Asp-Phe amide															44
Angiotensin II															
[Arg^8]angiotensin II					57 126		×			256–258	−62	Ac			315
[Asn^1,Val^5]angiotensin II															663

[Cha^8]angiotensin II					120 126 72		×			220 (dec)	−37	Ac		AA	316
[His^4-Tyr^6]angiotensin II					57 126		×			235–237	−80	Ac		AA	315
[Ile^5]angiotensin II	SP					DO-1	×							UV AA	157
[Ile^8]angiotensin II					57 126 40		×			238–240	−96	Ac		AA	315
p-[N,N-Bis(2-chloroethyl)amino]-phenylbutyryl-[Ile^8]angiotensin II	SP														465
[Leu^8]angiotensin II					81 126 72		×			253 (dec)	−66	Ac		AA	316
[Val^5]angiotensin II														AA	168
[Phe^4,Ala^8]angiotensin II					81 120 126		×			245–250 (dec)	−82	Ac		AA	316
[Suc^1,Phe^4,Tyr^8]angiotensin II					120 126 110a		×			214–218	−90	Ac		AA	316
[Sar^1,Ile^8]angiotensin II					81 126 72		×			215 (dec)	−36	DMF		AA	316
[Val^8]angiotensin II					81 126 72		×			247 (dec)	−70	Aq		AA	316
[Val^5,methyl-Ala^7]angiotensin II	SP			Sephadex G-25							−48	Aq		AA	17

(*continued*)

Table 8 Octapeptides (continued)

Compound	Method	Solvent	% Yield	Chromatography: Col	Tl	Ion	Paper	CCD	Recrys. Sol.	M.P. °C	$[\alpha]_D$	Solvent Rotation	Elect.	Other	Ref.
Cysteine															
S-Bz-Z-Cys-S-Tos-Lys-Asn-Gly-Gln-Thr-Asn-S-Bz-Cys-OMe	DCCI-NHS	DMF	39	Al_2O_3					Ac-Aq	180–185	−19	Ac			558
Z-Cys-N$^{\epsilon}$-Tos-Lys-Asn-Gly-Gln-Thr-Asn-Cys										190–194 (dec)					124
(Z-Cys-D-Phe-Leu-Gly-OMe)-(Z-Cys-D-Phe-Leu-Gly-OMe)									EtAc-An	245.5–246	−88				188
(Decanoyl-Cys-D-Phe-Leu-Gly-OMe)-(decanoyl-Cys-D-Phe-Leu-Gly-OMe)									Me	235–236	−49				188
(Propionyl-Cys-D-Phe-Leu-Gly-OMe)-(propionyl-Cys-D-Phe-Leu-Gly-OMe)									DMF-Eth	262–263	−50				188
S-Bzl-Tos-Cys-Tyr-Phe-Gln-Asn-S-Bzl-Cys-Pro-N$^{\beta}$-Z-α,β-diaminopropionyl-Gly amide									DMF-Aq	215–217	−17	DMF			730
Cystine															
Z-Hemi-Cyst-N$^{\epsilon}$-Tos-Lys-Asn-Gly-Gln-Thr-Asn-hemi-Cyst									Ac	212–214	−56	DMF		Mol. wt.	124
Glutamic acid															
γ-OBu^t-Z-Glu-Trp-Pro-Ile-Leu-Ser-Ser-Gly Boc-hydrazide	1 + 7 OSU	DMF-THF	84							160	−59	Me			83
γ-OBu^t-Glu-Trp-Pro-Ile-Leu-Ser-Ser-Gly Boc-hydrazide acetate									Alc-Eth		−60	Me		AA	83

Pyro-Glu-Tyr-(-O-sulfate)-Thr-Gly-Trp-Met-Asp-Phe amide															44
Glutamine															
Boc-Gln-Arg(HBr)-Leu-Leu-Gln-Gly-Leu-Val amide	1 +7 ONP	DMF HMPA	88		54		×		DMF-EtAc	225 (dec)	−50	Ac-Aq		AA	710
Gln-Arg(HBr)-Leu-Leu-Gln-Gly-Leu-Val amide HBr					54		×								710
DL-$(Gln)_8$										225					320
Glycine															
Z-Gly-Gly-Gly-Gly-Val-His-Pro-Phe-ONBzl	2 + 6 OSU	DMF	50						Me-EtAc-Eth	132–138 (dec)	−34	DMF			480
Gly-Gly-Gly-Gly-Val-His-Pro-Phe									Aq-Alc-Eth	204–208 (dec)	−44	Aq		AA	480
Z-Gly-Gly-Gly-Tyr-Val-His-Pro-Phe-ONBzl									Me-EtAc-Eth	138–141 (dec)	−21	DMF			480
Gly-Gly-Gly-Tyr-Val-His-Pro-Phe				Seph-adex G-10						208–212	−27	Aq		AA	480
Z-Gly-Gly-Val-Tyr-Val-His-Pro-Phe-ONBzl	2 + 6 OSU	DMF	68						Me-Eth	186–193 (dec)	−25	DMF			480
Gly-Gly-Val-Tyr-Val-His-Pro-Phe				Seph-adex G-10						202–204 (dec)	−38	Aq		AA	480
Gly-Ile-Gly-Ala-Val-Leu-Lys-Val	SP		49			DO-1 × 2					−91	Ac		AA	130
Z-Gly-Phe-Phe-Tyr-Pro-N^{ϵ}-Tos-Lys-Thr-OMe									Me-Aq	150–157	−34	DMF			297
Histidine															
Z-His-γ-OBu^t-Gln-β-OBu^t-Asp-Gly-γ-OBu^t-Gln-Thr-Val-Gly Boc-hydrazide	1 + 7 A	EtAc	75						DMF-EtAc	182–183 (dec)	−29	Me			39

(*continued*)

Table 8 Octapeptides (continued)

Compound	Method	Solvent	% Yield	Chromatography: Col	Tl	Ion	Paper	CCD	Recrys. Sol.	M.P. °C	$[\alpha]_D$	Solvent Rotation	Elect.	Other	Ref.
His-γ-OBu^t-Gln-β-OBu^t-Asp-Gly-γ-OBu^t-Gln-Thr-Val-Gly Boc-hydrazine diacetate			90								−28	Me		AA	39
Z-His-Leu-Cys-Gly-OMe (dimer)	1 + 7 A	EtAc	78						Me-Aq	198–200	−116	DMF			618
Z-His-N$^\epsilon$-For-Lys-Tyr-Asp-Asp-Tyr-γ-Bu^t-Gln-Gly Boc-hydrazide	1 + 7 A	DMF	63							215–217 (dec)	−29	DMF			39
His-N$^\epsilon$-For-Lys-Tyr-Asp-Asp-Tyr-γ-Bu^t-Gln-Gly Boc-hydrazide diacetate											−21	Ac-Aq		AA	39
Leucine															
Z-Leu-Ile-Leu-Asn-S-Bzl-Cys-Pro-Leu-Gly amide						×				264.5–266 (dec)	−86	Ac			242
Lysine															
N$^\alpha$,N$^\epsilon$-Bis-Boc-Lys-Ala-Pro-Ser-Gly-Arg-Val-Ser hydrazide acetate					54 137					121–126	−83	Aq		AA	58
N$^\alpha$,N$^\epsilon$-Bis-Boc-Lys-Ala-Pro-Ser-Gly-Arg-Val-Ser-OMe acetate	5 + 3 A	DMF			54 137	DO-1 × 8				135–140	−60	Me		AA	58
N$^\epsilon$-Tos-Boc-Lys-α-OBzl-Glu-Gln-N$^\epsilon$-Tos-Lys-OBzl-Tyr-Ser-Phe-Leu										205–209	−30	DMF			656
Lys-His-Pro-Pro-His-Leu-Ser-Phe (Lys)$_8$					126						−54	FM		AA	498 726

N^{ϵ}-Boc-Z-Lys-Pro-Val-Gly-N^{δ}-Boc-Orn-N^{δ}-Boc-Orn-N^{δ}-Boc-Orn-N^{δ}-Boc-Orn amide					9 24			4		202	−33	Me			613
N^{ϵ}-Boc-Z-Lys-Pro-Val-Gly-N^{δ}-Boc-Orn-N^{δ}-Boc-Orn-N^{δ}-Boc-Orn-N^{δ}-Boc-Orn-OMe	4 + 4 DCCI-NHS	DMF			9 24					145–155	−45	Me			613
Norleucine															
Boc-Nle-Phe-Glu-Arg-Gln-His-Met-Asp-D-sulfoxide monoacetate dihydrate	1 + 7 OSU	DMF-Aq	74			AG-1-× 2					−28			AA	238
Nle-Phe-Glu-Arg-Gln-His-Met-Asp-D-sulfoxide											−21	Aq		AA	238
Z-Nle-Phe-γ-OBu^t-Glu-Arg-Gln-His-Nle-Asp-Di-OBu^t diacetate	1 + 7 OSU	DMF-Aq	61								−20	DMF			238
Nle-Phe-Glu-Arg-Gln-His-Nle-Asp Tfa											−26	Aq		AA	238
Ornithine															
N^{δ}-Boc-Z-Orn-Pro-Val-Gly-N^{δ}-Boc-Orn-N^{δ}-Boc-Orn-N^{δ}-Boc-Orn-N^{δ}-Boc-Orn amide					9 24			4		198	−33	Me			613
N^{δ}-Boc-Z-Orn-Pro-Val-Gly-N^{δ}-Boc-Orn-N^{δ}-Boc-Orn-N^{δ}-Boc-Orn-N^{δ}-Boc-Orn-OMe					9 24					140–145	−41	Me			613
Phenylalanine															
Z-Phe-β-Bu^t-Asp-Phe-Ser-Val-Ser-Ser-Pro Boc-hydrazide	1 + 7 OSU	DMF-THF	88						Me-Aq	214–217 (dec)	−36	DMF			83
(Decanoyl-D-Phe-Leu-Gly-Cys-OMe)-(decanoyl-D-Phe-Leu-Gly-Cys-OMe)									EtAc-He	202–203	−45				188
Boc-$(Phe)_2$-$(Pro)_2$-$(Phe)_2$-$(Pro)_2$															676

(*continued*)

Table 8 Octapeptides (continued)

Compound	Method	Solvent	% Yield	Chromatography Col	Tl	Ion	Paper	CCD	Recrys. Sol.	M.P. °C	$[\alpha]_D$	Solvent Rotation	Elect.	Other	Ref.
N-Boc-Phe-Val-Asn-Glu-N^{im}-Trt-His-Leu-S-Bzh-Cys-Gly-OMe and hydrazide	2 + 6 DCCI-HBT	DMF		Seph-adex LH-20			×							AA	69
Z-Phe-Val-Asn-Gln-His-Leu-Cys-Gly-OMe									DMF-Eth	243	−19	Ac		AA	618
Boc-Phe-Val-Asn-Gln-N^{im}-Trt-His-Leu-S-Bzh-Cys-Gly-OMe															69
Proline															
Pro-Phe-His-Leu-Leu-Val-Tyr-Ser															497
Boc-Pro-Pro-Gly-Phe-OBzl-Ser-Pro-Phe-N^G-nitro-Arg-OPic	DCCI	Acn	91								−63				540
Boc-Pro-Pro-Gly-Phe-Ser-Pro-Phe-nitro-Arg OBzl	1 + 7 DCCI 1 + 7 OSU	DMF	55						An-Pe	125–129	−39	DMF			279
CHla-Pro-Pro-Gly-Phe-Ser-Pro-Phe-Arg															151
N^G-Nitro-Pro-Pro-Gly-Phe-Ser-Pro-Phe-N^G-nitro-Arg-OBzl										152–157	−51				279
4-(Nitroguanidino) valeryl-Pro-Pro-Gly-Phe-Ser-Pro-Phe-nitro-Arg-OBzl			79							135–137	−46	DMF			279
Pro-Pro-Gly-Phe-Ser-Pro-Phe-N^G-nitro-Arg-OBzl HCl										142–145	−50				279

Aoc-Pro-(Pro)$_6$-Pro															417
Aoc-Pro(Pro)$_6$-Pro-OBzl	4 + 4 DCCI	Chf	59						Me-EtAc						417
Boc-Pro-(Pro)$_6$-Pro	4 + 4 ONP		78							280	−256	Chf			530
Pro-(Pro)$_6$-Pro										260	−408	Aq			530
Serine															
OBzl-Boc-Ser-N$^\delta$-Z-Orn-OBzl-Ser-D-Phe-Pro-OBzl-Ser-N$^\delta$-Z-Orn-OBzl-Ser-ONp										110–112					261
Threonine															
OBu^t-Z-Thr-Phe-Pro-Gln-OBu^t-Thr-Ala-Ile-Gly			98						Acn-Aq	223–225	−18	DMF			574
OBu^t-Z-Thr-Phe-Pro-Gln-OBu^t-Thr-Ala-Ile-Gly-OMe	DCCI-NHS	DMF	69						Me	231–234	−57	Me			574
OBu^t-Thr-Phe-Pro-Gln-OBu^t-Thr-Ala-Ile-Gly			95												574
Tyrosine															
Tyr(-O-sulfate)-Ala-Ala-Gly-Trp-Met-Asp-Phe amide															44
OBzl-Z-Tyr-Ile-N^4-methyl-Asn-Asn-S-Bzl-Cys-Pro-Leu-Gly amide	1 + 7 ONP	DMF								262–264	−53	Fm-Aq			224
S-Bzl-β-mercaptoproprionyl-Tyr-Ile-N^4-methyl-Asn-Asn-S-Bzl-Cys-Pro-Leu-Gly amide										243–247	−57	Fm-Aq			224
OBzl-Z-Tyr-Ile-Gln-Asn-S-Bzl-Cys-Pro-Leu-Gly amide										245–247 (dec)	−41	DMF			62
Z-Tyr-Ile-Gln-Asn-S-(γ-OMe-carbonylpropyl)Cys-Pro-Leu-Gly amide									DMF-Aq	219–222	−38	DMF			285

(*continued*)

Table 8 Octapeptides (continued)

Compound	Method	Solvent	% Yield	Chromatography Col	Tl	Ion	Paper	CCD	Recrys. Sol.	M.P. °C	[α]D	Solvent Rotation	Elect.	Other	Ref.
β-(S-Bzl-mercapto)-β,β-diethyl-propionyl-OBzl-Tyr-Ile-Gln-Asn-S-Bzl-Cys-Pro-Leu-Gly amide					×					235	−43	DMF		AA	635
S-Bzl-mercaptodimethyl-Ac-Tyr(Bzl)-Ile-Gln-Asn-S-Bzl-Cys-Pro-Leu-Gly amide	SP									217–218	−33	DMF			635
β(S-Bzl-mercapto)-α,α-dimethyl-propionyl-OBzl-Tyr-Ile-Gln-Asn-S-Bzl-Cys-Pro-Leu-Gly amide	SP				×					225	−32	DMF		AA	635
S-Bzl-β-mercaptopropionyl-D-Tyr-D-Ile-D-Gln-D-Asn-S-Bzl-D-Cys-D-Pro-D-Leu-Gly amide										238–240	+42	DMF			148
OBu^t-Boc-Tyr-Ile-Gln-Asn-S-(Boc-hydrazinocarbonylpropyl)Cys-Pro-Leu-Gly amide				Sephadex LH-20					Me	228–230	−38	DMF			285
OBu^t-Boc-Tyr-Ile-Gln-Asn-S-β-carbosyethyl-Cys-Pro-Leu-Gly amide	1 + 7 OSU	DMF							DMF-Eth	219–220	−40	DMF			287
OBu^t-Boc-Tyr-Ile-Gln-Asn-S-β-carboxyethyl-homo-Cys-Pro-Leu-Gly amide	1 + 7 OSU	DMF							DMF-Eth	211–213	−35	DMF			289
OBu^t-Boc-Tyr-Ile-Gln-Asn-S-(γ-carboxypropyl)-Cys-Pro-Leu-Gly amide	1 + 7 OSU	DMF	68						DMF-Eth	213–214	−34	DMF			285

OBu^t-Boc-Tyr-Ile-Gln-Asn-S-γ-carboxypropyl-homo-Cys-Pro-Leu-Gly amide	1 + 7 OSU	DMF	82		64 76 125				DMF-Eth	209–211	−29	DMF		Mol. wt.	502
OBu^t-Boc-Tyr-Ile-Gln-Asn-S-(γ-ONp-carbonylpropyl)Cys-Pro-Leu-Gly amide									Me	225–227	−34	DMF			285
Tyr-Ile-Gln-Asn-S-Bzl-Cys-Pro-Leu-Gly amide															463
Tyr-Ile-Gln-Asn-S-β-carboxyethyl-Cys-Pro-Leu-Gly amide lactam					64 71 125			6	Me-Eth		−73	Ac		AA	287
Tyr-Ile-Gln-Asn-S-β-carboxyethyl-homo-Cys-Pro-Leu-Gly amide lactam								6			−85	Ac	×	AA	289
Tyr-Ile-Gln-Asn-S-(γ-carboxypropyl)Cys-Pro-Leu-Gly amide						DO-50 DO-1 × 2									285
Tyr-Ile-Gln-Asn-S-γ-carboxypropyl-Cys-Pro-Leu-Gly amide								6	Me-Eth		−58	Aq		AA	285
OBzl-Z-Tyr-Ile-Gln-Asn-S-Bzl-Cys-Pro-N$^\epsilon$-Tos-Lys-Gly	1 + 7 ONP	DMF								222 (dec)	−31	DMF			513
S-Bzl-β-mercaptopropionyl-Tyr-Ile-Gln-Asn-S-Bzl-Cys-Pro-N$^\epsilon$-Tos-Lys-Gly									DMF-EtAc	228 (dec)	−35	DMF			513
OBzl-Boc-Tyr-Ile-Gln-N$^\gamma$-Pht-Dbu-S-Bzl-Cys-Pro-Leu-Gly amide	1 + 7 ONP	DMF	86							236–240 (dec)	−54	Ac			221
OBzl-Boc-Tyr-Leu-N$^\epsilon$-Tfa-Lys-N$^\epsilon$-Tfa-Lys-Ala-Thr-Asn-Glu-Di-OBzl	1 + 7 OSU	DMF	87							243–244	−11	DMF		AA	689
OBzl-Tyr-Leu-N$^\epsilon$-Tfa-Lys-N$^\epsilon$-Tfa-Lys-Ala-Thr-Asn-Glu-Di-OBzl Tfa dihydrate							×			202	−8.7	DMF			689

(*continued*)

Table 8 Octapeptides (continued)

Compound	Method	Solvent	% Yield	Chromatography Col	Tl	Ion	Paper	CCD	Recrys. Sol.	M.P. °C	$[\alpha]_D$	Solvent Rotation	Elect.	Other	Ref.
O-Bzl-Z-Tyr-Phe-Gln-Asn-S-Bzl-Cys-Pro-homo-N$^{\epsilon}$-Tos-Lys-Gly amide										215–219	−40	DMF		AA	61
Valine															
[Boc-Val-Cys-Gly-Glu(OBu^t)$_2$]$_2$									EtAc-He		−68	Me			293
Boc-Val-(N$^{\delta}$-Z)-Orn-Leu-D-Phe-Pro-Val-(N$^{\delta}$-Z)-Orn-Leu-NHCH$_2$-CH$_2$OH	5 + 3 DCCI-NHS	Di-DMF	94							156–158	−91				384
Val-N$^{\delta}$-Z-Orn-Leu-D-Phe-Pro-Val-N$^{\delta}$-Z-Orn-Leu-NHCH$_2$-CH$_2$OH						DO-1				113–117	−107				384
Z-Val-Ser-Thr-Ala-Gln-Ala-Ala-Gly Boc-hydrazide	2 + 6 A	DMF-THF	65						DMF-Me	281–283 (dec)	−13	DMSO			39
Val-Ser-Thr-Ala-Gln-Ala-Ala-Gly-Boc-hydrazide										280–285 (dec)	−15	DMSO		AA	39
Boc-Val-OBzl-Thr-Val-Leu-Thr(Bzl)-Ala-Leu-Gly-OMe	SP			Seph-adex LH-20	54 77 52										36
Z-Val-Tyr-Ser-Gly-Pro-Gly-Ser-Gly Boc-hydrazide	2 + 6 A	DMF-THF	73						Me-Aq	215–217 (dec)	−23	DMF			83
Val-Tyr-Ser-Gly-Pro-Gly-Ser-Gly Boc-hydrazide acetate											−13	Me			83

Chapter 10

Nonapeptides

When plasma globulin is treated with trypsin or a snake venom protease, the biologically active nonapeptide bradykinin (Chart 8) is released. This substance has served as a useful model compound for various peptide synthetic methods including the original Merrifield solid phase procedure. Two more such model syntheses have just appeared. The new N-isobornyloxycarbonyl (IBOC) protecting group mentioned in Chapter 2 was employed with ONP and DCCI-HBT (last step) coupling procedures.[161] The IBOC group was removed at each stage by means of trifluoroacetic acid (room temperature) over 30 min and the N^G-nitro protecting group of arginine was removed by hydrogenolysis to yield pure bradykinin with all its usual biological activity. Nine analogs of bradykinin have been prepared by the 4-picolyl ester method.[146]

In addition to the new truncated analogs of oxytocin and vasopressin (Chart 8) mentioned in Chapter 7 can be added the following nonapeptide analogs and physical studies. A synthesis of [1-(L-2-hydroxy-3-mercaptopropanoic acid)] oxytocin has been completed and the substance was found to possess three times the oxytocic activity of oxytocin on the isolated rat uterus.[654] On the other hand, [*cyclo*-1-β-Ala-6-Asp] oxytocin proved to be a weak antagonist of these neurohypophyseal hormones in some systems.[220] Also, bromoacetyloxytocin was found to be a specific irreversible inhibitor of neurohypophyseal hormone action, but upon labeling may prove useful for locating hormone-receptor molecules.[651] Walter has also continued to study the conformation of oxytocin by nuclear magnetic resonance measurements. By means of Fourier-transformed carbon-13 magnetic resonance (at 25.16 MHz) measurements he and his colleagues have found no major conformational differences for oxytocin in dimethylformamide or water.[125] A comparison of the oxytocin conformation with that of Lys vasopressin has been made by proton magnetic resonance assignments.[182,183,648]

The preceding new advances in nonapeptide chemistry and biology were all received after the following tabular survey of nonapeptides was completed. However, one contribution read prior to July 1972, which did not lend itself to the tabular treatment of this chapter, involved a structure–activity study of previously described oxytocin analogs.[48] With the exception of Cys and Thr for which analog substitution data were not available, it appeared that the remaining 18 common protein amino acids could be placed in the following order of expected activity: Asp, Glu, Gly, Pro, Asn, Gln, Orn, Lys, Arg, Ser, Ala, Val, Leu, Ile, Phe, Tyr, Trp, and His. In some cases it now appears possible to predict what activity a particular analog may have even where double and triple substitutions have been made in the basic oxytocin structure.

Chart 8

161

Arg-Pro-Pro-Gly-Phe-Ser-Pro-Phe-Arg

Bradykinin

729

Cys-Tyr-X-Gln-Asn-Cys-Pro-Y-Gly-NH_2
(S————————S, disulfide bridge between the two Cys residues)

Oxytocin	X = Ile, Y = Leu
Arginine vasopressin	X = Phe, Y = Arg
Lysine vasopressin	X = Phe, Y = Lys

Table 9 Nonapeptides

Compound	Method	Solvent	% Yield	Chromatography Col	Tl	Ion	Paper	CCD	Recrys. Sol.	M.P. °C	$[\alpha]_D$	Solvent Rotation	Elect.	Other	Ref.
Alanine															
Z-Ala-$(Ala)_7$-Ala-morpholine amide	3 + 6 MA		40						TFA-Eth	>275 (dec)	−122	Tfa			186
2-OMe-[2-OEt-(2-OEt)]-Ac-Ala-$(Ala)_7$-Ala-morpholine	3 + 6 MA	DMSO	32						Hot TFE-Alc	>275 (dec)	−110	Tfa			186
Z-Ala-hemi-Cyst-N^ϵ-Tos-Lys-Asn-Gly-Gln-Thr-Asn-hemi-Cyst	1 + 8 OSU	DMF	56						DMF-EtAc	195–198	−8	DMF			124
Boc-$(Ala-Pro-Gly)_3$-OMe				Si		DO-50W-× 12			Me-EtAc	218–219				UV	367
Ala-Tyr-Leu-Lys-Lys-Ala-Thr-Asn-Glu triacetate				Seph-adex G-15			×		Eth		−480	Aq		AA	689
Ala-OBzl-Tyr-Leu-N^ϵ-Tfa-Lys-N^ϵ-Tfa-Lys-Ala-Thr-Asn-Glu-Di-OBzl-Tfa										223–224	−6.0	DMF			689
Boc-Ala-OBzl-Tyr-Leu-N^ϵ-Tfa-Lys-N^ϵ-Tfa-Lys-Ala-Thr-Asn-Glu-Di-OBzl	1 + 8 OSU	DMF	85						DMF-Aq	240	−15	DMF		AA	689
Arginine															
N^G-Tos-N^α-Z-Arg-Gly-Phe-Phe-Tyr-Thr-Pro-N^ϵ-Tos-Lys-Thr-OMe	1 + 8 WRK	Acn-DMF	80						Me-Eth	156–161	−12	DMF		AA	297
Arg-Pro-Pro-Gly-Phe-Ser-Pro-Phe-Arg Tri-Ac													×	AA	540

(*continued*)

Table 9 Nonapeptides (continued)

Compound	Method	Solvent	% Yield	Chromatography				CCD	Recrys. Sol.	M.P. °C	$[\alpha]_D$	Solvent Rotation	Elect.	Other	Ref.
				Col	Tl	Ion	Paper								
N^G-Nitro-Z-Arg-Pro-Pro-Gly-Phe-OBzl-Ser-Pro-Phe-N^G-nitro-Arg-OPic	OPCP	DMF	85								−51				540
Boc-nitro-Arg-Pro-Pro-Gly-Phe-Ser-Pro-Phe-nitro-Arg-OBzl	1 + 8 DCCI	DMF	50							141–145	−47	DMF			279
CHla-Arg-Pro-Pro-Gly-Phe-Ser-Pro-Phe-Arg															151
N^G-Nitro-Boc-Arg-Pro-Pro-Gly-Phe-Ser-Pro-Phe-N^G-nitro-Arg-OMe	SP		46												581
Aspartic acid															
β-OBu^t-Z-Asp-Val-Ser-Thr-Ala-Gln-Ala-Ala-Gly Boc-hydrazide monohydrate	1 + 8 OSU	DMF-Aq	78						DMF-Aq	284–286 (dec)	−15	DMSO			39
β-OBu^t-Asp-Val-Ser-Thr-Ala-Gln-Ala-Ala-Gly Boc-hydrazide										279–280 (dec)	−15	DMSO		AA	39
β-OBu^t-Z-Asp-Val-Tyr-Ser-Gly-Pro-Gly-Ser-Gly Boc-hydrazide	1 + 8 OSU	DMF-THF	82						Me	190–191 (dec)	−23	DMF		AA	83
Cysteine															
S-Bzl-Z-Cys-Leu-Ile-Leu-Asn-S-Bzl-Cys-Pro-Leu-Gly amide	1 + 8 ONP	DMF								257–259	−67	Ac			242
S-Bzl-Cys-OBzl-Tyr-Leu-Leu-Asn-S-Bzl-Cys-Pro-Leu-Gly amide	SP									225–228	−40	DMF		AA	681
S-Bzl-Z-Cys-Tyr-Ile-N^4-methyl-Asn-Asn-S-Bzl-Cys-Pro-Leu-Gly amide										246–248	−58	Fm-Aq			224

S-Bzl-Z-Cys-Tyr-Ile-Gln-Asn-S-Bzl-Cys-Pro-Leu-Gly amide										245–248 (dec)	−51 −65	DMF Ac			62
S-Bzl-Z-Cys-OBzl-Tyr-Ile-Gln-Dbu-S-Bzl-Cys-Pro-Leu-Gly amide Ac					55	CM-Seph-adex (H^+)					−46	DMF			221
S-Bzl-Z-Cys-OBzl-Tyr-Ile-Gln-N^{γ}-Pht-Dbu-S-Bzl-Cys-Pro-Leu-Gly amide									EtAc	265–268 (dec)	−43	DMSO			221
S-Bzl-Boc-Cys-OBzl-Tyr-Ile-N^{ϵ}-Boc-Lys-Asp-S-Bzl-Cys-Pro-Ile-Gly amide	SP													AA	388
S-Bzl-Boc-Cys-OBzl-Tyr-Ile-Pro-Asp-S-Bzl-Cys-Pro-Glu-Gly amide	SP													AA	388
S-Bzl-Boc-Cys-OBzl-Tyr-Ile-Pro Asp-S-Bzl-Cys-Pro-Ile-Gly amide														AA	388
S-Bzl-Boc-Cys-OBzl-Tyr-Ile-Pro-Asp-S-Bzl-Cys-Pro-Leu-Gly amide	SP													AA	388
S-Bzl-Cys-Tyr(Etoc)-Phe-Gln-Asn-S-Bzl-Cys-Pro-N^G-nitro-Arg-Gly hydrazide	SP														685
S-Bzl-N-Tos-Cys-Tyr-Phe-Gln-Asn-S-Bzl-Cys-Pro-N^G-Tos-Arg-Gly	SP			Seph-adex LH-20					Ac-Alc	230–231	−22	DMF			411
S-Bzl-N-Tos-Cys-Tyr-Phe-Gln-Asn-S-Bzl-Cys-Pro-N^G-Tos-Arg-Gly resin	SP	EtAc													411
S-Bzl-Z-Cys-O-Bzl-Tyr-Phe-Gln-Asn-S-Bzl-Cys-Pro-N^{ϵ}-Tos-Lys-Gly-NHMe										216–219	−36	Ac			179

(*continued*)

Table 9 Nonapeptides (continued)

Compound	Method	Solvent	% Yield	Chromatography				CCD	Recrys. Sol.	M.P. °C	$[\alpha]_D$	Solvent Rotation	Elect.	Other	Ref.
				Col	Tl	Ion	Paper								
S-Bzl-Z-Cys-O-Bzl-Tyr-Phe-Gln-Asn-S-Bzl-Cys-Pro-N^{ϵ}-Tos-Lys-Gly-NMe_2	SP								Ac-Alc	204–208	−40	Ac			179
S-Bzl-Z-Cys-Tyr-Phe-Gln-Asn-S-Bzl-Cys-Pro-homo-N^{ϵ}-Tos-Lys-Gly amide					×					218–221	−43	DMF			61
S-Bzl-N-Tos-Cys-Tyr-Phe-Gln-Asn-S-Bzl-Cys-Pro-N^{δ}-Pht-D-Orn-Gly amide										225–227					730
Glutamic acid															
Pyro-Glu-Trp-Ser-Tyr-Gly-Leu-Arg-Pro-Gly amide															634
Glutamine															
Pyr-Gln-Asp-Tyr(-O-sulfate)-Thr-Gly-Trp-Met-Asp-Phe amide															44
DL-Nona-Gln										220					320
Leucine															
Boc-Leu-Gln-Arg(HBr)-Leu-Leu-Gln-Gly-Leu-Val amide	1 + 8 OSU	DMF												AA	710
Leu-Gln-Arg(HBr)-Leu-Leu-Gln-Gly-Leu-Val amide HBr															710
Z-Leu-His-γ-OBu^t-Gln-β-OBu^t-Asp-Gly-γ-OBu^t-Gln-Thr-Val-Gly Boc-hydrazide	1 + 8 OSU	DMF	83						DMF-Aq	209–211 (dec)	−33	Me		AA	39

Leu-His-γ-OBu^t-Gln-β-OBu^t-Asp-Gly-γ-OBu^t-Gln-Thr-Val-Gly Boc-hydrazide diacetate											−29	Bu-Ac-Aq (3:1:1)		AA	39
Lysine															
$(Lys)_9$															726
N^ε-CHla-Lys-Pro-Pro-Gly-Phe-Ser-Pro-Phe-Arg															151
N^ε-Z-Boc-Lys-Pro-Val-Gly-N^ε-Boc-Lys-N^ε-Boc-Lys-Arg-Arg-Pro HCl															550
Lys-Pro-Val-Gly-Lys-Lys-Arg-Arg-Pro													×	AA	358
N^ε-Tos-Lys-Pro-Val-Gly-N^ε-Tos-Lys-N^ε-Tos-Lys-N^G-Tos-Arg-N^G-Tos-Arg-Pro	SP				55 116 126			7						AA	358
N^ε-Z-Boc-Lys-Pro-Val-Gly-N^ε-Boc-Lys-N^ε-Boc-Lys-N^ε-Boc-Lys-N^ε-Boc-Lys-Pro-NH-$(CH_2)_{15}CH_3$															77
N^ε-Boc-Lys-Pro-Val-Gly-N^ε-Boc-Lys-N^ε-Boc-Lys-N^ε-Boc-Lys-N^ε-Boc-Lys-Pro-NH-$(CH_2)_{15}CH_3$															77
N^ε-Boc-Z-Lys-Pro-Val-Gly-N^ε-Boc-Lys-Lys-Nle-Nle-Pro amide										164–170					78
N^ε-Boc-Lys-Pro-Val-Gly-N^ε-Boc-Lys-N^ε-Boc-Lys-Nle-Nle-Pro amide					23					165–169 (dec)	−71	Me			78
N^ε-Boc-Z-Lys-Pro-Val-Gly-N^ε-Boc-Lys-N^ε-Boc-Lys-Nva-Nva-Pro amide					27					188–191	−75	Me			72
N^ε-Boc-Lys-Pro-Val-Gly-N^ε-Boc-Lys-N^ε-Boc-Lys-Nva-Nva-Pro amide										171–175 (dec)	−72	Me			72

(*continued*)

Table 9 Nonapeptides (continued)

Compound	Method	Solvent	% Yield	Chromatography Col	Tl	Ion	Paper	CCD	Recrys. Sol.	M.P. °C	$[\alpha]_D$	Solvent Rotation	Elect.	Other	Ref.
Phenylalanine															
N-Tfa-Phe-Leu-Ile-Ile-Leu-Val-Pro-Pro-Phe-OBu^t															671
Z-Phe-Val-Asn-Gln-N^{im}-Bzl-His-Leu-S-Bzl-Cys-Gly-Ser hydrazide									DMSO-Aq	241–243	−37	Ac		AA	300
Z-Phe-Val-Asn-Gln-N^{im}-Bzl-His-Leu-S-Bzl-Cys-Gly-Ser-OMe									DMSO-Aq	232–234	−25	DMSO			300
Proline															
Z-Pro-His-N$^\epsilon$-For-Lys-Tyr-Asp-Asp-Tyr-γ-OBu^t-Gln-Gly Boc-hydrazide	1 + 8 OSU	DMF								207–208 (dec)	−38	DMF			39
Pro-His-N$^\epsilon$-For-Lys-Tyr-Asp-Asp-Tyr-γ-OBu^t-Gln-Gly Boc-hydrazide diacetate			94								−33	Ac-Aq		AA	39
Boc-Pro-(Pro)$_7$-Pro	3 + 6 ONP		85							280	−264				530
Pro-(Pro)$_7$-Pro					104					260	−421				530
Serine															
Z-Ser-Ala-Glu-Ala-Phe-Pro-Leu-Glu-Phe-OBu^t															575
Ser-Ala-Glu-Ala-Phe-Pro-Leu-Glu-Phe-OBu^t															575

Tyrosine															
N,O-Di-Z-Tyr-γ-OBu^t-Glu-Trp-Pro-Ile-Leu-Ser-Ser-Gly Boc-hydrazide	1 + 8 OTCP	DMF-THF	96		×				EtAc-Eth		−50	Me			83
Tyr-γ-OBu^t-Glu-Trp-Pro-Ile-Leu-Ser-Ser-Gly Boc-hydrazide acetate									Alc-Eth		−66	Me		AA	83
Valine															
Z-Val-[Asp(β-OBzl)]$_4$-N^ε-Z-Lys-Ile-Val-Gly-OBzl			92	Sephadex LH-20					DMF-EtAc	241–243	+28	HMPA			727
Boc-Val-[Asp(β-OBu^t)]$_4$-N^ε-Boc-Lys-Ile-Val-Gly-OBu^t				Sephadex LH-20						219–221 (dec)	−27	DMF			727
Val-[Asp]$_4$-Lys-Ile-Val-Gly		Ac-Me-Aq				×				292–295 (dec)	−36	Aq	×		727
4-Guanidino-Val-Pro-Pro-Gly-Phe-Ser-Pro-Phe-Arg								8			−51	DMF	×	AA	279
Bradykinin															
Arg-Pro-Pro-Gly-Phe-Ser-Pro-Phe-Arg tri-HCl (bradykinin)											−94	Aq	×	AA	279
Bradykinin															387
[5,8-Bis(thienyl-Ala)]bradykinin	SP														133
[5-β-Cyclohexyl-Ala]bradykinin															81
[5,8-β-Cyclohexyl-Ala]bradykinin															81
[8-β-Cyclohexyl-Ala]bradykinin															81
[5-Thienyl-Ala]bradykinin	SP														133
Mesotocin															
[4-Lys]mesotocin													×	AA	388
[4-Pro]mesotocin													×	AA	388
Oxytocin															
Oxytocin	SP									192–200 (dec)	−37	DMF		AA	62 134

(*continued*)

Table 9 Nonapeptides (continued)

Compound	Method	Solvent	% Yield	Chromatography: Col	Tl	Ion	Paper	CCD	Recrys. Sol.	M.P. °C	$[\alpha]_D$	Solvent Rotation	Elect.	Other	Ref.
[1-β-Mercaptopropionic acid, 4-N[4]-methyl-Asn]oxytocin				Sephadex G-25										AA	224
[4-N[4]-Methyl-Asn]oxytocin					121 57 123 58		×						×	AA	224
[5-Dbu]oxytocin					57 55 62	Sephadex G-25					−34	Ac		AA	221
[9-Glycinamide-1-^{14}C]oxytocin	SP														653
(9-Gly-amide-1-^{14}C)-oxytocin															649
[2,4-Di-Leu]oxytocin				Sephadex G-25							−43	Ac		AA	242
[3,4-Di-Leu]oxytocin				Sephadex G-25							−23	Ac		AA	681
Lys[4]-oxytocin															305
[4-Pro]oxytocin													×	AA	388
N$^{\alpha}$-Ac-2-OMe-Tyr-oxytocin					125 55 70						−105	Ac		AA	288
[4-Pro-8-Gln]oxytocin														AA	388

Deamino-D-oxytocin											+104	Ac		AA	148
([Deamino-homolanthionine[1,6]]oxytocin)					64 76			6	Me-Eth		−78	Ac		AA	502
Deaminotocinamide															243
[1-Deaminopenicillamine-d_8]oxytocin				Sephadex G-25							−52	1 *N* Ac			639
Deamino-6-seleno-oxytocin															187
Deamino-dicarba-oxytocin															721
Deamino-diseleno-oxytocin															187
Deamino-oxytocin				Sephadex G-25											187
[1-Mercaptodimethyl-acetic acid]-oxytocin				Sephadex G-25							+18	DMF			635
[1-β-Mercapto-α,α-dimethylpropionic acid]oxytocin											−45	DMF			635
[1-β-Mercapto-β,β-diethylpropionic acid]oxytocin					×						−38	DMF			635
Tocinamide															243
Vasopressin															
[1,6-Aminosuberic acid]-8-Arg-vasopressin				Sephadex G-25		A-IR-45	×				−66	Ac		AA	219
[8-Diaminopropionyl]vasopressin											−40	1 *N* Ac		AA	730
[8-D-Diaminopropionyl]vasopressin											−9.4	1 *N* Ac		AA	730
[8-Arg]vasopressin															411
[1,6-Aminosuberic acid]-8-Lys-vasopressin				Sephadex G-25		CM-Sephadex G-25 (H^+)	×						×	AA	219

(*continued*)

Table 9 Nonapeptides (continued)

Compound	Method	Solvent	% Yield	Chromatography Col	Tl	Ion	Paper	CCD	Recrys. Sol.	M.P. °C	[α]$_D$	Solvent Rotation	Elect.	Other	Ref.
(Bis-S-methyl)Lys-vasopressin[S-methyl-Cys]															409
1-Deamino-8-homo-Lys-vasopressin					57	DO-1 Seph-adex	×				−55	Ac		UV AA	361
8-Homo-Lys-vasopressin							×				−23	1 *N* Ac	×	AA	61
Lys-vasopressinoic acid methyl-amide						A-IRC-50					−25	1 *N* Ac			179
Lys-vasopressinoic acid dimethyl-amide						A-IRC-50								AA	179
[8-D-Orn]vasopressin											−21	1 *N* Ac			730
(9-Gly-amide-1-^{14}C)-Arg-vasopressin															649
(9-Gly-amide-1-^{14}C)-Lys-vasopressin															649
Vasotocin															
[1,6-Aminosuberic acid]-8-Arg-vasotocin				Seph-adex G-25			×				−75	Ac		AA	219
[1,6-Aminosuberic acid]-8-Lys-vasotocin							×				−79	1 *N* Ac		AA	219
[1-Deamino]Lys-vasotocin											−66	6.5 *N* Ac		AA	513

Chapter 11

Decapeptides

From 240,000 porcine hypothalami Schally and colleagues[399,542] were able to isolate 11.4 mg of the decapeptide (Chart 9) shown to be the luteinizing hormone-releasing factor/follicle-stimulating hormone-releasing factor (LH-RH/FSH-RH). In humans, 1 μg of the decapeptide is sufficient to stimulate the pituitary to synthesize and release LH and FSH. The medical possibilities for this discovery are impressive in respect to overcoming some cases of female sterility or as a new approach to contraception by disrupting the menstrual cycle or by means of analogs of the hormone which might inhibit the pituitary from releasing LH and FSH. For these reasons, syntheses of the decapeptide hormone and related substances stimulated considerable preparative efforts and have accounted for the major recent thrust in decapeptide chemistry.

Syntheses of LH-RH/FSH-RH (or more simply, LRH) quickly followed the structural elucidation and a typical example contributed by Folkers and colleagues[579] has been delineated in Chart 9. Other total syntheses of LRH by the fragmentation approach or by solid phase techniques have been reported within a few months of one another.[172,398,519,576] LRH has been synthesized employing the SP approach with a benzhydrylamine polymer[419] and both [2-Gly]LRH and des-2-His-LRH have been prepared by the same general method.[420] A description of the polymer and general experimental approach has been summarized in the first part of Chart 13 in Chapter 14.

The 36 analogs of LRH which were described after the table of this chapter was prepared constituted a similarly impressive advance in knowledge of this hypothalamus hormone (directed at the anterior pituitary gland). Conclusions, to date, indicate that the complete amino acid sequence of LRH may be necessary for maximum biological activity[541] and that each amino acid may have an important role.[159] An exception to these conclusions may involve the terminal Gly-NH_2 but the total chain length may be necessary for binding to its active site in the pituitary.[160] The *in vitro* biosynthesis of LRH from ^{14}C-Glu using hypothalamic tissue has been recently demonstrated.[266] As suggested by the discussions of hypothalamus-releasing hormones in Chapter 4 (TRH) and here, these new peptide hormones (and related members yet to be discovered) deserve considerable present and future synthetic endeavors. No doubt this will continue to be an exciting field for some time to come.

Chart 9

579

LRH Synthesis

Pyro-Glu-His —[(1) Trp-OBzl, DMF, DCCI-HBT, 3 days, room temp. (2) Recrys. Alc-EtAc]→ Pyro-Glu-His-Trp-OBzl
72%
d. p. 235°-238°

—[5% Pd/C, H_2; Me-Ac, 5 hr, room temp.]→ Pyro-Glu-His-Trp
~100%
d. p. 188°-190°

Part A

OBzl-N-Boc-Tyr —[(1) Gly-OBzl, THF, CH_3CH_2OCOCl, Et_3N, 1 hr ~0° (2) 16 hr, room temp. (3) Recrys. Me-He]→ OBzl-N-Boc-Tyr-Gly-OBzl
93%
m. p. 101°-104°

—[(1) Tfa, 15 min, room temp. (2) MCA, OBzl-N-Boc-Ser as above (3) Recrys. EtAc-Pe]→ OBzl-N-Boc-Ser-OBzl-Tyr-Gly-OBzl
83%
m. p. 92°-93°

—[(1) Tfa, 10 min, ~0° (2) Recrys. Me-Eth]→ OBzl-Ser-OBzl-Tyr-Gly-OBzl Tfa salt
88%
d. p. 143°-145°

—[(1) DCCI-HBT, DMF, Et_3N, 2 hr, ~0° (2) Part A, 36 hr, room temp. (3) Recrys. Alc]→ Pyro-Glu-His-Trp-OBzl-Ser-OBzl-Tyr-Gly-OBzl
69%
d. p. 197°-200°

Chart 9 (contd.)

(1) 8 *N* NaOH, Alc, 20 min, room temp.; (2) Ac, Recrys. Alc → Pyro-Glu-His-Trp-OBzl-Ser-OBzl-Tyr-Gly

84%
d. p. 171°-175° <u>Part B</u>

Z-Leu-OSu + N^G-Di-Adoc-Arg —[(1) DMF-Et_3N, 16 hr, room temp.; (2) Recrys. Me-Aq]→ Z-Leu-N^G-Di-Adoc-Arg

87%
m. p. 93°-96°

(1) DCCI-HBT, DMF, 2 hr, room temp.; (2) Pro-Gly-NH_2, 2 days, room temp.; (3) Prep. TLC → Z-Leu-N^G-Di-Adoc-Arg-Pro-Gly-NH_2

45%
d. p. 125°-130°

(1) 5% Pd/C, H_2, Alc, 1 hr, room temp.; (2) Part B, DMF, DCCI-HBT, 16 hr, room temp.; (3) Chromat. SiO_2; (4) Recrys. Alc →

Pyro-Glu-His-Trp-OBzl-Ser-OBzl-Tyr-Gly-Leu-N^G-Di-Adoc-Arg-Pro-Gly-NH_2

43%
d. p. 165°-170°

(1) 5% Pd-$BaSO_4$, Me-Ac, 25 hr, room temp.; (2) TFa, ~0°, 15 min, room temp.; (3) AGI-X8 (OH^-); (4) CMC Column → Pyro-Glu-His-Trp-Ser-Tyr-Gly-Leu-Arg-Pro-Gly-NH_2

98 mg. (42%), $\alpha]_D$ −34° (Me), LRH

Table 10 Decapeptides

Compound	Method	Solvent	% Yield	Chromatography				CCD	Recrys. Sol.	M.P. °C	$[\alpha]_D$	Solvent Rotation	Elect.	Other	Ref.
				Col	Tl	Ion	Paper								
Alanine															
Z-Ala-hemi-Cyst-N$^{\epsilon}$-Tos-Lys-Asn-Gly-Gln-Thr-Asn-hemi-Cyst-OBzl-Tyr-ONBzl									Me-DMF-Aq	215–217 (dec)	−15	DMF			124
β-Alanine															
Z-β-Ala-Tyr-Ser-Met-γ-Bzl-Glu-His-Phe-Arg-Trp-Gly							×				−19	DMF			249
α-Aminobutyric acid															
Z-Abu-OBzl-Tyr-Ser-Met-γ-Bzl-Glu-His-Phe-Arg-Trp-Gly			82		×				Me-Alc		−30	DMF			249
Arginine															
Arg(HBr)-Leu-Gln-Arg-(HBr)Leu-Leu-Gln-Gly-Leu-Val amide HBr					54				Me-EtAc		−51	80% Ac			710
Arg(HBr)-Leu-Gln-Arg(HBr)-Leu-Leu-Gln-Gly-Leu-Val amide HBr					54				Me-EtAc		−50	80% Ac			710
N$^{\alpha}$,N$^{\delta}$,N$^{\omega}$-Tris-Z-Arg-Leu-Gln-Arg-(HBr)Leu-Leu-Gln-Gly-Leu-Val amide	3 + 7 DCCI-HBT	DMF	70		49						−35	80% Ac			710
N$^{\alpha}$,N$^{\delta}$,N$^{\omega}$-Tris-Z-Arg-Leu-Gln-Arg-(HBr)Leu-Leu-Gln-Gly-Leu-Val amide	1 + 9 OSU	DMF													710
Cysteine															
S-Bzl-N-Z-Cys-octa-Gly-S-Bzl-Cys OBzl	5 + 5 A		95							289–290 (dec)	−36	DMSO			214

α,γ-Diaminobutyric acid															
Boc-Dbu(Meo)-αThr-N^ε^-Z-Dbu-αThr-N^ε^-Z-Dbu-N^ε^-Z-Dbu-D-Leu-Ile-N^ε^-Z-Dbu-N^ε^-Z-Dbu										228–229					22
Boc-Dbu(Meo)-αThr-N^ε^-Z-Dbu-αThr-N^ε^-Z-Dbu-N^ε^-Z-Dbu-D-Leu-Ile-N^ε^-Z-Dbu-N^ε^-Z-Dbu-OMe									DMF-Aq	206–209					22
Glutamic acid															
OBu^t-Z-Glu-N^G-Tos-Arg-Gly-Phe-Phe-Tyr-Thr-Pro-N$^\epsilon$-Tos-Lys-Ala-OMe	1 + 9 ONP	DMF	70							206–209	−25	DMF			299
γ-OBu^t-Z-Glu-N^G-Tos-Arg-Gly-Phe-Phe-Tyr-Thr-Pro-N$^\epsilon$-Tos-Lys-Thr-OMe	1 + 9 ONP	DMF	74						Me	200	−23	DMF		AA	297
Pyro-Glu-Gly-Trp-Ser-Tyr-Gly-Leu-Arg-Pro-Gly amide															634
N-Z-pyro-Glu-N^{im}-Tos-His-Trp-OBzl-Ser-OBzl-Tyr-Gly-Leu-N^G-nitro-Arg-Pro-Gly amide	SP		33						Me					AA	579
Pyro-Glu-His-Trp-OBzl-Ser-OBzl-Tyr-Gly-Leu-N$^\delta$,N$^\omega$-Adoc$_2$-Arg-Pro-Gly amide	6 + 4 DCCI-HBT	DMF	43	Si	120 99 7				Alc	165–170 (dec)	−11	Ac			91
Pyro-Glu-His-Trp-OBzl-Ser-OBzl-Tyr-Gly-Leu-N^G-nitro-Arg-Pro-Gly amide	3 + 7 DCCI-HBT	DMF	45		120 99 7									AA	579
Pyro-Glu-His-Trp-Ser-Tyr-Gly-Leu-Arg-Pro-Gly amide					120 7 136						−34	Me		AA	91
Pyro-Glu-His-Trp-Ser-Tyr-Gly-Leu-Arg-Pro-Gly amide	SP														519

(continued)

Table 10 Decapeptides (continued)

Compound	Method	Solvent	% Yield	Chromatography: Col	Tl	Ion	Paper	CCD	Recrys. Sol.	M.P. °C	$[\alpha]_D$	Solvent Rotation	Elect.	Other	Ref.
Pyro-Glu-His-Trp-Ser-Tyr-Gly-Leu-Arg-Pro-Gly amide					120 99 7			×						AA	579
Pyro-Glu-His-Trp-Ser-Tyr-Gly-Leu-Arg-Pro-Gly amide					120 7			×						AA	579
Pyro-Glu-His-Trp-OBzl-Ser-OBzl-Tyr-Gly-Leu-N^{ϵ}-Z-Lys-Pro-Gly amide	6 + 4 DCCI-HBT	DMF	47		120 99 7						−28	Me			91
Pyro-Glu-His-Trp-Ser-Tyr-Gly-Leu-Lys-Pro-Gly amide					120 7 136						−28	Me			91
Z-Glu-Ser-Ala-Glu-Ala-Phe-Pro-Leu-Glu-Phe-OBu^t															575
Glu-Ser-Ala-Glu-Ala-Phe-Pro-Leu-Glu-Phe-OBu^t															575
Glutamine															
Z-Gln-His-Leu-Cys-Gly-OMe (dimer)									DMF-EtAc	215–216	+10	Ac			618
Glycine															
Z-$(Gly)_9$-S-Bzl-Cys-OBzl										313–315 (dec)					214
$(Gly)_9$-S-Bzl-Cys-OBzl										315–318 (dec)	−18	DMSO			214
$(Gly)_9$-S-Bzl-Cys-OBzl HBr										297–300	−17	DMSO			214

Isoleucine															
Boc-Ile-Ala-OBzl-Tyr-Leu-N$^{\epsilon}$-Tfa-Lys-N$^{\epsilon}$-Tfa-Lys-Ala-Thr-Asn-Glu-Di-OBzl tetrahydrate	1 + 9 OSU	DMF								232–234	−12	DMF		AA	689
Ile-Ala-OBzl-Tyr-Leu-N$^{\epsilon}$-Tfa-Lys-N$^{\epsilon}$-Tfa-Lys-Ala-Thr-Asn-Glu-Di-OBzl-Tfa hydrate										217–220	−5	DMF			689
Leucine															
Boc-Leu-γ-OBu^{t}-Glu-Asn-Tyr-Cys-Asn-OBu^{t} | Boc-Val-Cys-Gly-Glu(OBu^{t})					×										293
[Boc-Leu-Val-Cyst-Gly-Glu(OBu^{t})$_2$]											−12	DMF			293
[Trt-Leu-Val-Cyst-Gly-Glu(OBu^{t})$_2$]$_2$									Me-EtAc	224–225	−25	Chf			293
Lysine															
(Lys)$_{10}$															726
Methionine															
Boc-Met-Glu-N$^{\epsilon}$-Boc-Lys-Phe-γ-OBu^{t}-Glu-S-*p*-methoxy-Bzl-Cys-Thr-Leu-S-*p*-methoxy-Bzl-Cys-Gly hydrazide									Me	248 (dec)	−31	DMF			10
Boc-Met-Glu-N$^{\epsilon}$-Boc-Lys-Phe-γ-OBu^{t}-Glu-S-*p*-methoxy-Bzl-Cys-Thr-Leu-S-*p*-methoxy-Bzl-Cys-Gly-OMe	5 + 5 DCCI-NHS	DMF	64							244–246 (dec)	−31	N,N′-DMF			10
Phenylalanine															
Boc-(Phe)$_2$-(Pro)$_2$-(Phe)$_2$-Val-(Pro)$_2$-Gly				Seph-adex LH-20										AA	676
Boc-(Phe)$_2$-(Pro)$_2$-(Phe)$_2$-Val-(Pro)$_2$-Gly-OMe	1 + 9 OSU	THF	67												676
	6 + 4 MCA	THF	90												

(*continued*)

Table 10 Decapeptides (continued)

Compound	Method	Solvent	% Yield	Chromatography				CCD	Recrys. Sol.	M.P. °C	$[\alpha]_D$	Solvent Rotation	Elect.	Other	Ref.
				Col	Tl	Ion	Paper								
$(Phe)_2$-$(Pro)_2$-$(Phe)_2$-Val-$(Pro)_2$-Gly															676
Boc-Phe-$(Pro)_2$-$(Phe)_2$-Ala-$(Pro)_2$-Ala-Phe-OMe										136–139					676
Phe-$(Pro)_2$-$(Phe)_2$-Ala-$(Pro)_2$-Ala-Phe															676
Boc-Phe-$(Pro)_2$-$(Phe)_2$-Ala-$(Pro)_2$-Gly-Phe-OMe										133–136					676
Phe-$(Pro)_2$-$(Phe)_2$-Ala-$(Pro)_2$-Gly-Phe															676
Boc-Phe-$(Pro)_2$-$(Phe)_2$-Ala-$(Pro)_2$-Val-Phe-OMe															676
Phe-$(Pro)_2$-$(Phe)_2$-Ala-$(Pro)_2$-Val-Phe														AA	676
Phe-Pro-Pro-Phe-Phe-Abu-Pro-Pro-Ala-Phe	SP														674
Phe-Pro-Pro-Phe-Phe-Abu-Pro-Pro-Abu-Phe	SP														674
Boc-Phe-$(Pro)_2$-$(Phe)_2$-Gly-$(Pro)_2$-Ala-Phe-OMe										138–142					676
Phe-$(Pro)_2$-$(Phe)_2$-Gly-$(Pro)_2$-Ala-Phe															676
Boc-Phe-$(Pro)_2$-$(Phe)_2$-Gly-$(Pro)_2$-Gly-Phe-OMe										134					676

Phe-(Pro)$_2$-(Phe)$_2$-Gly-(Pro)$_2$-Gly-Phe										151–157					676
Boc-Phe-(Pro)$_2$-(Phe)$_2$-Ile-(Pro)$_2$-Ala-Phe-OMe				Sephadex LH-20										AA	676
Phe-(Pro)$_2$-(Phe)$_2$-Ile-(Pro)$_2$-Ala-Phe															676
Boc-Phe-(Pro)$_2$-(Phe)$_2$-Leu-(Pro)$_2$-Ala-Phe-OMe										130–134				AA	676
Phe-(Pro)$_2$-(Phe)$_2$-Leu-(Pro)$_2$-Ala-Phe														AA	676
Boc-Phe-(Pro)$_2$-(Phe)$_3$-(Pro)$_2$-(Phe)$_2$-OMe													×		676
Phe-(Pro)$_2$-(Phe)$_3$-(Pro)$_2$-(Phe)$_2$															676
Boc-Phe-Pro-Pro-Phe-Phe-Val-Pro-Pro-D-Ala-Phe				Sephadex LH-20	69										677
Boc-Phe-Pro$_2$-Phe$_2$-Val-Pro$_2$-D-Ala-Phe-OMe				Sephadex LH-20	69									AA	677
Phe-(Pro)$_2$-(Phe)$_2$-Val-(Pro)$_2$-Ala-Phe	SP													AA	52
Phe-Pro-Pro-Phe-Phe-Val-Pro-Pro-Ala-Phe				Sephadex LH-20										AA	677
Phe-Pro-Pro-Phe-Phe-Val-Pro-Pro-Abu-Phe	SP														674
Boc-Phe-(Pro)$_2$-(Phe)$_2$-Val-(Pro)$_2$-(Phe)$_2$-OMe															676
Phe-(Pro)$_2$-(Phe)$_2$-Val-(Pro)$_2$-(Phe)$_2$														AA	676
For-D-Phe-Pro-Val-N$^\delta$-Z-Orn-Leu-D-Phe-Pro-Val-N$^\delta$-Z-Orn-Leu-NHCH$_2$CH$_2$OH	2 + 8 DCCI-NHS		86	DO-50 DO-1						132–135	−81			Mol. wt.	384

(*continued*)

Table 10 Decapeptides (continued)

Compound	Method	Solvent	% Yield	Chromatography Col	Tl	Ion	Paper	CCD	Recrys. Sol.	M.P. °C	$[\alpha]_D$	Solvent Rotation	Elect.	Other	Ref.
For-D-Phe-Pro-Val-Orn-Leu-D-Phe-Pro-Val-Orn-Leu-$NHCH_2$-CH_2OH 2 HCl			82						Me-Eth	204–205	−113			AA	384
Z-(OMe)-D-Phe-Pro-Val-δ-Z(NO_2)-Orn-Leu-D-Phe-Pro-Val-N^{δ}-Z(NO_2)-Orn-Leu hydrazide							×							AA	453
D-Phe-Pro-Val-N^{δ}-Z(NO_2)-Orn-Leu-D-Phe-Pro-Val-N^{δ}-Z(NO_2)-Orn-Leu hydrazide HCl			100							212–216					453
Proline															
Boc-Pro-$(Pro)_8$-Pro	4 + 6 ONP		71							280	−267				530
Pro-$(Pro)_8$-Pro					104					260	−454				530
Serine															
OBzl-Boc-Ser-N^{δ}-Z-Orn-OBzl-Ser-D-Phe-Pro-OBzl-Ser-N^{δ}-Z-Orn-OBzl-Ser-D-Phe-Pro									Me-Aq	113–115	−39	Ac			261
OBzl-Boc-Ser-N^{δ}-Z-Orn-OBzl-Ser-D-Phe-Pro-OBzl-Ser-N^{δ}-Z-Orn-OBzl-Ser-D-Phe-Pro-OMe				Si							−48	Ac			261
OBzl-Boc-Ser-OBzl-Tyr-OBzl-Ser-Met-Gln-N^{im}-Bzl-His-Phe-N^{G}-Tos-Arg-Trp-Gly hydrazide														AA	357

N^{α}-Boc-Ser-Tyr-Ser-Met-Gln-His-Phe-Arg-Trp-Gly hydrazide					79 126	A-IRC-50							×	AA	357
Ser-Tyr-Ser-Met-Gln-His-Phe-Arg-Trp-Gly hydrazide					79 126								×	AA	357
Threonine															
8-Deutero-actinocinyl-(α)-[O-Ac-Thr-D-Val-Pro-Sar-N-methyl-Val-OMe]-(β)-[cyclo-(Thr-D-Val-Pro-Sar-N-methyl-Val-O-Thr)]															344
Tryptophan															
Z(OMe)-Trp-D-Trp-Asn-Gln-Tyr-Val-N^{δ}-Z-Orn-Leu-D-Phe-Pro	DCCI-NHS	EtAc	68						DMF-Eth	208–209 (dec)	−24	DMF			342
Z(OMe)-Trp-D-Trp-Asn-Gln-Tyr-Val-N^{δ}-Z-Orn-Leu-D-Phe-Pro-ONp															342
Tyrosine															
OBzl-Boc-Tyr-Pro_2-Phe_2-Val-Pro_2-Ala-Phe														AA	678
OBzl-Boc-Tyr-Pro_2-Phe_2-Val-Pro_2-Ala-Phe-OMe															678
OBzl-Tyr-Pro_2-Phe_2-Val-Pro_2-Ala-Phe															678
N,O-Di-Z-Tyr-Tyr-γ-Bu^{t}-Glu-Trp-Pro-Ile-Leu-Ser-Ser-Gly Boc-hydrazide	1 + 9 OTCP	DMF-THF	95						Alc-Eth		−43	Me			83
Valine															
Z(OMe)-Val-N^{δ}-Z-Orn-Leu-D-Ala-Pro-Val-N^{δ}-Z-Orn-Leu-D-Ala-Pro	A	DMF	88						Me-Eth-Pe	185–187	−60	Me			349
Z(OMe)-Val-N^{δ}-Z-Orn-Leu-D-Phe-Pro-Val-N^{δ}-Z-Orn-Leu-D-Phe-Gly	A	DMF	69		124 25				Me-Pe	128–129	−21	DMF			401

(*continued*)

Table 10 Decapeptides (continued)

Compound	Method	Solvent	% Yield	Chromatography Col	Tl	Ion	Paper	CCD	Recrys. Sol.	M.P. °C	$[\alpha]_D$	Solvent Rotation	Elect.	Other	Ref.
Ac-Val-N^δ-Z-Orn-Leu-D-Phe-Pro-Val-N^δ-Z-Orn-Leu-D-Phe-Pro amide									EtAc-Pe	230–232					385
Ac-Val-N^δ-Z-Orn-Leu-D-Phe-Pro-Val-N^δ-Z-Orn-Leu-D-Phe-Pro-$NHCH_2CH_2OH$										230–233					385
Ac-Val-N^δ-Z-Orn-Leu-D-Phe-Pro-Val-N^δ-Z-Orn-Leu-D-Phe-Pro-$NHCH_2CH_2$-O_2CH_3									EtAc-Pe	179–181					385
Ac-Val-Orn-Leu-D-Phe-Pro-Val-Orn-Leu-D-Phe-Pro amide 2 HCl									Me-Eth	245–246					385
Ac-Val-Orn-Leu-D-Phe-Pro-Val-Orn-Leu-D-Phe-Pro-$NHCH_2CH_2O_2CH_2$ 2 HCl										220–222					385
Ac-Val-Orn-Leu-D-Phe-Pro-Val-Orn-Leu-D-Phe-Pro-$NHCH_2CH_2OH$ 2 HCl										226–227					385
Boc-Val-N^δ-Z-Orn-Leu-D-Phe-Pro-Val-N^δ-Z-Orn-Leu-D-Phe-Pro	5 + 5 DCCI-NHS	Di-DMF	95			DO-50			Me-Eth	173–175					385
Boc-Val-N^δ-Z-Orn-Leu-D-Phe-Pro-Val-N^δ-Z-Orn-Leu-D-Phe-Pro amide	5 + 5 DCCI-NHS	Di-DMF				DO-50 DO-1							×		385

Boc-Val-N$^{\delta}$-Z-Orn-Leu-D-Phe-Pro-Val-N$^{\delta}$-Z-Orn-Leu-D-Phe-Pro-NHCH$_2$CH$_2$OH	5 + 5 DCCI-NHS	Di-DMF	89							127–130					385
For-Val-N$^{\delta}$-Z-Orn-Leu-D-Phe-Pro-Val-N$^{\delta}$-Z-Orn-Leu-D-Phe-Pro amide									EtAc-Pe	215–216					385
For-Val-N$^{\delta}$-Z-Orn-Leu-D-Phe-Pro-Val-N$^{\delta}$-Z-Orn-Leu-D-Phe-Pro-NHCH$_2$CH$_2$OH						DO-50				148–150					385
For-Val-N$^{\delta}$-Z-Orn-Leu-D-Phe-Pro-Val-N$^{\delta}$-Z-Orn-Leu-D-Phe-Pro-NHCH$_2$CH$_2$O$_2$CH									EtAc-Pe	153–155					385
For-Val-Orn-Leu-D-Phe-Pro-Val-Orn-Leu-D-Phe-Pro-NHCH$_2$CH$_2$-O$_2$CH 2 HCl									Me-Eth	234–236					385
For-Val-Orn-Leu-D-Phe-Pro-Val-Orn-Leu-D-Phe-Pro-NHCH$_2$CH$_2$OH									Me-Eth	239–241					385
For-Val-Orn-Leu-D-Phe-Pro-Val-Orn-Leu-D-Phe-Pro-amide 2 HCl			94		×				Me-Eth	247–248					385
Val-N$^{\delta}$-Z-Orn-Leu-D-Phe-Pro-Val-N$^{\delta}$-Z-Orn-Leu-D-Phe-Pro amide						DO-1				199–200					385
Val-N$^{\delta}$-Z-Orn-Leu-D-Phe-Pro-Val-N$^{\delta}$-Z-Orn-Leu-D-Phe-Pro-OH HCl									Me-Pe	155–157	−132				385
Val-N$^{\delta}$-Z-Orn-Leu-D-Phe-Pro-Val-N$^{\delta}$-Z-Orn-Leu-D-Phe-Pro-NHCH$_2$CH$_2$OH						DO-1				114–116					385
Val-Orn-Leu-D-Phe-Pro-Val-Orn-Leu-D-Phe-Pro-OH 3 HCl										233–234					385
Z-Val-Phe-Gly-N^{G}-nitro-Arg-S-OBzl(OMe)-Cys-γ-Bzl-Glu-Leu-Ala-Ala-Ala Boc-hydrazide	3 + 7 DCCI-NHS	DMF	76						DMF-Alc	264–267	−27	DMF			196

(continued)

Table 10 Decapeptides (continued)

Compound	Method	Solvent	% Yield	Chromatography				CCD	Recrys. Sol.	M.P. °C	$[\alpha]_D$	Solvent Rotation	Elect.	Other	Ref.
				Col	Tl	Ion	Paper								
Z-Val-Phe-Gly-N^G-nitro-Arg-S-OBzl(OMe)-Cys-γ-OBzl-Glu-Leu-Ala-Ala-Ala hydrazide									DMF-Alc	256–260	−20	DMF			196

Chapter 12

Undecapeptides to Nonadecapeptides

The tantalizing objective of developing a very simple experimental procedure for rapidly obtaining pure peptides for biological studies in the approximate ten-amino acid unit range has encouraged the introduction of a reaction-processing technique designated the two-phase method.[547] The procedure is based on using methylene chloride as a water-immiscible solvent for OSU-coupling reactions. The side products from OSU esters are quite soluble in water. Therefore by extracting the reaction mixture with aqueous acetic acid followed by dilute potassium carbonate solution these materials and excess reactants are easily removed. The aqueous carbonate easily hydrolyzes OSU esters. The general technique has been used for preparing hexapeptides without characterizing intermediates as noted in Chart 10. A series of six dodecapeptides were next obtained by coupling appropriate hexapeptide fragments, and 102 tube countercurrent separation experiments with the dodecapeptides resulted in distribution curves which corresponded well with those calculated. A limitation of the two-phase approach is the present need for water-immiscible solvent systems for peptides of 8 to 9 or more units in length.

At this point, it should be mentioned that Sheehan and colleagues[565] earlier introduced the two-phase concept using the EDCI water-soluble carbodiimide coupling method with methylene chloride as solvent. However, the Schneider and Wirz[547] modification is a nice advance and, as these investigators point out, other coupling methods involving water-soluble reagents (and side products) such as the WRK procedure offer potentially useful alternatives. In summary, the two-phase method entails the use of two immiscible solvents in concert with appropriate protecting groups and coupling techniques such that no intermediate of a growing peptide chain requires isolation.

For applications such as those mentioned in the Schneider study[547] involving preparation of relatively small but carefully defined antigens, the two-phase method appears eminently suitable. These chemists have predicted that some small peptides may be only weakly immunogenic but important differences might be expected with this type of antigen and this should be evaluatable by the newer highly sensitive radioimmunological techniques. Toward this end the series of dodecapeptides (one is designated in Chart 10) were coupled with antigenic determinants and the products are undergoing immunological evaluation. One such attachment involved a penicilloyl group which was introduced following cleavage of the N^{ϵ}-Z-protecting groups with hydrogen bromide in glacial acetic acid.

A synthetic approach to pure peptides of 10 or more units is best achieved by condensing fragments of established structure and purity. The synthesis of ribonuclease T_1 unit 1–11 presented in the following chapter nicely exemplifies the practical approach to pure peptides of more than 10 units.[602] This new contribution by Hofmann and colleagues[602] adequately demonstrates a workable approach to synthetic peptides of the size covered in this chapter. Thus further introductory discussion will be limited to advances read too late to be included in the following table.

The kidney endopeptidase renin is known to cleave the decapeptide angiotensin I from a plasma globulin and thereby has a fundamental role in certain types of hypertension and salt balance. As a new approach to the assay of renin, Arg-Val-Tyr-Ile-His-Pro-Phe-His-Leu-Leu-Val-Tyr-Ser was synthesized and coupled through its N-terminal amino group to the carboxyl groups of N-acetyl-poly-Glu bearing a molecular weight of 60,000 to 80,000.[33] The Tyr unit was then labeled with ^{125}I and the action of renin on this product was shown to yield the radio-labeled tetrapeptide Leu-Val-[^{125}I]Tyr-Ser. By using this reaction sequence a method was developed to determine human renin content on an exceptionally small scale.

A biosynthetic approach to a linear gramicidin (pentadecapeptide) has been studied.[34] A valuable illustration of the fragment condensation approach to heptadecapeptide preparation appears in the synthesis of [15-Leu] human gastrin I by Wünsch and Deimer.[696,697] In this synthesis the Asp and Glu carboxyl groups were blocked with a Bu^t ester, Z-N protection was used for amino groups and OSu, OTCP, and MCA methods were employed for coupling reactions.

The 11-unit C-terminal section of a new vasoactive peptide isolated from porcine intestines has been synthesized (fragment condensation approach) by Bodanszky and colleagues.[60] The same group has also prepared the N-terminal 6-unit fragment of this intestinal peptide.[324]

Chart 10

547 Two-Phase Synthesis

N $^{\epsilon}$-Z-N $^{\alpha}$-Boc-Lys-OSu + Gly-OEt —[(1) MeCl-TEA ~0°; (2) 5 hr, room temp.; (3) Wash 3 × each with 1 *N* Ac, Aq. 0.5 *M* K_2CO_3 and Aq]→ N $^{\epsilon}$-Z-N $^{\alpha}$-Boc-Lys-Gly-OEt

11.9 g → 10.5 g (90%)

—[(1) TFA 40 min, room temp.; (2) Dowex-2 × 4 ($^{\ominus}$OAc) 50% Alc]→ N $^{\epsilon}$-Z-Lys-Gly-OEt Ac salt

9.4 g (98%)

—[(1) MeCl-DMF Boc-Leu-OSu, TEA 4 hr, room temp.; (2) Wash as above except 6 × with Aq to remove DMF; (3) Extract Aq with MeCl]→

Boc-Leu-N $^{\epsilon}$-Z-Lys-Gly-OEt —[Subsequent steps as before in 90+% yields]→ Boc-Leu-N$^{\epsilon}$-Z-Lys-Ala-Leu-N$^{\epsilon}$-Z-Lys-Gly-OEt

91%

—[(1) DMF-Di-Alc 0.5 *N* NaOH, Ppt. 90 min, 40°; (2) 1 *N* citric acid to; (3) Reppt. from DMF-Aq]→ Boc-Leu-N$^{\epsilon}$-Z-Lys-Ala-Leu-N$^{\epsilon}$-Z-Lys-Gly

92%
m. p. 177°-180°

—[(1) TFA, 30 min, room temp.; (2) Dowex-2 ($^{\ominus}$OH) Alc-DMF (5:1)]→ Leu-N$^{\epsilon}$-Z-Lys-Ala-Leu-N$^{\epsilon}$-Z-Lys-Gly-OEt

—[(1) DCCI-NHS, DMF 1 day, −15°; (2) 1 day, room temp.; (3) −15° for 2 hr to collect urea; (4) Ppt. DMF-Aq]→ Boc-Leu-N$^{\epsilon}$-Z-Lys-Ala-Leu-N$^{\epsilon}$-Z-Lys-Gly-Leu-N$^{\epsilon}$-Z-Lys-Ala-Leu-N$^{\epsilon}$-Z-Lys-Gly-OEt

50-70%
d. p. 282°-284°

Table 11 Undecapeptides

Compound	Method	Solvent	% Yield	Chromatography Col	Chromatography Tl	Chromatography Ion	Chromatography Paper	CCD	Recrys. Sol.	M.P. °C	$[\alpha]_D$	Solvent Rotation	Elect.	Other	Ref.
Asparagine															
N-Ac-Asn-Pro-N$^{\epsilon}$-Ac-Lys-N$^{\epsilon}$-Ac-Lys-Tyr-Ile-Pro-Gly-Thr-N$^{\epsilon}$-Ac-Lys-Met															693
Asn-Pro-Lys-Lys-Tyr-Ile-Pro-Gly-Thr-Lys-Met tetraacetate dihydrate						A-IRA-400 Sephadex					−25	Aq		AA	693
Boc-Asn-Pro-N$^{\epsilon}$-Tfa-Lys-N$^{\epsilon}$-Tfa-Lys-Tyr-Ile-Pro-Gly-Thr-N$^{\epsilon}$-Tfa-Lys-Met											−47	Me		AA	693
Aspartic acid															
Asp-Glu-Ser-Ala-Glu-Ala-Phe-Pro-Leu-Glu-Phe-OBu^t															575
Z-Asp-Glu-Ser-Ala-Glu-Ala-Phe-Pro-Leu-Glu-Phe-OBu^t															575
Cysteine															
S-Bzl-N-Z-Cys-(Gly)$_9$ S-Bzl-Cys-OBzl	5 + 6 SPh	AC	70							302–304 (dec)	−25	DMSO			214
	2 + 9 ONP	DMSO	51							302–304 (dec)	−25	DMSO			

Leucine															
Boc-Leu-Ile-Ala-OBzl-Tyr-Leu-N$^{\epsilon}$-Tfa-Lys-N$^{\epsilon}$-Tfa-Lys-Ala-Thr-Asn-Glu-Di-OBzl	1 + 10 OSU	DMF	85							245–246	−15	DMF			689
Leu-Ile-Ala-OBzl-Tyr-Leu-N$^{\epsilon}$-Tfa-Lys-N$^{\epsilon}$-Tfa-Lys-Ala-Thr-Asn-Glu Di-OBzl Tfa										237–238	−6.9	DMF			689
Leu-Ile-OBzl-Ser-OBzl-Tyr-Ile-N$^{\epsilon}$-Z-Lys-N^{G}-nitro-Arg-N$^{\epsilon}$-Z-Lys-N^{G}-nitro-Arg-Gln-Gln amide	SP														689
Lysine															
$(Lys)_{11}$															726
Lys-Val-Leu-Thr-Thr-Gly-Leu-Pro-Ala-Leu-Ile	SP					DO 1 × 2								AA	130
Proline															
Boc-Pro-$(Pro)_9$-Pro	5 + 6 ONP		50							300	−269	Chf			530
Pro-$(Pro)_9$-Pro					106					270	−464	Aq			530
Serine															
Z-Ser-Ser-β-Bu^{t}-Asp-Val-Ser-Thr-Ala-Gln-Ala-Ala-Gly Boc-hydrazide	2 + 9 A	DMF	64							270–277 (dec)	−16	DMSO			39
Ser-Ser-β-Bu^{t}-Asp-Val-Ser-Thr-Ala-Gln-Ala-Ala-Gly Boc-hydrazide acetate						×								AA	39
Z-Ser-Tyr-Pro-His-N$^{\epsilon}$-For-Lys-Tyr-Asp-Asp-Tyr-γ-Bu^{t}-Gln-Gly Boc-hydrazide	2 + 9 A	DMF	99						DMF-EtAc	210–211 (dec)	−26	DMF			39
Ser-Tyr-Pro-His-N$^{\epsilon}$-For-Lys-Tyr-Asp-Asp-Tyr-γ-Bu^{t}-Gln-Gly Boc-hydrazide diacetate											−39	Ac-Aq		AA	39

(*continued*)

Table 11 Undecapeptides (continued)

Compound	Method	Solvent	% Yield	Chromatography				CCD	Recrys. Sol.	M.P. °C	$[\alpha]_D$	Solvent Rotation	Elect.	Other	Ref.
				Col	Tl	Ion	Paper								
Tyrosine															
Z-Tyr-Gln-Leu-His-γ-Bu^t-Gln-β-Bu^t-Asp-Gly-γ-Bu^t-Gln-Thr-Val-Gly Boc-hydrazide	2 + 9 A	DMF	69							218–220 (dec)	−24	DMF			39

Table 12 Dodecapeptides

Compound	Method	Solvent	% Yield	Chromatography Col	Tl	Ion	Paper	CCD	Recrys. Sol.	M.P. °C	$[\alpha]_D$	Solvent Rotation	Elect.	Other	Ref.
Alanine															
12-β-(Pyrazolyl-3)-Ala-S-peptide$_{3-14}$											−47	Aq		AA	238
Boc-(Ala-Pro-Gly)$_4$-OMe									Me-Aq	235–236 (dec)				UV	367
Asparagine															
Z-Asn-Gln-His-Leu-Cys-Gly-OMe (dimer)									DMF-EtAc	223–224	−10	Ac			618
Asn-N$^\epsilon$-Boc-Lys-Phe-His-O-Bu^t-Thr-Phe-Pro-Gln-O-Bu^t-Thr-Ala-Ile-Gly									Me	218–221 (dec)	−26	DMF			574
Aspartic Acid															
Z-Asp-Ser-Tyr-Pro-His-N$^\epsilon$-For-Lys-Tyr-Asp-Asp-Tyr-γ-Bu^t-Gln-Gly Boc-hydrazide	1 + 11 OTCP	DMF	94							208–209 (dec)	−31	DMF			39
Asp-Ser-Tyr-Pro-His-N$^\epsilon$-For-Lys-Tyr-Asp-Asp-Tyr-γ-Bu^t-Gln-Gly Boc-hydrazide diacetate			92								−38	Ac-Aq		AA	39
Cysteine															
Boc-Val-γ-OBu^t-Glu-Ala-Leu-Tyr-Leu-Val-Cys-Gly-Glu(OBu^t)$_2$ | Boc-Cys-Asn-OBu^t					×										293

(continued)

Table 12 Dodecapeptides (continued)

Compound	Method	Solvent	% Yield	Chromatography				CCD	Recrys. Sol.	M.P. °C	$[\alpha]_D$	Solvent Rotation	Elect.	Other	Ref.
				Col	Tl	Ion	Paper								
Glutamic acid															
Z-Glu-Asp-Glu-Ser-Ala-Glu-Ala-Phe-Pro-Leu-Glu-Phe-OBu^t															575
Glu-Asp-Glu-Ser-Ala-Glu-Ala-Phe-Pro-Leu-Glu-Phe-OBu^t															575
Leucine															
Boc-Leu-Ala-Ala-Leu-N$^\epsilon$-Z-Lys-Gly-Leu-Ala-Ala-Leu-Ala-Gly-OEt	6 + 6 DCCI-NHS	DMF	50–70		57					290–293 (dec)	−11	DMF			547
Boc-Leu-Ala-Ala-Leu-N$^\epsilon$-Z-Lys-Gly-Leu-Ala-Ala-Leu-N$^\epsilon$-Z-Lys-Gly-OEt					61					281–284	−10	DMF			547
Boc-Leu-Ala-Ala-Leu-N$^\epsilon$-Z-Lys-Gly-Leu-N$^\epsilon$-Z-Lys-Ala-Leu-Ala-Gly-OEt					61					285–288	−14	DMF			547
Boc-Leu-Ala-Ala-Leu-N$^\epsilon$-Z-Lys-Gly-Leu-N$^\epsilon$-Z-Lys-Ala-Leu-N$^\epsilon$-Z-Lys-Gly-OEt					63					277–280 (dec)	−13	DMF			547
(Leu-Ala)$_6$	SP														473
Boc-Leu-N$^\epsilon$-Z-Lys-Ala-Leu-Ala-Gly-Leu-N$^\epsilon$-Z-Lys-Ala-Leu-Ala-Gly-OEt					63					274–276 (dec)	−11	DMF			547

Boc-Leu-N$^{\epsilon}$-Z-Lys-Ala-Leu-N$^{\epsilon}$-Z-Lys-Gly-Leu-N$^{\epsilon}$-Z-Lys-Ala-Leu-N$^{\epsilon}$-Z-Lys-Gly-OEt					63					282–284 (dec)	−16	DMF			547
Lysine															
N$^{\alpha}$,N$^{\epsilon}$-Di-Boc-Lys-γ-OBu^{t}-Glu-Thr-Ala-Ala-Ala-N$^{\epsilon}$-Boc-Lys-cyclopentyl-Gly-γ-Bu^{t}-Glu-N$^{\delta}$-Boc-Orn-Gln-His hydrazide									DMF	219–220 (dec)	−27	Ac			67
N$^{\alpha}$,N$^{\epsilon}$-Di-Boc-Lys-γ-OBu^{t}-Glu-Thr-Ala-Ala-Ala-N$^{\epsilon}$-Boc-Lys-cyclopentyl-Gly-γ-Bu^{t}-Glu-N$^{\delta}$-Boc-Orn-Gln-His-OMe															67
N$^{\alpha}$,N$^{\epsilon}$-Di-Boc-Lys-γ-Bu^{t}-Glu-Thr-Ala-Ala-Ala-N$^{\epsilon}$-Boc-Lys-Phe-γ-Bu^{t}-Glu-N$^{\delta}$-Boc-Orn-Glu-N$^{\delta}$-Boc-Orn-OMe	5 + 7 A	DMF	85						DMF-Eth	230–232	−14	DMF		AA	68
N$^{\alpha}$,N$^{\epsilon}$-Di-Boc-Lys-γ-Bu^{t}-Glu-Thr-Ala-Ala-Ala-N$^{\epsilon}$-Boc-Lys-Phe-γ-Bu^{t}-Glu-N$^{\delta}$-Boc-Orn-Gln-N$^{\delta}$-Boc-Orn hydrazide										220–222 (dec)	−25	Ac			68
(Lys)$_{12}$															726
Phenylalanine															
(Phe-Ala)$_{6}$	SP														472
Proline															
Boc-Pro-(Pro)$_{10}$-Pro	6 + 6 ONP		68							300	−272	Chf			530
Pro-(Pro)$_{10}$-Pro					106					280	−472	Aq			530
Serine															
Z-Ser-Ser-Ser-β-Bu^{t}-Asp-Val-Ser-Thr-Ala-Gln-Ala-Ala-Gly Boc-hydrazide	1 + 11 A	DMF	83								−13	DMSO		AA	39

(*continued*)

Table 12 Dodecapeptides (continued)

Compound	Method	Solvent	% Yield	Chromatography				CCD	Recrys. Sol.	M.P. °C	$[\alpha]_D$	Solvent Rotation	Elect.	Other	Ref.
				Col	Tl	Ion	Paper								
Z-Ser-Ser-Ser-β-OBu^t-Asp-Val-Ser-Thr-Ala-Gln-Ala-Ala-Gly Boc-hydrazide						×								AA	39
Z-Ser-Ser-Ser-γ-OBu^t-Asp-Val-Ser-Thr-Ala-Gln-Ala-Ala-Gly Tfa														AA	39

Table 13 Tridecapeptides

Compound	Method	Solvent	% Yield	Chromatography Col	Tl	Ion	Paper	CCD	Recrys. Sol.	M.P. °C	$[\alpha]_D$	Solvent Rotation	Elect.	Other	Ref.
Alanine															
12-β-(Pyrazolyl-3)-Ala-S-peptide$_{2-14}$											−48	Aq		AA	238
Ala-Glu-Asp-Glu-Ser-Ala-Glu-Ala-Phe-Pro-Leu-Glu-Phe-OBu^t															209
Z-Ala-Glu-Asp-Glu-Ser-Ala-Glu-Ala-Phe-Pro-Leu-Glu-Phe-OBu^t															209
Glutamic acid															
γ-OBu^t-Z-Glu-Thr-Ala-Ala-Ala-N$^\epsilon$-Boc-Lys-Phe-Glu-Arg-Gln-His-Nle-Asp														AA	238
Lysine															
N$^\epsilon$,N$^\alpha$-Boc-Lys-Glu-Thr-Ala-Ala-Ala-Lys-Phe-Glu-Arg-Gln-His-Leu-OMe	8 + 5 A	DMF		Sephadex LH-20						227 (dec)	−12	DMF			643
N$^\epsilon$,N$^\alpha$-Boc-Lys-Glu-Thr-Ala-Ala-Ala-Lys-Phe-Glu-Arg-Gln-His-Leu-OMe	7 + 6 A	DMF	47												643
Lys-Glu-Thr-Ala-Ala-Ala-Lys-Phe-Glu-Arg-Gln-His-Leu														AA	643
Lys-Glu-Thr-Ala-Ala-Ala-Lys-Phe-Glu-Arg-Gln-His-Leu				Sephadex G-15		A-IR-45	×				−72 ±2	Aq	×		643
Lys-Glu-Thr-Ala-Ala-Ala-Lys-Phe-Glu-homo-Arg-Gln-His-Leu			80	Ce		A-IR-45	×				−72		×	AA	642

(*continued*)

Table 13 Tridecapeptides (continued)

Compound	Method	Solvent	% Yield	Chromatography Col	Tl	Ion	Paper	CCD	Recrys. Sol.	M.P. °C	$[\alpha]_D$	Solvent Rotation	Elect.	Other	Ref.
Lys-Glu-Thr-Ala-Ala-Ala-Lys-Phe-Glu-Cit-Gln-N^{im}-Bzl-His-Leu	SP			Seph-adex 15			×						×	AA	641
Lys-Glu-Thr-Ala-Ala-Ala-Lys-Phe-Glu-Cit-Gln-His-Leu		Ac-Aq	65	Ce			×					Aq	×	AA	641
$(Lys)_{13}$															726
Serine															
Z-Ser-Asn-Ser-Tyr-Pro-His-For-Lys-Tyr-Asn-Asn-Tyr-γ-OBu^t-Glu-Gly Boc-hydrazide														AA	39
Z-Ser-Asp-Ser-Tyr-Pro-His-N^ϵ-For-Lys-Tyr-Asp-Asp-Tyr-γ-Bu^t-Gln-Gly Boc-hydrazide	1 + 12 A	DMF	79							201–202 (dec)	−26	DMF			39
Threonine															
Z-Thr-Ala-Ala-Ala-N^ϵ-Boc-Lys-Phe-Ala-Glu-Arg-Gln-His-Nle-Asp														AA	238

Table 14 Tetrapeptides

Compound	Method	Solvent	% Yield	Chromatography: Col	Chromatography: Tl	Chromatography: Ion	Chromatography: Paper	CCD	Recrys. Sol.	M.P. °C	$[\alpha]_D$	Solvent Rotation	Elect.	Other	Ref.
Arginine															
Arg(HBr)-Asp(β-OBu^t)-OBu^t-Ser-Ala-Arg(HBr)-Leu-Gln-Arg(HBr)-Leu-Leu-Gln-Gly-Leu-Val amide HBr					54				Me-EtAc		−32	80% Ac		AA	711
N$^\alpha$,N$^\delta$,N$^\omega$-Tris-Z-Arg-β-OBu^t-Asp-OBu^t-Ser-Ala-Arg(HBr)-Leu-Gln-Arg(HBr)-Leu-Leu-Gln-Gly-Leu-Val amide	4 + 10 DCCI-NHS	DMF	76		54		×				−31	80% Ac		AA	711
Cysteine															
S-Bzl-N-Z-Cys-(Gly)$_{12}$ S-Bzl-Cys-OBzl	7 + 7 ONP	DMSO	93							330–333 (dec)	−21 −18	DMSO FM			214
	4 + 10 SPh	DMSO	66							314–316 (dec)	−20	DMSO			
Glutamine															
[Z-Gln-Leu-γ-OBu^t-Glu-Asn-O-Bu^t-Tyr-Cys-Asn-OBzl]$_2$										245–250 (dec)	−80	DMF			141
[Gln-Leu-Glu-Asn-Tyr-Cys-Asn]$_2$				Sephadex G-25						235 (dec)	−40	Aq		UV AA	141
Glycine															
Z-Gly-Ala-Glu-Asp-Glu-Ser-Ala-Glu-Ala-Phe-Pro-Leu-Glu-Phe-OBu^t															575

(*continued*)

Table 14 Tetrapeptides (continued)

Compound	Method	Solvent	% Yield	Chromatography Col	Chromatography Tl	Chromatography Ion	Chromatography Paper	CCD	Recrys. Sol.	M.P. °C	$[\alpha]_D$	Solvent Rotation	Elect.	Other	Ref.
Gy-Ala-Glu-Asp-Glu-Ser-Ala-Glu-Ala-Phe-Pro-Leu-Glu-Phe-OBu^t															575
Lysine															
N^α,N^ϵ-Boc-Lys-γ-bu^t-Glu-Thr-Ala-Ala-Ala-N^ϵ-Boc-Lys-Phe-Glu-Arg-Gln-Ser-Met-Asp-D-sulfoxide														AA	238
N^α,N^ϵ-Boc-Lys-γ-OBu^t-Glu-Thr-Ala-Ala-Ala-N^ϵ-boc-Lys-Phe-Glu-Arg-Gln-His-Nle-Asp														AA	238
N^α,N^ϵ-boc-Lys-γ-bu^t-Glu-Thr-Ala-Ala-Ala-Nle-Phe-Glu-Arg-Gln-His-Met-Asp-D-sulfoxide														AA	238
N^α,N^ϵ-Boc-Lys-γ-bu^t-Glu-Thr-Ala-Ala-Ala-Nle-Phe-Glu-Arg-Gln-His-Nle-Asp														AA	238
$(Lys)_{14}$															726
N^ϵ-Z-Boc-Lys-Pro-Val-Gly-N^ϵ-Boc-Lys-N^ϵ-Boc-Lys-Arg-Arg-Pro-Val-DNS-Lys-Val-Tyr-Pro-OBu^t 2 HCl															550
N^ϵ-Boc-Lys-Pro-Val-Gly-N^ϵ-Boc-Lys-N^ϵ-Boc-Lys-Arg-Arg-Pro-Val-DNS-Lys-Val-Tyr-Pro-OBu^t 2 HCl															550

Lys-Pro-Val-Gly-Lys-Lys-Arg-Arg-Pro-Val-DNS-Lys-Val-Tyr-Pro-OH 5 Ac															550
Norleucine															
7-Nle-S-peptide$_{1-14}$											−36			AA	238
13-Nle-S-peptide$_{1-14}$														AA	238
7,13-Di-Nle-S-peptide$_{1-14}$														AA	238
Serine															
12-Ser-S-peptide$_{1-14}$														AA	238
Valine															
Val-Phe-Gly-Arg-S-acetamido-methyl-Cys-Glu-Leu-Ala-Ala-Ala-Met-Lys-Arg-His-OMe											−89	Aq	×	AA	196

Table 15 Pentadecapeptides

Compound	Method	Solvent	% Yield	Chromatography				CCD	Recrys. Sol.	M.P. °C	$[\alpha]_D$	Solvent Rotation	Elect.	Other	Ref.
				Col	Tl	Ion	Paper								
Alanine															
Boc-(Ala-Pro-Gly)$_5$-OMe						DO-50W			Tfe-Aq	>250 (dec)				UV	367
Asparagine															
Asn-Gly-Ala-Glu-Asp-Glu-Ser-Ala-Glu-Ala-Phe-Pro-Leu-Glu-Phe-OBu^t															575
Z-Asn-Gly-Ala-Glu-Asp-Glu-Ser-Ala-Glu-Ala-Phe-Pro-Leu-Glu-Phe-OBu^t															575
Leucine															
Z-Leu-Arg(HBr)-β-OBu^t-Asp-OBu^t-Ser-Ala-Arg(HBr)-Leu-Gln-Arg(HBr)-Leu-Leu-Gln-Gly-Leu Val amide	1 + 14 OSU	DMF-DMA	85								−33	80% Ac		AA	711
Leu-Arg(HBr)-β-OBu^t-Asp-OBu^t-Ser-Ala-Arg(HBr)-Leu-Gln-Arg(HBr)-Leu-Leu-Gln-Gly-Leu-Val amide HBr											−33	80% Ac		AA	711
Leu-Arg(HBr)-OBu^t-Asp-OBu^t-Ser-Ala-Arg(HBr)-Leu-Gln-Arg(HBr)-Leu-Leu-Gln-Gly-Leu-Val amide HBr															711

Lysine														
Lys-His(DNp)-Pro-Pro-His(DNp)-Leu-Ser-Phe-Met-Ala-Ile-Pro-Pro-Lys-Lys-4-HBr					126								AA	498
Lys-His-Pro-Pro-His-Leu-Ser-Phe-Met-Ala-Ile-Pro-Pro-Lys-Lys				Seph-adex G-25	126 53 131					−122	Fm		AA	498
Proline														
$(Pro-Pro-Gly)_5$	SP													319
Boc-Pro-$(Pro)_{13}$-Pro	5 + 10 ONP		58						300	−280	Chf			530
Pro-$(Pro)_{13}$-Pro					106				290	−495	Aq			530
Serine														
Ser-Asp-Asn-Asn-Gln-Gln-Gly-Lys-Ser-Ala-Gln-Gln-Gly-Gly-Tyr amide	SP													476

Table 16 Hexadecapeptides

Compound	Method	Solvent	% Yield	Chromatography: Col	Chromatography: Tl	Chromatography: Ion	Chromatography: Paper	CCD	Recrys. Sol.	M.P. °C	$[\alpha]_D$	Solvent Rotation	Elect.	Other	Ref.
Arginine															
Tri-Z-Arg-Leu-Arg(HBr)-OBu^t-Asp-OBu^t-Ser-Ala-Arg(HBr)-Leu-Gln-Arg(HBr)-Leu-Leu-Gln-Gly-Leu-Val amide					54		×				+32	80% Ac	×	AA	711
Tri-Z-Arg-Leu-Arg(HBr)-OBu^t-Asp-OBu^t-Ser-Ala-Arg(HBr)-Leu-Gln-Arg(HBr)-Leu-Leu-Gln-Gly-Leu-Val amide				Sephadex G-25 and G-15	130		×				−28	80% Ac		AA	711
Tri-Z-Arg-Leu-Arg(HBr)-β-OBu^t-Asp-OBu^t-Ser-Ala-Arg(HBr)-Leu-Gln-Arg(HBr)-Leu-Leu-Gln-Gly-Leu-Val amide	2 + 14 DCCI-HBT	DMF	87		130						−25	80% Ac		AA	711
Leucine															
Z-Leu-Tyr-Leu-Val-S-Bzl-Cys-Gly-γ-OBu^t-Glu-N$^\omega$-Tos-Arg-Gly-Phe-Phe-Tyr-Thr-Pro-N$^\epsilon$-Tos-Lys-Ala-OMe	6 + 10 A	DMF	73				×		DMF-Eth	257–259	−30	DMF		AA	299
Z-Leu-Tyr-Leu-Val-S-Bzl-Cys-Gly-γ-bu^t-Glu-N$^\omega$-Tos-Arg-Gly-Phe-Phe-Tyr-Thr-Pro-N$^\epsilon$-Tos-Lys-Thr-OMe	6 + 10 A	DMF	92						DMF-Eth	255–256	−31	DMF		AA	297

Phenylalanine															
(Z-D-Phe-Leu-Gly-Cys-OMe)$_4$									Alc-Eth	182–183	−33	DMF			188
Tyrosine															
[OBu^t-Z-Tyr-Gln-Leu-γ-OBu^t-Glu-Asn-OBu^t-Tyr-Cys-Asn-OBzl]$_2$										235 (dec)	−57	DMF			141
[Tyr-Gln-Leu-Glu-Asn-Tyr-Cys-Asn]$_2$										253 (dec)	−31	Aq		AA S-S de-term.	141

Table 17 Heptadecapeptides

Compound	Method	Solvent	% Yield	Chromatography Col	Tl	Ion	Paper	CCD	Recrys. Sol.	M.P. °C	$[\alpha]_D$	Solvent Rotation	Elect.	Other	Ref.
Alanine															
Z-Ala-Leu-Tyr-Leu-Val-S-Bzl-Cys-Gly-γ-Bu^t-Glu-N$^\omega$-Tos-Arg-Gly-Phe-Phe-Tyr-Thr-Pro-N$^\epsilon$-Tos-Lys-Ala-OMe	7 + 10 CDI	DMF	54				×		DMF-Aq	256–259	−34	DMF		AA	299
β-Ala-Tyr-Ser-Met-Glu-His-Phe-Arg-Trp-Gly-Lys-Pro-Val-Gly-Lys-Lys-Lys-NH-$(CH_2)_4$-NH_2	10 + 7 DCCI-HBT	DMF									−69	1% Ac		AA	171
Cysteine															
S-Bzl-N-Z-Cys-$(Gly)_{15}$-S-Bzl-Cys-OBzl	7 + 10 ONP	DMSO	83							367 (dec)	−21	Fm			214
Glycine															
Gly-Ile-Gly-Ala-Val-Leu-Lys-Val-Leu-Thr-Thr-Gly-Leu-Pro-Ala-Leu-Ile				Cell SE Sephadex G-15							−97	Ac		AA	130

Table 18 Octadecapeptides

Compound	Method	Solvent	% Yield	Chromatography Col	Tl	Ion	Paper	CCD	Recrys. Sol.	M.P. °C	$[\alpha]_D$	Solvent Rotation	Elect.	Other	Ref.
Alanine															
Boc-(Ala-Pro-Gly)$_6$-OMe					78	DO-50W								UV	367
β-Alanine															
β-Ala-Tyr-Ser-Met-Glu-His-Phe-Arg-Trp-Gly-Lys-Pro-Val-Gly-Lys-Lys-Arg-Arg amide						CM-Ce					−58	0.1 *N* Ac	×	AA	249
α-Aminoisobutyric acid															
Abu-Tyr-Ser-Met-Glu-His-Phe-Arg-Trp-Gly-Lys-Pro-Val-Gly-Lys-Lys-Arg-Arg amide						A-CG-400 CM-Cell					−58	0.1 *N* Ac	×	AA	249
Arginine															
Z-[N^G-nitro-Arg-Pro-Pro-Gly-Phe-OBzl-Ser-Pro-Phe-N^G-nitro-Arg]$_2$-OPic															81
Proline															
Pro-Hyp-Gly-(Pro-Pro-Gly)$_5$															319
Aoc-(Pro-Pro-Gy)$_6$	3 +15 ONP	Py-Aq	48										×		319
Serine															
Ser-Tyr-Ser-Met-Glu-His-Phe-Arg-Trp-Gly-Lys-Pro-Val-Gly-Orn-Orn-Orn-Orn amide															613

(*continued*)

Table 18 Octadecapeptides (continued)

Compound	Method	Solvent	% Yield	Chromatography				CCD	Recrys. Sol.	M.P. °C	$[\alpha]_D$	Solvent Rotation	Elect.	Other	Ref.
				Col	Tl	Ion	Paper								
D-Ser-Tyr-Ser-Met-Glu-His-Phe-Arg-Trp-Gly-Lys-Pro-Val-Gly-Orn-Orn-Orn-Orn amide															613
Ser-Tyr-Ser-Met-Glu-His-Phe-Arg-Trp-Gly-Orn-Pro-Val-Gly-Orn-Orn-Orn-Orn amide															613
D-Ser-Tyr-Ser-Met-Glu-His-Phe-Arg-Trp-Gly-Orn-Pro-Val-Gly-Orn-Orn-Orn-Orn amide															613
Threonine															
OBu^t-Z-Thr-OBu^t-Tyr-OBu^t-Thr-Gln-OBu^t-Asp-Phe-Asn-N^ϵ-Boc-Lys-Phe-His-OBu^t-Thr-Phe-Pro-Glu-OBu^t-Thr-Ala-Ile-Gly	A	DMF	74											AA	574

Table 19 Nonadecapeptides

Compound	Method	Solvent	% Yield	Chromatography Col	Tl	Ion	Paper	CCD	Recrys. Sol.	M.P. °C	$[\alpha]_D$	Solvent Rotation	Elect.	Other	Ref.
Glycine															
Z-Gly-Pro-Hyp-Gly-(Pro-Pro-Gly)$_5$	1 + 18 ONP	Py-Aq	68	Sephadex LH-20									×	AA	319
Serine															
Ser-Tyr-Ser-Met-Gln-His-Phe-Arg-Trp-Gly-Lys-Pro-Val-Gly-Lys-Lys-Arg-Arg-Pro					55 126	IRC-50 CM-Ce								AA	358
Boc-D-Ser-Tyr-Ser-Met-γ-OBu^t-Glu-His-Phe-Arg-Trp-Gly-N$^\epsilon$-Boc-Lys-Pro-Val-Gly-N$^\epsilon$-Boc-Lys-N$^\epsilon$-Boc-Lys-N$^\epsilon$-Boc-Lys-N$^\epsilon$-Boc-Lys-Pro-NH-$(CH_2)_{15}$-CH_3	10 + 9 DCCI-NHS	DMF	78												77
D-Ser-Tyr-Ser-Met-Glu-His-Phe-Arg-Trp-Gly-Lys-Pro-Val-Gly-Lys-Lys-Lys-Lys-Pro-NH$(CH_2)_{15}$-CH_3															77
Boc-D-Ser-Tyr-Ser-Met-OBu^t-Glu-His-Phe-Arg-Trp-Gly-Boc-Lys-Pro-Val-Gly-Boc-Lys-Boc-Lys-Nle-Nle-Pro amide								1							78

(*continued*)

Table 19 Nonadecapeptides (continued)

Compound	Method	Solvent	% Yield	Chromatography				CCD	Recrys. Sol.	M.P. °C	$[\alpha]_D$	Solvent Rotation	Elect.	Other	Ref.
				Col	Tl	Ion	Paper								
D-Ser-Tyr-Ser-Met-Glu-His-Phe-Arg-Trp-Gly-Lys-Pro-Val-Gly-Lys-Lys-Nle-Nle-Pro amide													×		78
Boc-D-Ser-Tyr-Ser-Met-OBu^t-Glu-His-Phe-Arg-Trp-Gly-Boc-Lys-Pro-Val-Gly-Boc-Lys-Boc-Lys-Nva-Nva-Pro amide								2							78
D-Ser-Tyr-Ser-Met-Glu-His-Phe-Arg-Trp-Gly-Lys-Pro-Val-Gly-Lys-Lys-Nva-Nva-Pro amide													×		78

Chapter 13

Peptides of 20 to 188 Units

Important achievements reviewed during the period covered by the table of this section include revised structures (for the human and porcine hormone, Asn for Asp at position 25 and Glu for Gln at position 30) for the 39-unit adrenocorticotropic hormone nicely deduced by Riniker and colleagues[515] and a masterful total synthesis of the human hypocalcemic hormone calcitonin M by the same Riniker, Rittel, and Sieber group[574] at CIBA-GEIGY, Basel. Synthesis of the 32-unit human calcitonin M provides an excellent illustration of the fragment condensation approach to complex peptides. In the final coupling step the 1–10- and 11–32-unit segments were condensed using the DCCI-NHS procedure and the OBu^t- and N-Boc-protected hormone was purified by careful countercurrent distribution to provide a 65% yield of pure peptide. Acid catalyzed cleavage of the protecting groups, even under optimum conditions, led to side reactions, namely, alkylation of Met and N → 0 acyl migration at the Ser or Thr units. But these minor side products were completely removed by a countercurrent distribution procedure using the system Bu-Ac-Aq (4:1:5) to afford the pure hormone free of diastereoisomers and exhibiting the biological activity of highly purified natural calcitonin M.

Syntheses of human insulin B-chain S-sulfonate,[298] the destripeptide B of human insulin, and the destripeptide B of bovine insulin[303] were described by Katsoyannis and collaborators. Two advances noted which were not suitable for incorporation in the table concerned the complete amino acid sequence of ribonuclease T_1 based on chymotryptic digests,[607] and the photochemical modification of Gly-containing proteins such as collagen lysozyme and ribonuclease by photoalkylation with 1-butene.[592]

After the tabular section was completed, Wünsch and co-workers described an impressive total synthesis of the 27-unit peptide hormone secretin employing the following protection and peptide bond-forming methods: Arg as the monohydrobromide, N^{α},N^{im}-Adoc for His, N-Z or Nps for Thr, and OBu^t for Asp, Glu, Ser, and Thr with DCCI-NHS and DCCI-HBT for coupling.[695,701,709] The synthetic secretin was purified by ion-exchange chromatography on SP-Sephadex C-25 and found to possess the full biological activity of the natural product. Protection of the arginine guanidino group by protonation represents a particularly interesting aspect of this total synthesis.

Recent progress in insulin chemistry has included a synthesis of [59-For-Lys]-bovine $proinsulin_{31-60}$ corresponding to the entire sequence of the connecting peptide portion,[724] syntheses of the destetrapeptide A of porcine and sheep insulin,[301] and a chemical conversion of porcine insulin to human insulin.[532] The latter interconversion offers an inexpensive route to human insulin and the possibility of treating diabetes mellitus patients with the actual human hormone thereby avoiding the immunologic intolerance which develops through current use of insulin from animal sources. As suggested by Ruttenberg[532] this immunologic response may play a role in the pathological changes that are seen with long-term insulin therapy. Here is but one facet of a vast area where the peptide chemist can greatly contribute to medicine.

Great challenges also await the bioorganic chemist in devising synthetic approaches to other small proteins. Eight peptide units corresponding to the amino acid sequence of yeast iso-1-cyatochrome c (108 amino acid units) have been synthesized for subsequent fragment condensation by the A method.[421] The fragment condensation approach is presently being applied to the synthesis of ribonuclease T_1 (Chart 11) by Hofmann and colleagues and the 1-46 sequence has recently been completed.[602] Salient features of the synthesis (Chart 12) comprise generation of hydrazides by TFA-catalyzed cleavage of Z-protected hydrazides (instead of exposing the peptide to hydrazine treatment) and protection of the Cys thiol with the ethylcarbamoyl group (Eac). The Eac-S group of Cys was found stable to trifluoroacetic acid and to hydrogen bromide in the same reagent. However, it can be easily cleaved by aqueous alkali or by a mercury(II) or silver(I) salt in methanol as shown at the end of Chart 12. Completion of this synthetic route to ribonuclease T_1 should unambiguously settle the total synthesis of this enzyme which has already, albeit in low yield, been obtained by SP and NCA procedures.[202]

Chart 11

1 2 3 4 5 6 7 8 9 10 11 12 13 14 15 16 17
Ala - Cys - Asp - Tyr - Thr - Cys - Gly - Ser - Asn - Cys - Tyr - Ser - Ser - Ser - Asp - Val - Ser -

18 19 20 21 22 23 24 25 26 27 28 29 30 31 32 33 34
Thr - Ala - Gln - Ala - Ala - Gly - Tyr - Gln - Leu - His - Glu - Asp - Gly - Glu - Thr - Val - Gly -

35 36 37 38 39 40 41 42 43 44 45 46 47
Ser - Asn - Ser - Tyr - Pro - His - Lys - Tyr - Asn - Asn - Tyr - Glu - Gly - Phe - Asp - Phe - Ser -

Val - Ser - Ser - Pro - Tyr - Tyr - Glu - Trp - Pro - Ile - Leu - Ser - Ser - Gly - Asp - Val - Tyr -

Ser - Gly - Pro - Gly - Ser - Gly - Ala - Asp - Arg - Val - Val - Phe - Asn - Glu - Asn - Asn - Gln -

Leu - Ala - Gly - Val - Ile - Thr - His - Thr - Gly - Ala - Ser - Gly - Asn - Asn - Phe - Val - Glu -

104
Cys - Thr

607 Ribonuclease T_1

Chart 12

602 Ribonuclease T_1 1-47

$CH_3CH_2NHC(=O)-S-CH_2CH(NH_2)CO_2H$ —— (1) Boc-N_3, TEA, DMF-Di-Aq, 1 day, room temp. (2) Recrys. Acn ——→ S-Eac-N-Boc-Cys
62%
m.p. 139°-140°

(1) 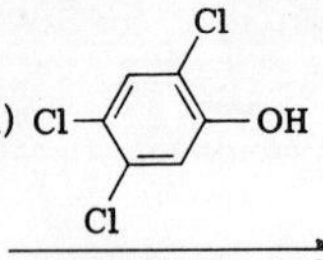
—— DCCI, THF, −10°, 4°, 40 hr (2) Recrys. EtAc ——→ S-Eac-N-Boc-Cyr-OTcp
69%
m.p. 152°-154°

—— THF, −10°, 44 hr, 4°; Tyr–NHNHZ Ac salt ——→
70% yield

N-Boc-Tyr —— (1) Z-$NHNH_2$, DCCI, −10°, 18 hr, room temp. (2) Ppt. EtAc-Pe ——→ N-Boc-Tyr-NHNH-Z
99%
m.p. 136°-139°

N-Boc-Tyr-NHNH-Z —— (1) 90% TFA (2) AGl-x2 ($^{\ominus}$OAc) ——→ Tyr–NHNHZ Ac salt

S-Eac-N-Boc-Cys-Tyr-NHNH-Z (residues 10, 11)
60%
m.p. 142°-144°
(ppt. from EtAc-Pe)

—— (1) 90% TFA → 80% yield TFA salt (2) N-Boc-Asp-OTcp, THF, TEA, −10°, 30 min, 1 day, room temp. ——→

N-Boc-Asp-S-Eac-Cys-Tyr-NHNH-Z (residues 9, 10, 11)
55%

—— (1) 90% TFA → 94% yield TFA salt (2) N-Boc-Ser-$NHNH_2$, + ONO, 6.9 *M* HCl-Di, −20°, 20 min (3) −40°, TEA-DMF to pH 7.5–8.0 (4) 3 days, 4°, ppt. Me-Pe ——→

N-Boc-Ser-Asp-S-Eac-Cys-Tyr-NHNH-Z
82% (crude)

—— (1) 90% TFA (2) 15 A ion-exchange resin, 0.1 to 2.0 *M* pyridinium acetate gradient elution ——→

Ser-Asp-S-Eac-Cys-Tyr-NHNH-Z Ac salt
59%

Part A

Chart 12 (cont.)

S-Eac-N-Boc-Cys-OTcp + Gly-NHNH-Z Tfa salt

→ (1) DMF, −10°
(2) TEA, ~ 0°, 3 hr
3 days, room temp.
(3) Recrys. EtAc

6 7
S-Eac-N-Boc-Cys-Gly-NHNH-Z
92%
m.p. 164°-166°

→ (1) 90% TFA
(2) N-Boc-Thr-OSu, DMF
−10°
(3) TEA, ~ 0°, 3 hr
40 hr, room temp.
(4) Ppt. EtAc-Pe

N-Boc-Thr-S-Eac-Cys-Gly-NHNH-Z
97%
m.p. 100°-105°

→ (1) 90% TFA
(2) DMF, −10°
N-Boc-Tyr-OTcp
(3) TEA, ~ 0°, 3 hr
3 days, room temp.
(4) Recrys. EtAc-Eth

4 5 6 7
N-Boc-Tyr-Thr-S-Eac-Cys-Gly-NHNH-Z
69%

→ (1) As above
β-OBu^t-N-Boc-Asp-OTcp
(2) Ppt. Me-Eth

3 4 5 6 7
β-OBu^t-N-Boc-Asp-Tyr-Thr-S-Eac-Cys-Gly-NHNH-Z

→ (1) 90% TFA, 40 min
(2) As above
pyridinium acetate
ion exchange
→ 83% yield
(3) DMF, ~ 0°
S-Eac-N-Boc-Cys-OTcp
18 hr, room temp.
1 day, 40°
(4) AG 1-X2
gradient elution
0.03 to 0.09 *M* Ac

S-Eac-N-Boc-Cys-Asp-Tyr-Thr-S-Eac-Cys-Gly-NHNH-Z
74%

→ (1) 90% TFA
(2) N-For-Ala-OSu
DMF, −10°
4 hr, ~ 0°, 40 hr,
room temp.
(3) AG 1-X2
as above

Chart 12 (cont.)

1 2 3 4 5 6 7
N-For-Ala-S-Eac-Cys-Asp-Tyr-Thr-S-Eac-Cys-Gly-NHNH-Z

66%

(1) HBr-TFA
35 min, ~0°, 35 min,
room temp.
(2) 6.9 *M* HCl-Di
DMF, ~10°
→
+ ONO, 84 hr, 4°
(3) Part A in DMF
TEA to pH 7.5-8.0
(4) AG 1-X2
as above

1 11
N-For-Ala-S-Eac-Cys-Asp-Tyr-Thr-S-Eac-Cys-Gly-Ser-Asp-S-Eac-Cys-Tyr-NHNH-Z

64%

(1) HBr-TFA
(2) + ONO
→ Ribonuclease T_1 1-23 (61%) → Repeat with fragment 24-27 Z-hydrazide → Ribonuclease T_1 1-47 (55%)
(3) Fragment 12-23
Z-hydrazide
(4) AG 1-X-2
All as above

S-Eac-N-Z-Cys

(1) $Hg(OAc)_2$, Me
(2) 30 min, room temp.
(3) Ac, H_2S, 4 hr
→
(4) As below

64%
m.p. 189°-190°

(1) 1 *N* NaOH, 2 hr,
room temp.
(2) 1 *N* HCl to pH 8
(3) Air, 18 hr, room temp.
→ N, N′-Z-Cystine DCHA salt
(4) HCl to pH 2
(5) DCHA-Eth
(6) Recrys. Me-Eth

78%
m.p. 189°-190°

S-Eac-N-Z-Cys-Gly-OEt

(1) $AgNO_3$, Me
(2) TEA, Δ
(3) DMF-HCl, 3 hr,
→ 62% m.p. 122°-123°
room temp.
(4) Recrys. EtAc-Pe

N-Z-Cys-Gly-OEt

(1) $Hg(OAc)_2$, 80% Ac
1 hr, room temp.
→
(2) H_2S, 4 hr, room temp.
(3) Recrys. EtAc-Pe

79%
m.p. 121°-122°

Table 20 20- to 188-Unit Peptides

Compound	Method	Solvent	% Yield	Chromatography Col	Tl	Ion	Paper	CCD	Recrys. Sol.	M.P. °C	$[\alpha]_D$	Solvent Rotation	Elect.	Other	Ref.
20-Unit Peptides															
Isoleucine															
Ile-Lys-Lys-Lys-Thr-Glu-Arg-Glu-Asp-Leu-Ile-Ala-Tyr-Leu-Lys-Lys-Ala-Thr-Asn-Glu				Sephadex G-25	81 128		×				−64	Aq		AA	688
N-Ac-Ile-Val-Glu-Gln-S-sulfo-Cys-S-sulfo-Cys-Ala-Gly-Val-S-sulfo-Cys-Ser-Leu-Tyr-Gln-Leu-Glu-Asn-Tyr-S-sulfo-Cys-Asn				Sephadex G-50										AA	302
Lysine															
Lys-Asp-Glu-Glu-Glu-Glu-Val-Glu-Ser-Phe-Ser-Gly-Pro-Asp-Ala-Pro-Leu-Pro-Ala-Gly															501
Lys-Glu-Thr-Ala-Ala-Ala-Lys-Cyclopentyl-Gly-Glu-Orn-Gln-His-Met-Asp-Ser-Ser-Thr-Ser-Ala-Ala				Sephadex G-25		A-CG-50					−85	Aq		AA	67
Lys-Glu-Thr-Ala-Ala-Ala-Lys-Phe-Glu-Orn-Gln-Orn-Met-Asp-Ser-Ser-Thr-Ser-Ala-Ala				Sephadex G-25							−78	Aq	×	AA	68
$1^{\epsilon},7^{\epsilon},10^{\delta}$-Triguanidino[cyclopentyl-$Gly^{8}Orn^{10}$]-S-peptide													×		67
A-Chain of human insulin															328

(*continued*)

Table 20 20- to 188-Unit Peptides (continued)

Compound	Method	Solvent	% Yield	Chromatography: Col	Tl	Ion	Paper	CCD	Recrys. Sol.	M.P. °C	$[\alpha]_D$	Solvent Rotation	Elect.	Other	Ref.
Arginine															
N^{G}-Nitro-Arg-A-chain S-sulfonate				Sephadex G-25											569
21-Unit Peptides															
Histidine															
N^{im}-Bzl-His-Leu-Val-Glu-Ala-Leu-Tyr-Leu-Val-S-Bzl-Cys-Gly-Glu-N^{ω}-Tos-Arg-Gly-Phe-Phe-Tyr-Thr-Pro-N^{ϵ}-Tos-Lys-Ala HBr									DMSO-EtAc	267–268	−23	DMF		AA	299
									DMSO-EtAc	265–268	−24	DMF		AA	299
N^{α}-Z-N^{im}-Bzl-His-Leu-Val-γ-Bu^{t}-Glu-Ala-Leu-Tyr-Leu-Val-S-Bzl-Cys-Gly-Glu-N^{ω}-Tos-Arg-Gly-Phe-Phe-Tyr-Thr-Pro-N^{ϵ}-Tos-Lys-Ala	5 + 16 A	DMF	87						DMSO-Aq	257–259	−25	DMSO		AA	299
N^{im}-Bzl-His-Leu-Val-Glu-Ala-Leu-Tyr-Leu-Val-S-Bzl-Cys-Gly-Glu-N^{ω}-Tos-Arg-Gly-Phe-Phe-Tyr-Thr-Pro-N^{ϵ}-Tos-Lys-Thr Tri-HBr	5 + 16 A	DMF	91				×			267–269	−23	DMSO		AA	297
Lysine															
Lys-N^{G}-nitro-Arg-A chain S-sulfonate	1 + 20 OSU	Aq-Di											×		569

22-Unit Peptides															
Threonine															
O-Bu^t-Thr-O-Bu^t-Tyr-O-Bu^t-Thr-Gln-O-Bu^t-Asp-Phe-Asn-N^ϵ-Boc-Lys-Phe-His-O-Bu^t-Thr-Phe-Pro-Gln-O-Bu^t-Thr-Ala-Ile-Gly-Val-Gly-Ala-Pro amide					×					220 (dec)	−35	Chf-MeAq			574
Thr-Tyr-Thr-Gln-Asp-Phe-Asn-Lys-Phe-His-Thr-Phe-Pro-Gln-Thr-Ala-Ile-Gly-Val-Gly-Ala-Pro amide														AA	574
24-Unit Peptides															
Serine															
Boc-Ser-Tyr-Ser-Met-γ-OBu^t-Glu-His-Phe-Arg-Trp-Gly-N^ϵ-Boc-Lys-Pro-Val-Gly-N^ϵ-Boc-Lys-N^ϵ-Boc-Lys-Arg-Arg-Pro-Val-Lys-Val-Tyr-Pro-OBu^t, 3 HCl															550
Ser-Tyr-Ser-Met-Glu-His-Phe-Arg-Trp-Gly-Lys-Pro-Val-Gly-Lys-Lys-Arg-Arg-Pro-Val-DNS-Lys-Val-Tyr-Pro-6 Ac															550
Tyrosine															
Z-Tyr-Gln-Leu-His-γ-OBu^t-Glu-β-OBu^t-Asp-Gly-γ-OBu^t-Glu-Thr-Val-Gly-Ser-Asn-Ser-Tyr-Pro-His-For-Lys-Tyr-Asn-Asn-Tyr-γ-OBu^t-Glu-Gly Boc-hydrazide						AG 1 × 2									39
25-Unit Peptides															
Proline															
Pro-Pro-Gln-N^ϵ-Z-Lys-N^G-nitro-Arg-A-chain S-sulfonate	4 + 21 A	DMF-THF	Sephadex G-25												569

(*continued*)

Table 20 20- to 188-Unit Peptides (continued)

Compound	Method	Solvent	% Yield	Chromatography				CCD	Recrys. Sol.	M.P. °C	$[\alpha]_D$	Solvent Rotation	Elect.	Other	Ref.
				Col	Tl	Ion	Paper								
26-Unit Peptides															
Glycine															
Gly-Ile-Gly-Ala-Val-Leu-Lys-Val-Leu-Thr-Thr-Gly-Leu-Pro-Ala-Leu-Ile-Ser-Tyr-Ile-Lys-Arg-Lys-Arg-Gln-Gln amide (19-tyrosine melittin)	SP														416
28-Unit Peptides															
E. coli acyl carrier protein	SP														207
30-Unit Peptides															
Phenylalanine															
Phe-Val-Asn-Gln-His-Leu-S-sulfo-Cys-Gly-Ser-His-Leu-Val-Glu-Ala-Leu-Tyr-Leu-Val-S-sulfo-Cys-Gly-Glu-Arg-Gly-Phe-Phe-Tyr-Thr-Pro-Lys-Ala (sheep insulin B-chain S-sulfonate)														AA	304
Phe-Val-Asn-Gln-His-Leu-S-sulfo-Cys-Gly-Ser-His-Leu-Val-Glu-Ala-Leu-Tyr-Leu-Val-S-sulfo-Cys-Gly-Glu-Arg-Gly-Phe-Phe-Tyr-Thr-Pro-Lys-Thr-(human insulin B-chain S-sulfonate)														AA	298
Sheep B-chain S-sulfonate															304

32-Unit Peptides															
Cysteine															
S———S Boc-Cys-Gly-Asn-Leu-OBu^t-Ser-OBu^t-Thr-Cys-Met-Leu-Gly-OBu^t-Thr-OBu^t-Tyr-OBu^t-Thr-Gln-OBu^t-Asp-Phe-Asn-N$^\epsilon$-Boc-Lys-Phe-His-OBu^t-Thr-Phe-Pro-Gln-OBu^t-Thr-Ala-Ile-Gly-Val-Gly-Ala-Pro amide								10		250 (dec)	−49	Chf-Me-Aq			574
S———S Cys-Gly-Asn-Leu-Ser-Thr-Cys-Met-Leu-Gly-Thr-Tyr-Thr-Gln-Asp-Phe-Asn-Lys-Phe-His-Thr-Phe-Pro-Gln-Thr-Ala-Ile-Gly-Val-Gly-Ala-Pro amide					×						−77	10% Ac	×	AA	574
33-Unit Peptides															
Human growth hormone ditriacontapeptide 122–153	SP														97
36-Unit Peptides															
Serine															
Z-Ser-Ser-Ser-Asp-Val-Ser-Thr-Ala-Gln-Ala-Ala-Gly-Tyr-Gln-Leu-His-γ-OBu^t-Glu-β-OBu^t-Asp-Gly-γ-OBu^t-Glu-Thr-Val-Gly-Ser-Asn-Ser-Tyr-Pro-His-For-Lys-Tyr-Asn-Asn-Tyr-γ-OBu^t-Glu-Gly Boc-hydrazide	13 + 24 A	DMF	48			AG-1 × 2								AA	39

(continued)

Table 20 20- to 188-Unit Peptides (continued)

Compound	Method	Solvent	% Yield	Chromatography Col	Tl	Ion	Paper	CCD	Recrys. Sol.	M.P. °C	$[\alpha]_D$	Solvent Rotation	Elect.	Other	Ref.
39-Unit Peptides															
Serine															
Ser-Tyr-Ser-Met-Glu-His-Phe-Arg-Trp-Gly-Lys-Pro-Val-Gly-Lys-Lys-Arg-Arg-Pro-Val-Lys-Val-Tyr-Pro-Asn-Gly-Ala-Glu-Asp-Glu-Ser-Ala-Glu-Ala-Phe-Pro-Leu-Glu-Phe (human ACTH)															575
42-Unit Peptides															
Monotetracontapeptide 81–121 of human growth hormone	SP														97
48-Unit Peptides															
Destripeptide B^{28-30} porcine (human) insulin															303
49-Unit Peptides															
Phage fd B protein															326
51-Unit Peptides															
Insulin															
Arg-insulin				Seph-adex G-75											661
Beef insulin															73
Des-Al-Gly-insulin															74

Al,B29-Bis-Boc-des-Phe^{BL}-insulin				Seph-adex G-50									×		338
Al,B29-Bis-Boc-[Boc-Phe(J)^{B1}]insulin															338
[Phe(J)^{B1}]insulin															338
Porcine proinsulin															436
Acetamidino-insulin															509
Bic(Boc) insulin															173
Des-Al-amino-insulin															74
Destripeptide B^{28-30} bovine insulin															303
58-Unit Peptides															
Bovine basic pancreatic trypsin inhibitor	SP														450
188-Unit Peptides															
Human pituitary growth hormone	SP														359

Chapter 14

Solid Phase Synthesis

A variety of equipment changes and experimental modifications for the general Merrifield solid phase (SP) procedure[394] have been suggested over the past two years. The original Merrifield synthesis of bradykinin has been modified by Anfinsen and co-workers to permit preparation of the fully active peptide in less than 5 hr.[103] The synthesis was carried out in a jacketed glass column with provision for mixing by bubbling nitrogen from the bottom and a vacuum source at the top. To remove reagents, the unit was equipped to easily switch to vacuum at the bottom and nitrogen entry at the top. Cleavage of Boc groups was performed in two 5-min stages with 50% TFA in methylene chloride, and coupling reactions were performed twice using the Boc-amino acid and DCCI each in tenfold excess with respect to the polymer amino component. A complete cycle of cleavage and coupling was effected in approximately 30 min. Peptide bond-forming reactions with DCCI for 2 or more hours was considered unnecessarily long as kinetic studies have shown that coupling is actually completed in 1 or 2 min for most amino acids.

Usual attachment of the first amino acid unit to the chloromethylated polystyrene-divinylbenzene resin has been improved, for example, with Boc-Asn, by condensing the amino acid (2.0 mmoles) in DMF with triethylamine (1.8 mmoles) during one day at room temperature.[389] Cleavage of the peptide end product from the polymer as the amide derivative has been accomplished using ammonia in DMF in a sealed reaction vessel at room temperature for 5 days.[476] A careful study of several peptide bond-forming reagents and polymer supports in respect to racemization has revealed the DCCI procedure to be far inferior to DCCI-NHS or EEDQ.[346] With the model tripeptide Gly-Ala-Leu (racemization assessment by means of the Izumiya procedure), the product was found to contain Gly-D-Ala-Leu in the amount of 24.6% with DCCI, 2.8% with DCCI-NHS, and 6.0% with EEDQ.

Both Ile and Val as Boc derivatives were found to react at a far slower rate than other Boc amino acids presumably due to steric hindrance.[506] A careful study of N-protecting groups for SP syntheses has demonstrated that the Bpoc group and its cleavage by dilute Tfa in methylene chloride should prove particularly useful in future solid phase syntheses.[505] A suggested improvement for SP syntheses with Trp-containing peptides depends on the use of hydrogen chloride in formic acid for cleavage of the Boc-protecting group. Utility of this method was believed due to formylation of the indole nitrogen thereby protecting the pyrrole ring from oxidation. Treatment with 0.1 *N* aqueous piperidine was used to remove the N-formyl group.[454,455] The N-trimethylsilyl method of protection has been suggested for SP synthesis to take advantage of the milder conditions for deprotection, namely, 1% Ac in Alc.[381] The use of EEDQ in Alc-MeCl has been recommended as a substitute for DCCI.[581,712] In addition to the new reactor mentioned by Anfinsen and co-workers,[103] another new reaction vessel has been described.[52]

In an attempt to synthesize the octapeptide [Me-Phe^8]-angiotensin II, the C-terminal N-methyl-Phe unit led to premature cleavage from the polymer.[317] Some of the more subtle but quite important aspects of the SP method have been uncovered by Brunfeldt and co-workers[80] who found by using ^{14}C-labeled acetic acid that the resin and particularly the Teflon parts of the reactor absorbed enough acetic acid to result in some N-acetyl peptides during the coupling step. Such failure sequences were assessed by mass spectrometry techniques. Error sequences have also been evaluated by means of ^{19}F NMR.[36] A method for the elimination of truncated peptides resulting from failure sequences has been proposed[685] and is based on the use of 3-sulfopropionic acid anhydride (Chart 13). The N-3-sulfopropionyl derivatives can then be removed by filtration through a weakly basic ion exchanger (acetate form). Since the latter step is frequently employed in the final isolation of peptides as acetate derivatives the new procedure is reasonably convenient. Evidence for doubling of peptide chains during SP synthesis due to interchain aminolysis has been compiled by Beyerman and collaborators.[50] For the final amino acid analysis and monitoring of peptide composition during the SP procedure, hydrolysis with propionic acid-hydrochloric acid at 130° has been recommended.[670] An advantage of this procedure is that the peptide need not be cleaved from the resin prior to the hydrolysis step and total preparation for the amino acid analysis requires only 2 hr.

The design and application of new polymer supports for peptide synthesis remains a rapidly developing field. The use of glass beads coated with the polystyrene chloromethylated resin in a packed bed has been studied in detail.[551] With a model tetrapeptide the procedure was developed to a point where it became ten times faster than the usual Merrifield method but the average yield of each amino acid attachment was about 98.8% and the tetrapeptide

Leu-Ala-Gly-Val was obtained in an average purity of 96%. The prospects for large scale synthesis seemed promising by this approach.[551]

A macroporous polystyrene bead has undergone initial evaluation and may have some specialized utility because of its large nonswelling internal cavities and compatibility with water and other polar solvents.[616] Application of a dinitrophenylene group for bidirectional extension of peptide chains from the polymer has been described and an advantage appears in mild cleavage of the peptide at the penultimate stage using simply 2-mercaptoethanol in DMF.[180] Another approach to mild peptide cleavage has been termed the "safety catch principle." In this method the relatively stable peptide–polymer bond is made labile by an appropriate chemical reaction.[313] An example has been summarized in Chart 13 where treatment with diazomethane affords the N-methyl secondary sulfonamide which is quite labile to base.[313] Limitations of the method so far observed include partial racemization of the C-terminal unit and transpeptidation reactions with Asp and Glu residues protected by esters.

Sodium borohydride (in diglyme) reduction of a benzoyl polystyrene has been used to obtain the corresponding benzhydryl polystyrene.[588] The chloro and bromo derivatives were prepared by passing the respective hydrogen halide through a suspension of the polymer in MeCl. The corresponding *p*-methoxy derivatives have also been synthesized. The chemistry here is quite reminiscent of that earlier developed by Geiger for protecting amino acid amides with a benzhydryl group.

A quite new and interesting approach to polymer supports has been introduced by Bayer[37] and Parr[472,473] which is based on binding to an inorganic matrix such as glass. A previous limitation of this approach involving the easy removal of peptide from the silicate polymer seems to have been overcome by constructing a short aliphatic chain between the 3-dimensional silicate network and the benzyl group as shown in Chart 13.[472] The fundamental simplicity of this technique using porous glass beads seems attractive but no doubt will require further investigation.

A reverse approach to solid phase synthesis that is attaching the coupling reagent to a polymer thereby leaving the peptide in solution has been explored with EEDQ attached to a polymer[75] (see Chart 13). In terms of racemization the polymeric reagent is slightly less effective than the usual technique with monomeric EEDQ. The next logical step in polymer support methodology would be to have all reactants and products in solution, and such soluble phase techniques have been explored along several avenues in the recent past. Presumably peptide synthesis in solution can be brought more easily to the brink of quantitative yields than can heterogeneous reactions typical of the solid phase method. Toward this end Bayer[38,434] has deployed a methylene chloride soluble polyethylene glycol for this purpose as shown in Chart 13. Yields of better than 99% have been obtained for each coupling step and no racemization was detected by a gas chromatographic method. Separation of excess reagents was achieved by membrane filtration. The method offers the possibility of continuing a fragment condensation synthesis without cleaving the C-terminal polymeric protecting group as this part of the molecule can provide a valuable solubilizing effect for larger peptides. Further elaboration here with soluble polymers of varying molecular weights should nicely augment other workable polymer support techniques. Hopefully, the better solid and soluble polymer methods will be developed to the point where we will have a select list of various procedures, each most suitable for certain specialized problems such as the various reactions of organic chemistry have evolved.

A series of practical applications of solid phase peptide synthesis were included in the review of LRH decapeptide syntheses in Chapter 11. Related work viewed while preparing the following table concerned SP syntheses of human growth hormone position 17–20 with Bpoc-protected His,[722] an angiotensin II analog,[316] the 19-Tyr analog of melittin the hemolytic and surface-active peptide component of honey bee venom,[416] and the 74-unit *E. coli* acyl carrier protein.[207] Papers received after the closing date for this table reported SP syntheses of desglycinamide Lys vasopressin which was found to be active *in vivo*,[657] bradykinin analogs with Phe replaced by β-cyclohexyl-Ala,[136] a series of human growth hormone segments,[95] oligo-Pro peptides of up to 20 units,[529] and a synthesis of human thyrocalcitonin using the benzhydrylamine polymer.[517]

The enticing concept of solid phase synthesis has already begun to influence other areas of organic chemistry and is already well represented in the nucleotide discipline.[486] A poly-*p*-styryldiphenylalkylidenephosphorane polymer for performing the Wittig olefin synthesis was utilized to obtain good carbonyl → olefin conversion[406] (Chart 13). Synthesis of oligosaccharides was investigated using a polymeric support bearing an allyl alcohol side chain coupled through an acetal bond to the first carbohydrate unit. The oligomer was removed from the resin by oxidative cleavage of the allyl olefin.[149,150] A polymer support procedure has also been applied to the otherwise cumbersome task of selectively protecting one alcohol group of a symmetrical diol (Chart 13).[356]

Chart 13

685

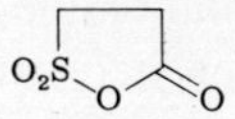

419

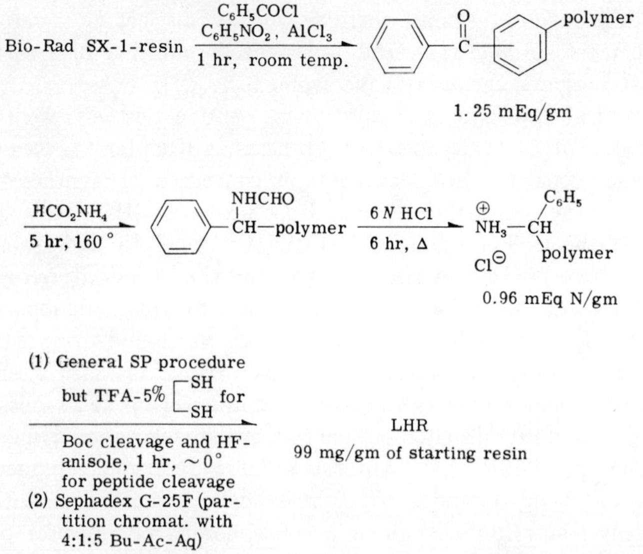

313

Polystyrene $\xrightarrow[\text{(2) } NH_4OH]{\text{(1) } HOSO_2Cl}$ Polymer-SO_2NH_2 $\xrightarrow[\text{2 days, room temp.}]{\text{Boc-Gly-OTcp, DMF, TEA}}$ Boc-Gly-$NHSO_2$-polymer

0.7 mmole Gly/gm

$\xrightarrow[\text{(2) Chf, Chf-5\% TEA, Chf-1\% Ac, Chf, An}]{\text{(1) TFA or HBr-Ac}}$ $\overset{\oplus}{N}H_3CH_2\overset{O}{\overset{\|}{C}}-\overset{\ominus}{N}-SO_2-$polymer $\xrightarrow[\text{repeated with 2,6-lutidine as base}]{\text{Sequence}}$

Chart 13 (contd.)

e.g., Boc-Thr-Ala-Ile-Gly-Val-Gly-$NHSO_2$-polymer $\xrightarrow[\text{(2) 0.5 } N \text{ NaOH or 0.5 } M \text{ } NH_3 \text{ in Di or } NH_2NH_2 \text{ in Me}]{\text{(1) } CH_2N_2 \text{ Eth-An}}$

[R—N(CH_3)SO_2-polymer] ⟶ Boc-Thr-Ala-Ile-Gly-Val-Gly

57%
(overall yield after recrys.)

472

Br—C_6H_4—Br $\xrightarrow[\text{Mg}]{CH_2\ CHCH_2Br}$ Br—C_6H_4—$CH_2CH{=}CH_2$ $\xrightarrow[\text{(2) HCHO}]{\text{(1) Mg}}$

$HOCH_2$—C_6H_4—$CH_2CH{=}CH_2$ $\xrightarrow[\text{6 hr, room temp.}]{HCl_g}$ $ClCH_2$—C_6H_4—$CH_2CH{=}CH_2$

$\xrightarrow[\text{0.1 } M \text{ } H_2PtCl_6 \text{, 3 hr, } \Delta]{HSiCl_3}$ $Cl_3SiCH_2CH_2CH_2$—C_6H_4—CH_2Cl (44%)

(1) Shake with porous glass beads, 1 day, room temp.
(2) 50% Alc to hydrolyze remaining Si—Cl bonds
(3) Polymerized by 1 day at 100° in vac.
(4) Wash Be, THF

glass—Si—O—Si($-$O$-$Si—)($-$O$-$Si—)—$CH_2CH_2CH_2$—C_6H_4—CH_2Cl $\xrightarrow[\text{TEA-Di}]{\text{Boc-Ala}}$

0.3 to 0.8 mmole Cl/gm

Boc-Ala-$NHCH_2$—C_6H_4—$(CH_2)_3$-glass $\xrightarrow[\text{HBr-TFA for cleavage } (\sim 100\%)]{\text{(1) General SP procedure and}}$ Phe-Ala$)_6$

0.24 mmole Ala/gm

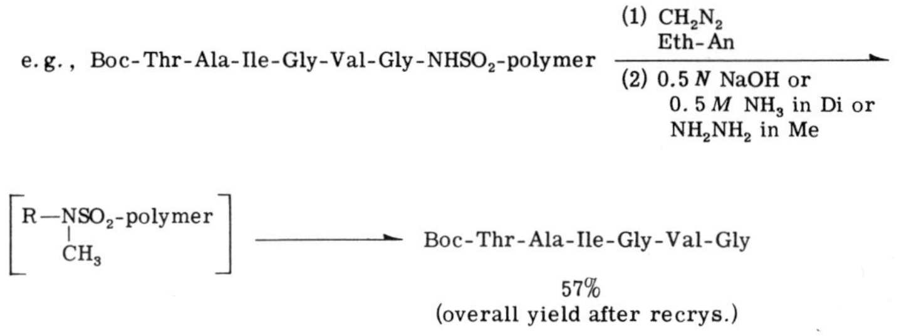

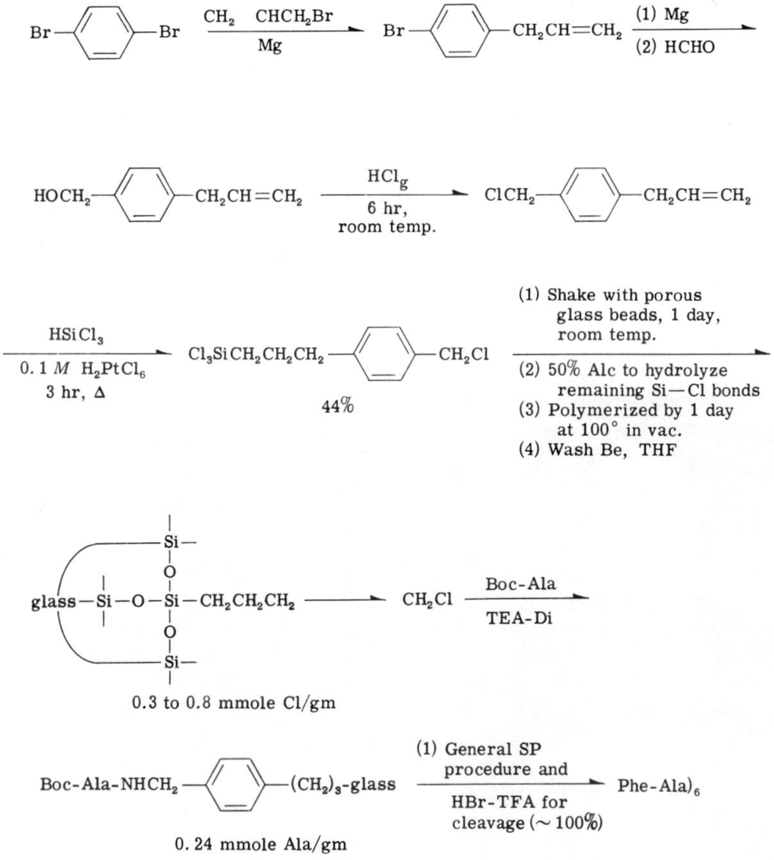

Chart 13 (cont.)

75

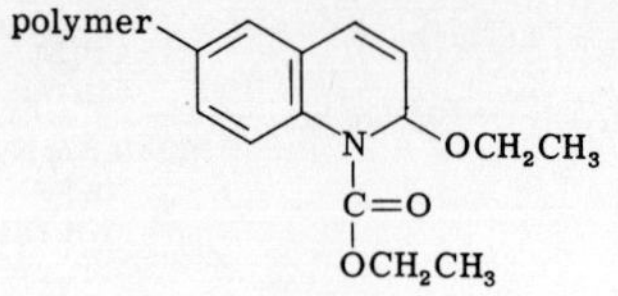

38, 434

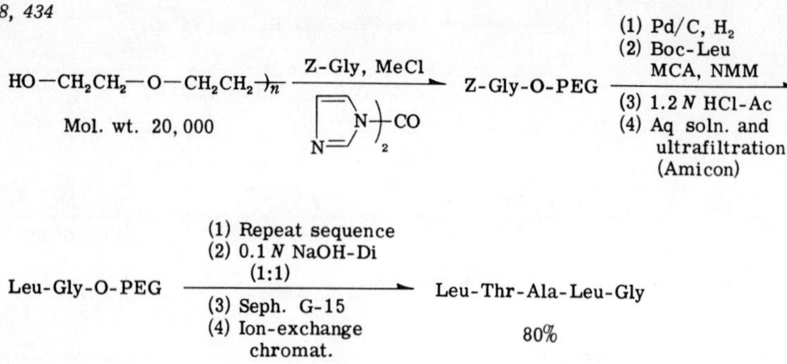

406

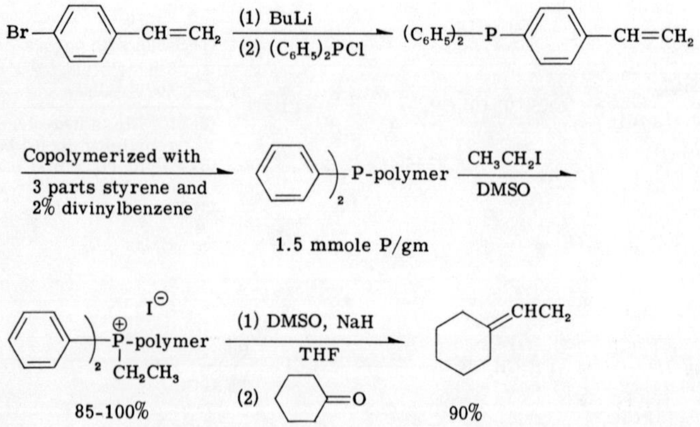

Chart 13 (cont.)

352

$$\underset{0.8 \text{ mmole Cl/gm}}{\text{Polymer-}CH_2COCl} \xrightarrow[\text{Py, 2 days, room temp. 1 day, } 105^\circ]{HOCH_2CH_2CH_2CH_2OH \text{ (large excess)}} \underset{0.2 \text{ mmole/gm}}{\text{polymer-}CH_2CO_2CH_2(CH_2)_2CH_2OH}$$

$$\xrightarrow[\text{(2) } NH_4OH\text{-Di}]{\text{(1) } (C_6H_5)_3CCl\text{, Py}} \underset{37\%}{HOCH_2(CH_2)_2CH_2{-}OC(C_6H_5)_3}$$

Table 21 Solid Phase

Compound	Method	Solvent	% Yield	Chromatography				CCD	Recrys. Sol.	M.P. °C	$[\alpha]_D$	Solvent Rotation	Elect.	Other	Ref.
				Col	Tl	Ion	Paper								
Boc-Ala-polymer															450
															620
Boc-Gln-polymer															416
Boc-Gly-polymer															383
Boc-N^{im}-Bzl-His-polymer															317
Z(OMe)-Leu-polymer															453
Boc-Leu-polymer															643
Boc-D-Leu-polymer															423
															732
N^{ϵ}-Z-Boc-Lys-polymer															498
Boc-N-methyl-Phe-polymer															317
Boc-Phe-polymer															17
															524
															664
Boc-Pro-polymer															518
N-Boc-Pro-polymer															49
N-Boc-OBzl-Tyr-polymer															474
Ala-Gly-polymer															370
Z(OMe)-Ala-Gly-polymer															712
γ-OBzl-Z(OMe)-Asp-Gly-polymer														IR	712
Boc-Cys-Gly-O-polymer | CH_2COOH															376

Boc-Gln-Gln-copolymer														416
Z-Gly-Phe hydrazide-polymer														677
Gly-Val-polymer														382
Z-Ile-Ala-polymer														505
Z(OMe)-Ile-Ala-polymer														505
Boc-Ile-Ala-polymer														505
Bpoc-Ile-Ala-polymer														505
Nps-Ile-Ala-polymer														505
Leu-Gly-polymer														382
Lys-His-copolymer														483
N^{α}-Boc-N^{ϵ}-Z-Lys-D-Leu-polymer														732
Boc-Val-Phe-O-polymer														203
N-Z-β-(1,3-Dioxy-4,4,5,5-tetramethyldihydroimidazol-2-yl)-Ala-Pro-Phe-polymer														664
N^{G}-Nitro-Boc-Arg-Gln-Gln-copolymer														416
PTC-Gly-Leu-Phe-polymer amide														382
N^{im}-Bzl-Boc-His-Pro-Phe-polymer												×		283
Boc-Cys-Gly-O-polymer | Boc-Cys-Gly-O-polymer														376
N^{ϵ}-Z-N^{α}-Boc-Lys-N^{G}-nitro-Arg-Gln-Gln-polymer														416
OBzl-Boc-Thr-Pro-N^{ϵ}-Z-Lys-Ala-O-polymer	SP													389
Ala-N^{im}-Boc-His-N^{G}-nitro-Arg-Leu-N^{im}-Boc-His-Gln-Leu-polymer														722
Boc-Ala-His-N^{G}-nitro-Arg-Leu-His-Gln-Leu-polymer														722
Boc-Ala-His-N^{G}-Tos-Arg-Leu-His-Gln-Leu-polymer														722

(*continued*)

Table 21 Solid Phase (continued)

Compound	Method	Solvent	% Yield	Chromatography Col	Tl	Ion	Paper	CCD	Recrys. Sol.	M.P. °C	$[\alpha]_D$	Solvent Rotation	Elect.	Other	Ref.
Bpoc-Ala-N^{im}-Boc-His-N^G-nitro-Arg-Leu-N^{im}-Boc-His-Gln-Leu-polymer													×		722
β-OBzl-Z-Asp-Arg(Tos)-Val-Tyr-(Bzl)-Ile-His(Tos)-Pro-Phe-polymer															157
Boc-β-OBzl-Asp-N^G-nitro-Arg-Val-O-Bzl-Tyr-Val-N^{im}-Bzl-His-N-methyl-Ala-polymer															17
Boc-Leu-N^6-Z-Lys-N^6-Z-Lys-Thr-N^2-nonanoyl-Lys-N^6-Z-Lys-D-Leu-polymer															732
Z(OMe)-Val-N^ϵ-Z-Lys-Val-Tyr-Pro-Ala-Gly-polymer															712
Z(OMe)-Val-N^ϵ-Z-Lys-Val-Tyr-Pro-α-OBzl-Asp-Gly-polymer															712
Z-β-[α-(NBzl)Asp]-N^G-nitro Arg-Val-OBzl-Tyr-Val-N^{im}-Bzl-His-Pro-Phe-polymer															524
Phe-Val-Asn-Gln-N^{im}-Bzl-His-Leu-S-Ac-Cys-Gly-O-polymer															206
Phe-Val-Asn-Gln-N^{im}-Bzl-His-Leu-S-Bz-Cys-Gly-O-polymer															206

Phe-Val-Asn-Gln-N^{im}-Bzl-His-Leu-S-Z-Cys-Gly-O-polymer															206
Phe-Val-Asn-Gln-N^{im}-Bzl-His-Leu-S-Bzl-Cys-Gly-O-polymer															206
Phe-Val-Asn-Gln-N^{im}-Bzl-His-Leu-S-OBzl(OMe)-Cys-Gly-O-polymer															206
Phe-Val-Asn-Gln-N^{im}-Bzl-His-Leu-S-Btm-Cys-Gly-O-polymer															206
Phe-Val-Asn-Gln-N^{im}-Bzl-His-Leu-S-ethylcarbamoyl-Cys-Gly-O-polymer															206
Phe-Val-Asn-Gln-N^{im}-Bzl-His-Leu-S-diphenylmethyl-Cys-Gly-O-polymer															206
Phe-Val-Asn-Gln-N^{im}-Bzl-His-Leu-S-Trt-Cys-Gly-O-polymer															206
Phe-Val-Asn-Gln-N^{im}-Bzl-His-Leu-S-Thp-Cys-Gly-O-polymer															206
OBzl-Ser-Met-OBzl-Glu-OBzl-His-Phe-N^{G}-nitro-Arg-Trp-Gly-polymer															99
OBzl-Tyr-Gly-OBzl-Ser-Leu-Pro-Gln-N^{ϵ}-Z-Lys-Gly-polymer															99
O-Bzl-Boc-Tyr-Ile-Gln-Asn-S-Bzl-Cys-Pro-Leu-Gly-polymer															635
Boc-N^{ϵ}-Tos-Lys-Pro-Val-Gly-N^{ϵ}-Tos-Lys-N^{ϵ}-Tos-Lys-N^{G}-Tos-Arg-N^{G}-Tos-Arg-Pro-polymer															358
S-Bzl-Z-Cys-O-Bzl-Tyr-Phe-Gln-Asn-S-Bzl-Cys-Pro-N^{ϵ}-Tos-Lys-Gly-polymer															179

(continued)

Table 21 Solid Phase (continued)

Compound	Method	Solvent	% Yield	Chromatography				CCD	Recrys. Sol.	M.P. °C	$[\alpha]_D$	Solvent Rotation	Elect.	Other	Ref.
				Col	Tl	Ion	Paper								
Z(OMe)-D-Phe-Pro-Val-δ-*p*-nitro-Z-Orn-Leu-D-Phe-Pro-Val-$N^δ$-Z(NO_2)-Orn-Leu-polymer															453
OBzl-Boc-Ser-OBzl-Tyr-OBzl-Ser-Met-Gln-N^{im}-Bzl-His-Phe-N^G-Tos-Arg-Trp-Gly-polymer														AA	357
Ala-N^G-nitro-Arg-OBzl-Thr-OBzl-Thr-OBzl-His-OBzl-Thr-Gly-OBzl-Ser-Leu-Pro-Gln-$N^ε$-Z-Lys-Gly-polymer															99
OBzl-Ser-OBzl-Tyr-OBzl-Ser-Met-OBzl-Glu-OBzl-His-Phe-Tos-Arg-Trp-Gly-$N^ε$-Z-Lys-Pro-Val-polymer															99
$N^ε$-Z-Boc-Lys-His(Dnp)-Pro-Pro-His(Dnp)-Leu-OBzl-Ser-Phe-Met-Ala-Ile-Pro-Pro-$N^ε$-Z-Lys-$N^ε$-Z-Lys-polymer															498
N-Boc-Ser-Asn-Asn-Asn-Gln-Gln-Gly-$N^ε$-Z-Lys-OBzl-Ser-Ala-Gln-Gln-Gly-Gly-OBzl-Tyr-polymer															474
N-Boc-Ser-Asp-Asn-Asn-Gln-Gln-Gly-$N^ε$-Z-Lys-OBzl-Ser-Ala-Gln-Gln-Gly-Gly-OBzl-Tyr-polymer															474

Chapter 15

Polypeptides

Racemization free syntheses of polypeptides are becoming of increasing importance from the standpoint of obtaining model antigenic bio-polymers. Such polymers also continue to be required for physical measurements and the preparation of poly-(β-Asp) for ORD study provides a good illustration (Chart 14). The OSU coupling technique gave a polymer shown to be optically pure.[213] Analogous syntheses of poly-(γ-OMe-Glu)[9] and poly-(γ-OBzl-Glu) were completed.[215] An NCA method was recently utilized (Chart 14) to prepare a poly-(β-Asp) but the extent of racemization was not indicated.[425] A similar method was used to obtain poly-His (Chart 14).[152]

Synthetic poly-Gly-Pro-Ala was prepared as a model of the fibrous protein collagen and was found to be antigenic in guinea pigs and to display some immunological cross reactivity with collagen.[139] Jones and colleagues have continued their study of the collagens using experimental techniques briefly summarized in Chart 14.[107,108] The specimen of poly-(Glu-Ala-Ala) prepared as entered in Chart 14 gave evidence of being antigenic in guinea pigs.[107] Related interesting immunochemical investigations of antigenic linear polypeptides are being pursued by Johnson.[269,270,273]

The thermal preparation of polypeptides from amino acids of the type so far identified in meteor and lunar material has suggested that proteinlike polymers could exist extraterrestrially.[535] A series of physiochemical measurements of polypeptides, particularly by CD and ORD for conformational reasons, have been undertaken with copper(II) complexes of poly-His,[353] poly-(N^{ϵ}-Bzl-Lys),[719] random copolymers of Glu-Asp benzyl ester with D-Glu-Asp benzyl ester,[471] poly-β-(γ-Glu)-Asp,[522] and poly-(*cis*-5-Me-Pro).[461]

The direction of current endeavors in the polypeptide field toward elucidating the mechanism of antibody formation should produce knowledge of great medical significance. Also such studies based on specific antigenic determinants, animal genetic factors, and the secondary structures of polypeptides of known primary structure should enhance our overall understanding of protein chemistry. The recent contributions, for example, of Rydon,[213,215] and Jones,[107,139] abbreviated in Chart 14 provide useful synthetic models for future racemization free syntheses of specific polypeptides required for such fundamentally important research.

Chart 14

213

β-OSu-N-Z-Asp-OBzl —(7% HBr-Ac / Cl_2CHCO_2H)→ β-OSu-Asp-OBzl HBr salt (85%, m.p. 156°-157°) —((1) TEA-DMF, 1 day, room temp. (2) Ppt Aq (3) Sephadex G-15 with DMF)→

poly-(β-Asp-OBzl) (75%) —((1) 5% Pd/C, 5 atm DMF, 3 days, room temp. (2) Dialysis-Aq 1 day, room temp. (3) Lyophilize)→ $[-NHCH(CO_2H)CH_2CO-]$ (81%, Mol. wt. ~11,000 (by Sephadex G-75)) —(6 *N* HCl / 1 day, 100°)→

Asp
$\alpha]_D$ + 24.7° (+24.8° for L)
0% D

425

Asp-OEt —(ClCOCl)→ $HO_2CCH_2CH(NHCOCl)CO_2Et$ —(TEA)→

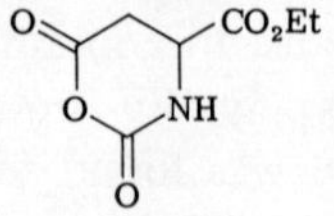

—(Δ, 2 hr)→ poly-(β-Asp) + CO_2

152

N^{im}-Boc-His-NCA —(TEA-Di)→ $(N^{im}\text{-Boc-His})_n$ (Mol. wt. ~21,000) —(HBr-TFA)→ poly-His

139

Z-Ala-Gly-Pro-OTcp —(HBr-Ac / 1 hr, room temp.)→ Ala-Gly-Pro-OTcp HBr salt (94%, m.p. 159°-161°)

Chart 14 (cont.)

(1) NMM, DMSO
3 days, room temp.
(2) Alc-centrifuge
→ poly-(Ala-Gly-Pro)
(3) 50% Ac soln.
dialized against Aq
(4) Lyophilize

50-70%
Mol. wt. ~12,000
(by gel chromat.)

107

γ-OBu^t-N-Z-Glu-Ala-Ala-OPh(OBzl) → [(1) 10% Pd/C, H_2, Ac, 3 hr, room temp. (2) TEA, DMSO, 2 days, room temp.]

poly-(γ-OBu^t-Glu-Ala-Ala) → poly-(Glu-Ala-Ala)

65%

(1) 90% TFA
1 hr, room temp.
(2) in 0.5 *N* NaOH
dialyzed against Aq
(3) HCl to pH 3
repeat dialysis
(4) Lyophilize

31%
Mol. wt. to ~6,000

Table 22 Polypeptides

Compound	Method	Solvent	% Yield	Chromatography Col	Tl	Ion	Paper	CCD	Recrys. Sol.	M.P. °C	$[\alpha]_D$	Solvent Rotation	Elect.	Other	Ref.
Poly-Ala															477
Poly-DL-α-Abu															464
Poly-Arg HCl											−72	Aq			451
							×				+23	6 *N* HCl			451
Poly-N^G-Tos-Arg	NCA	DMA	94								+7	DMF		Mol. wt.	451
Poly-β,N-Bzl-Asn	NCA	Di-DMF									+12	CH_3-SO_3H			225
Poly-β-N-(DL)-(1)-phenethyl-Asn	NCA	Di													225
Poly-β,N-(D)-(1)-phenethyl-Asn	NCA	Di									+24	CH_3-SO_3H			225
Poly-β,N-(L)-(1)-phenethyl-Asn	NCA	Di-DMF									−54	CH_3-SO_3H			225
Poly-Asp	NCA														452
Poly(β-Bzl-Asp)	NCA													Mol. wt.	212
Poly-β-Bzl-Asp															175
Poly-β-Bzl-OBu^t-*threo*-β-hydroxy-Asp														IR	580
Poly-β-Bzl-OBu^t-*threo*-β-hydroxy-DL-Asp	NCA	Be												Mol. wt.	580
Poly-β-Bzl-*threo*-β-hydroxy-DL-Asp															580

β-Poly-α-ethyl-Asp	NCA									253–256 (dec)	−2	Phenol-Aq		Mol. wt.	425
Poly-*threo*-β-hydroxy-DL-Asp														IR	580
Poly(γ-OBzl-Glu)															212
Poly-α-OBzl-Glu	OPCP	Be													72
Poly(γ-OBzl-Glu)	NCA	Di													13
Poly(γ-OBu^t-Glu)											+21	Chf			393
Poly-Glu	NCA	Di												Mol. wt.	452
Poly-DL-Glu														Mol. wt.	452
Poly-Gly															464
Poly(1-Bzl-His)															104
Poly-N^{im}-Boc-His											−8.2	DMF			152
Poly-N^{im}-DNp-His	NCA	Di					×				+18	DMF	×	Mol. wt.	153
Poly-His															483
									An		−34	Aq			153
Poly-His HBr											−19	Aq			152
Palladium-on-Poly-S-Leu															40
Poly-DL-Leu															464
Poly-DL-*n*-Leu															464
Poly-Lys															485
Poly-Lys HCl															483
Poly(N^δ-Z-Orn)	NCA	DMF													57
Poly-N^δ-Z-Orn	NCA														174
Poly-Pro															516
														Mol. wt.	121
															402
Poly(Tyr)															102
Palladium-on-Poly-S-Val															40

(*continued*)

Table 22 Polypeptides (continued)

Compound	Method	Solvent	% Yield	Chromatography Col	Tl	Ion	Paper	CCD	Recrys. Sol.	M.P. °C	$[\alpha]_D$	Solvent Rotation	Elect.	Other	Ref.
Poly-DL-Val															464
Poly-DL-*n*-Val															464
(DL-Ala)$_m$(Gly)$_n$															593
Poly(Ala-Gly)	OPCP	Be									−84	DCA		Mol. wt.	71
Poly(β-Ala-Phe)	OPhBzl	DMSO	39								−9.0	DCA			110
Poly-N-Z-α,γ-Di-Abu	NCA	Di	80												174
Copoly(Glu-Asp)	NCA	Di					×							Mol. wt.	452
Poly(γ-Bzl-Glu$^{m'}$Gly$^{n'}$	NCA	Di	85												13
Poly[N^5-(4-hydroxybutyl)-Glu^{m}Gly^n]															13
Poly(γ-OBzl-Glu-γ-Bzl-Glu)	OSU	Chf		Seph-adex G-150										Mol. wt.	212
	ONP	DMF													226
Poly(γ-OBzl-D-Glu-γ-OBzl-Glu)	OPCP	Be													226
Poly(γ-OBu^t-Glu-γ-OBu^t-Glu)	DCCI	Acn		Seph-adex G-15 and G-75										Mol. wt.	212
Poly(γ-OBu^t-Glu-γ-OBu^t-D-Glu)											+1.5	Chf			393
Poly(γ-OBu^t-D-Glu-γ-OBu^t-Glu)											−1.5	Chf			393

Poly(γ-OBzl-Glu-Leu)	OSU	THF												Mol. wt.	212
Poly(γ-OBzl-D-Glu-Leu)	OSU	THF													216
Poly-Gly-Cys															214
Poly(N^{ϵ}-Gly-Lys HBr)									Alc						12
Poly(N^{ϵ}-N-Z-Gly-Lys)														Mol. wt.	12
Poly(N^{ϵ}-N-Z-Leu-Lys)															12
Poly(N^{ϵ}-Leu-Lys HBr)															12
Poly(N^{ϵ},N-Z-Phe-Lys)															12
Poly(N^{ϵ}-Phe-Lys HBr)															12
Poly(Pro-Gly)	ONP														625
Poly(Ala_2-Gly)											−18	DCA		Mol. wt.	71
[Ala-Gly-Ala]$_n$															593
[Ala-Gly-D-Ala]$_n$															593
Poly(Ala-Gly_2)	OPCP													Mol. wt.	71
Poly(Ala-Pro-Gly)$_n$	ONP	DMSO	60–80											Mol. wt. AA	367
Poly(α-Bzl-Asp-S-Bzl-Cys-Gly)	OPCP	DMF	73												336
Poly(β-Bzl-Asp-S-Bzl-Cys-Gly)	OPCP														336
Cys-poly-Gly-Cys															214
Poly(γ-Bzl-Glu-S-Bzl-Cys-Gly)	OPCP														336
Poly(α-OBzl-Glu-S-Bzl-Cys-Gly)	OPCP	DMF													336
Poly-[Di-(γ-OBu^t-Glu)-γ-OBu^t-D-Glu]											−2.5	Chf			393
Poly[Di-(γ-OBzl-D-Glu)-Leu]															216
Poly(Gly-Gly-Phe)	OPhBzl	DMSO	45												110
Poly(Gly-Pro-Ala)	OPhBzl	DMSO	66												110
Poly(Lys-Ala-Ala)-Tfa															106

(continued)

Table 22 Polypeptides (continued)

Compound	Method	Solvent	% Yield	Chromatography Col	Tl	Ion	Paper	CCD	Recrys. Sol.	M.P. °C	$[\alpha]_D$	Solvent Rotation	Elect.	Other	Ref.
(DL-Lys)$_{80}$(Ala)$_{160}$(DL-Lys)$_{80}$															355
[Pro-Gly-Pro]$_n$															593
[Pro-Pro-Gly]$_n$															593
Poly(Ala$_2$-Gly$_2$)	OPCP	Be									−89	DCA		Mol. wt. AA	71
Poly(Ala-Gly$_3$)	OPCP	Be	60								−46	DCA		Mol. wt.	71
Poly[Tri-(γ-OBu^t-Glu)-γ-OBu^t-D-Glu]											+4.3	Chf			393
Poly[Tri-(γ-OBzl-D-Glu)-Leu]															216
Poly(Gly-Gly-Pro-Gly)	ONP														625
Poly(Ala$_2$-Gly$_3$)	OPCP	Be	52								−74	DCA		Mol. wt. AA	71
Poly(Ala-Glu-Ala-Gly)Gly-1-[^{14}C]OEt															274
Poly[tetra-(γ-OBzl-D-Glu)-Leu]	OSU	Be							Chf-Me	260–280 (dec)	−28	Chf			216
Poly(Glu-Tyr-Ala-Gly)Gly-1-[^{14}C]OEt															268
Poly(Phe-Glu-Ala-Gly)Gly-1-[^{14}C]OEt															273

Poly(Trp-Glu-Ala-Gly)Gly-OMe				Seph- adex G-10											278
Poly(OMe-Tyr-Glu-Ala-Gly)Gly-1-[^{14}C]OEt															272
Poly-(Tyr-Glu-Ala-Gly)Gly-1-[^{14}C]OEt															271
Poly(Tyr-γ-Glu-Ala-Gly)Gly-OMe															267
Poly(Tyr-Glu-Gly-Gly)Gly-OMe											−9.8	Tris buffer			277
Poly-(OBu^t-Tyr-γ-Bu^t-Glu-Val-Gly)[1-^{14}C]Gly-OEt															276
Poly-(Tyr-Glu-Val-Gly)[1-^{14}C]Gly-OEt															276
Poly(Ala$_3$-Gly$_3$)	OPCP	Be									−86	DCA		AA	71

Chapter 16

Cyclopeptides

The first synthesis of *cyclo*-Cys, by reacting the S-Eac precursor with iodine,[292] and a new synthesis of gramicidin S (Chart 15) represents some of the more noteworthy recent aspects of cyclopeptide preparation included in the table of this subdivision. After the tabular survey was closed a number of contributions were reviewed. The crystal structure of *cyclo*-(β-Ala-β-Ala) was determined.[673] Some *cyclo*-(Orn-Leu) peptides were prepared and screened for antibacterial activity (none found)[247] and syntheses of, for example, *cyclo*-(Phe-D-Pro)[373] and N-methylated cyclodipeptides (by methyl iodide, silver oxide, DMF)[403] were entered in the literature.

A dehydropeptide has been cyclized to an imidazole derivative to serve as a model for the biosynthesis of luciferin from a crustacean.[405] Unequivocal evidence for the lichen metabolite *cyclo*-(R-β-phenyl-β-Ala-Pro)$_2$ was provided by total synthesis.[63] The cyclization step (20% yield) was achieved by ONP peptide-bond formation in pyridine at high dilution (6 mmole to 3000 ml). Nine isomeric cyclohexapeptides have been synthesized as models for the study of molecular interactions.[53]

The cyclic decapeptide antibiotic gramicidin S belongs to a family of similar substances isolated from *Bacillus brevis*. These antibiotics are not generally used systemically but are valuable, for example, in certain eye-threatening infections. The final stages of a recent solid phase synthesis of gramicidin S were accomplished by the experimental methods outlined in Chart 15. The 90% conversion at the cyclization step seems especially remarkable.[453] Perhaps the spatial arrangement of this particular decapeptide sequence particularly favors cyclization.

A nicely executed synthesis of [2,7-Cys]gramicidin S has been described.[374] The Schwyzer route to this gramicidin analog was based on a fragment condensation procedure using DCCI-HBT for coupling reactions, and formation of the disulfide bridge by subjecting the S-Eac-protecting group to reaction with iodine. The 27% yield at the cyclization step with an ONP coupling reaction was more typical and the latter steps of the synthesis have been entered in Chart 15. In related studies the [6-Tyr]analog of Antamanide, the cyclic decapeptide of *Amanita phalloides*, has been obtained by Wieland[678] employing a fragment condensation approach.

Substances of the amphotericin D (a polyene), polyether (for example crown cyclic ethers), Antamanide (a peptide), and valinomycin (a depsipeptide) represent structural types capable of complexing with alkali metal ions and thereby promoting their dissolution in fairly nonpolar solvents. Such compounds are known as "ion carriers" and some display antibiotic properties which may in part reside with activity in natural membranes. In order to evaluate structural changes upon such interesting functions, Gisin and Merrifield[178] have synthesized a cyclododecapeptide (Chart 15) where the D-α-hydroxyisovaleric acid and L-lactic acid units of valinomycin were replaced respectively with D-Pro and L-Pro. In MeCl-Aq the valinomycin analog was found to exhibit a seven times greater affinity for potassium picrate (to form a 1:1 hydrophobic complex) than that of the parent depsipeptide.

The conformations of cyclic peptides continue to be of interest and the Schwyzer group[412] has studied a series of cyclic pentapeptides by PMR. With the examples investigated, replacement of Pro by D-Pro or substitution of Gly by Sar markedly changed the peptide conformation. Again the results were significant from the viewpoint of peptide complex formation with alkali metal and alkaline metal ions.

Chart 15

453 Gramicidin S

Z(OMe)-D-Phe-Pro-Val-N $^{\delta}$-Z-(NO_2)-Orn-Leu-D-Phe-Pro-Val-Z(NO_2)-Orn-Leu-polymer

(1) 100% NH_2NH_2, DMF 2 days, room temp.
(2) 3.5 *N* HCl-EtAc anisole
→ D-Phe-Pro-Val-N $^{\delta}$-Z(NO_2)-Orn-Leu-D-Phe-Pro-Val-Z(NO_2)-Orn-Leu-$NHNH_2$ HCl salt

~50%
estimated to be 98% pure

(1) 1 *N* HCl-DMF-Ac $NaNO_2$, −10°, 10 mm
(2) 500:1 Py added 3 hr, 0° and 36 hr, 4°
(3) Dowex-50 (H^+), Me-Aq
→ *Cyclo*-(D-Phe-Pro-Val-N $^{\delta}$-Z(NO_2)-Orn-Leu-D-Phe-Pro-Val-Z(NO_2)-Orn-Leu)

90%
m. p. 228°-230°
Mol. wt. (osmometer) 1499, calcd. 1492

(1) Pd, H_2, 0.2 *N* HCl 4 hr, room temp.
(2) Recrys. Alc-1 *N* HCl
→ *Cyclo*-(D-Phe-Pro-Val-Orn-Leu-D-Phe-Pro-Val-Orn-Leu)

50%
m. p. 275°-277°

374 (2, 7-Cys)-Gramicidin S

Boc-Val-S-Eac-Cys-Leu-Phe-Pro-Val-S-Eac-Cys-Leu-Phe-Pro-ONp

(1) TFA, 1 hr, room temp. (→92%)
(2) ~600:1 Py added DMF, 1 day, 55°
(3) Sephadex LH-20, Alc
(4) Recrys. Me-Eth
→ *Cyclo*-(Val-S-Eac-Cys-Leu-Phe-Pro)$_2$

27%
m. p. 181°-182°

(1) I_2, Me 2 hr, ~0° 5 hr, room temp.
(2) Recrys. Me-Eth
→ *Cyclo*-(Phe-Pro-Val-Cys-Leu-Phe-Pro-Val-Cys-Leu) (S———S bridge between the two Cys)

65%
d. p. 325, M^+ 1116

Chart 15 (cont.)

178 Valinomycin analog

$ClCH_2$-polymer —[(1) $KOCOCH_3$, Me cellosolve, 16 hr, 100°-110°; (2) Repeated 2 ×]→ $CH_3CO_2CH_2$-polymer —[$CH_3)_2NH$, MeCl, 1 day, Δ]→

$HOCH_2$-polymer —[(1) 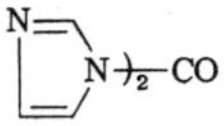 Boc-Pro-MeCl, 30 min, − 5°; (2) 3 days, room temp.]→ Boc-Pro-OCH_2-polymer

0.6 mmole Pro/gm

—[(1) Automatic SP synthesizer (→58%); (2) 30% HBr-Ac, 90 min, room temp.; (3) Sephadex LH-20, 10^{-3} *N* HCl]→ Val-D-Pro-D-Val-Pro-Val-D-Pro-D-Val-Pro-Val-D-Pro-D-Val-Pro HCl salt

43-46%

—[(1) TEA-Aq to pH 5.5; (2) Lyophilize; (3) WRK 800:1 MeCl added, 5 days, room temp.; (4) AG 501-X8; (5) Recrys. MeCl-Be]→ *Cyclo*-(Val-D-Pro-D-Val-Pro)$_3$

16%

d. p. >330°, M^+ 1176

Table 23 Cyclopeptides

Compound	Method	Solvent	% Yield	Chromatography: Col	Tl	Ion	Paper	CCD	Recrys. Sol.	M.P. °C	$[\alpha]_D$	Solvent Rotation	Elect.	Other	Ref.
N-*t*-Boc-*cyclo*-Leu			17						AcOEt-Hep	131–133					282
D-Ala-D-*cyclo*-Ser															352
Cyclo-(Cys-Cys) S — S									Me		+310	DMSO			292
Cyclo-(γ-OBzl-D-Glu-Leu)										175–177				Mass spec.	216
Cyclo-(His-His)															733
Cyclo-(Pro-methyl-Phe)					57 126 72					137				IR	317
Cyclo-(Ser-His)															733
Cyclo-(Tri-Gly)									Hot Aq						566
Cyclo-(Pro-Pro-Hyp)	ONP	Py	12							~350				Mol. wt.	122
Cyclo-(Pro-Pro-Hyp) acetate									An	206–210				Mol. wt.	122
Cyclo-(Pro-Pro-Hyp) benzoate										212–216				Mol. wt. IR	122
Cyclo-(Ala_3-Sar)															115
Cyclo-[β-Ala-Gly-(α-Tos-amino)-β-Ala-Gly]	A	Aq			×					269–273 (dec)	−30	DMSO		Mol. wt.	223

Cyclo-(β-Ala-Gly-β-Asp-Gly) hydrazide														Mol. wt.	223
Cyclo-(β-Ala-Gly-β-Asp-Gly)-OMe	MA									260–310 (dec)				Mol. wt.	223
Cyclo-di-(γ-OBzl-D-Glu)-Leu-di-(γ-OBzl-D-Glu)-Leu										240 (dec)					216
Cyclo-(Gly_2-Sar_2)															115
Cyclo-(Gly-Sar-Gly-Sar)															115
Cyclo-(Gly-Sar_3)															115
Cyclo-(Sar_4)															115
Cyclo-(S-S′-dehydro-Ile-D-Leu-Cys-Val-Cys)				Seph-adex LH-20							−98	TFE		AA	549
Cyclo-(S-S′-dehydro-Ile-D-Leu-D-Cys-Val-D-Cys)											−43	TFE		AA	549
Cyclo-(Ile-D-Leu-S-Bzl-Cys-Val-S-Bzl-Cys)	DCCI-NHS	DMF	52	Seph-adex LH-20							−65	TFA		AA	548
Cyclo-(Ile-D-Leu-S-Bzl-D-Cys-Val-S-Bzl-D-Cys)	DCCI-NHS	DMF	62	Seph-adex LH-20							+16	TFA			548
Cyclo-(S-S′-dehydro-D-Leu-Ile-D-Cys-Val-D-Cys)											+59	DMF		Mol. wt. AA	549
Cyclo-(D-Leu-Ile-S-Bzl-Cys-Val-S-Bzl-Cys)	DCCI-NHS	DMF		Seph-adex							−37	TFA		Mol. wt. AA	548
Cyclo-(D-Leu-Ile-S-Bzl-D-Cys-Val-S-Bzl-D-Cys)	DCCI-NHS	DMF	80	Seph-adex LH-20							+42	DMF			548

(*continued*)

Table 23 Cyclopeptides (continued)

Compound	Method	Solvent	% Yield	Chromatography: Col	Tl	Ion	Paper	CCD	Recrys. Sol.	M.P. °C	$[\alpha]_D$	Solvent Rotation	Elect.	Other	Ref.
Cyclo-(OBzl-Ser-N$^{\delta}$-Z-Orn-OBzl-Ser-D-Phe-Pro)			76	DO-1 DO-50	×					118–119	−93	Ac		Mol. wt.	261
Cyclo-(Ser-Orn-Ser-D-Phe-Pro) HCl										250 (dec)	−157	Ac			261
Cyclo-(Val-N$^{\delta}$-Z-Orn-Leu-D-Ala-Pro)										130–133	−65	Me			349
Cyclo-(Gly-S-Bzl-Cys-Leu-Gly-S-Bzl-Cys-Leu)	A	Aq	55						Ac-Aq	283–285			×		54
Cyclo-(Gly-S-Bzl-Cys-D-Leu-Gly-S-Bzl-Cys-D-Leu)	A	DMF-Aq	43						Ac-Aq	277–281					54
Cyclo-(Gly-Gly-D-Ala-D-Ala-Gly-Gly)															621
Cyclo-(Gly-Gly-β-Ala-Gly-Gly-β-Ala)										358 (dec)					264
Cyclo-(Gly-Gly-DL-Ala-Gly-Gly-DL-Ala)															264
Cyclo-(hexa-Gly)										>350					264
Cyclo-$(Gly_2$-Phe$)_2$	A		25			DO-1 DO-50				280–285 (dec)	−52	DMF			331
Cyclo-$(Gly_2$-Tyr$)_2$	ONP	DMF	9			A-IR4B DO-50	×		Me-Eth				×	Mol. wt.	331

Cyclo-(Gly-His-Gly-Ala-Tyr-Gly)	DCCI	DMF		Ce				11							332
Cyclo-(Gly-His-Gly-Tyr-Ala-Gly)	DCCI	Phenol	10					11							332
Cyclo-(Gly-Leu-Gly-Gly-Leu-Gly)														Mass spec.	332
Cyclo-(Gly-Tyr-Gly-d_2-Gly-Tyr-Gly-d_2)															332
Cyclo-(Pro-Ser-Gly-Pro-Ser-Gly)	ONP	DMF-Py	16		×				Me	308–313				Mol. wt. IR AA	626
Cyclo-(Ser-Pro-Gly-Ser-Pro-Gly)	ONP	DMF-Py	4						Me	315–319				IR Mol. wt.	627
Cyclo-(Val-Ala-Phe-Ala-Gly-Gly)	A	THF	10		69										677
Cyclo-⟨[Ala-D-Thr-Ala-yl-(β)]-[α-Hyp-Ala-Trp-Nva-(ind-2)] sulfide⟩				Sephadex LH-20			×						×	UV	138
Cyclo-⟨[Boc-Ala-D-Thr-Ala-yl-(β)]-[α-Hyp-Ala-Trp-Nva(OCH_3)-(ind-2)] sulfide⟩				Sephadex LH-20										AA	138
Cyclo-⟨[Boc-Ala-D-Thr-Ala-yl-(β)]-[α-Hyp-Ala-Trp-Nva-(ind-2)] sulfide⟩													×		138
Cyclo-⟨[*i*-Boc-α-Hyp-Ala-Trp-Nva-(ind-2)]-[Ala-D-Thr-Ala(NH_2)-yl-(β)] sulfide⟩				Sephadex LH-20										UV	138
Cyclo-[Val-Orn-Leu-δ-aminovaleryl-Val-Orn-Leu-D-Phe-Pro]															433
Cyclo-[Ala-$(Pro)_2$-Ala-$(Phe)_2$-$(Pro)_2$-Phe-Phe]	DCCI-OSU		45							174–178				Mol. wt. AA	676

(*continued*)

Table 23 Cyclopeptides (continued)

Compound	Method	Solvent	% Yield	Chromatography				CCD	Recrys. Sol.	M.P. °C	$[\alpha]_D$	Solvent Rotation	Elect.	Other	Ref.
				Col	Tl	Ion	Paper								
Cyclo-[Ala-(Pro)$_2$-Gly-(Phe)$_2$-(Pro)$_2$-Phe-Phe]	DCCI-OSU		49							181–183				Mol. wt. AA	676
Cyclo-[Ala-(Pro)$_2$-Val-(Phe)$_2$-(Pro)$_2$-Phe-Phe]														Mol. wt. AA	676
Cyclo-[Dbu(Meo)-aThr-N$^\delta$-Z-Dbu-aThr-N$^\delta$-Z-Dbu-N$^\delta$-Z-Dbu-D-Leu-Ile-N$^\delta$-Z-Dbu-N$^\delta$-Z-Dbu]						DO-50									22
Cyclo-[Dbu(Meo)-aThr-Dbu-aThr-Dbu-Dbu-D-Leu-Ile-Dbu-Dbu]							×			206–208			×		22
Cyclo-[Gly-(Pro)$_2$-Ala-(Phe)$_2$-(Pro)$_2$-Phe-Phe]	DCCI-OSU		49							180–182				Mol. wt. AA	676
Cyclo-[Gly-(Pro)$_2$-Gly-(Phe)$_2$-(Pro)$_2$-Phe-Phe]	DCCI-OSU		45						Alc-Aq	168–180				Mol. wt.	676
Cyclo-[Ile-(Pro)$_2$-Ala-(Phe)$_2$-(Pro)$_2$-Phe-Phe]	DCCI-OSU		48						Alc-Aq	164–167				Mol. wt. AA	676
Cyclo-[Leu-(Pro)$_2$-Ala-(Phe)$_2$-(Pro)$_2$-Phe-Phe]	DCCI-OSU		46						Alc-Aq	167–169				Mol. wt.	676
Cyclo-[Phe-(Pro)$_2$-Phe-(Phe)$_2$-(Pro)$_2$-Phe-Phe]															676

Cyclo-(D-Phe-Pro-Val-N^{δ}-*p*-nitro-Z-Orn-Leu-D-Phe-Pro-Val-N^{δ}-*p*-nitro-Z-Orn-Leu)						DO-50				228–230	−149	Me		AA	453
Cyclo-(D-Phe-Pro-Val-Orn-Leu-D-Phe-Pro-Val-Orn-Leu) 2 HCl									Alc-1 *N* HCl	275–277	−251	Alc			453
Cyclo-(OBzl-Ser-N^{δ}-Z-Orn-OBzl-Ser-D-Phe-Pro-OBzl-Ser-N^{δ}-Z-Orn-OBzl-Ser-D-Phe-Pro)	ONP	DMF	30	Sephadex LH-20		DO-1, DO-50			Alc-Aq	246–247	−185	Ac			261
Cyclo-(Ser-Orn-Ser-D-Phe-Pro-Ser-Orn-Ser-D-Phe-Pro) 2 HCl										320 (dec)	−126	Ac		AA	261
Cyclo-(Trp-D-Trp-Asn-Gln-Tyr-Val-N^{δ}-Z-Orn-Leu-D-Phe-Pro)						D-IR-50, D-IR-45			DMF-Eth	225–228	−36	Me			342
Cyclo-(Trp-D-Trp-Asn-Gln-Tyr-Val-Orn-Leu-D-Phe-Pro) HCl									Me-EtAc-Eth	218–220 (dec)	−84	Me		UV	342
Cyclo-[OBzl-Tyr-Pro_2-Phe_2-Val-Pro_2-Ala-Phe]									An-Aq	158–160				AA	678
Cyclo-[Tyr(OMe)-Pro_2-Phe_2-Val-Pro_2-Ala-Phe]									An-Aq						678
Cyclo-(Val-N^{δ}-Z-Orn-Leu-D-Ala-Pro)-(Val-N^{δ}-Z-Orn-Leu-D-Ala-Pro)	ONP	DMF-Py	23			DO-1, DO-50			Me-Eth	234–236 (dec)	−199	Me			349
Cyclo-(Val-Orn-Leu-D-Ala-Pro)-(Val-Orn-Leu-D-Ala-Pro)									Me-Eth-Pe	190–191 (dec)	−189	Me			349
Cyclo-(Val-N^{δ}-Z-Orn-Leu-D-Phe-Pro-Val-N^{δ}-Z-Orn-Leu-D-Phe-Gly)	ONP	Py	31							156–157	−94	Ac		Mol. wt.	401
Cyclo-[Val-$(Pro)_2$-Gly-$(Phe)_2$-$(Pro)_2$-Phe-Phe]														Mol. wt.	676
Cyclo-[Val-$(Pro)_2$-Phe-$(Phe)_2$-$(Pro)_2$-Phe-Phe]														Mol. wt. AA	676

(*continued*)

Table 23 Cyclopeptides (continued)

Compound	Method	Solvent	% Yield	Chromatography				CCD	Recrys. Sol.	M.P. °C	$[\alpha]_D$	Solvent Rotation	Elect.	Other	Ref.
				Col	Tl	Ion	Paper								
Cyclo-(Val-Orn-Leu-D-Phe-Pro-Val-Orn-Leu-D-Phe-Gly) Di-HCl ([Gly^5]gramicidin S · 2 HCl)				×	124 25				Me-Eth-Pe	205–207 (dec)	−89	Alc	×	AA	401
Cyclo-(Tyr-Pro_2-Phe_2-Val-Pro_2-Ala-Phe) (Tyr^6-Antamanid)															678
[4-5-δ-Aminovaleric acid]gramicidin S															433

Chapter 17

Depsipeptides and Cyclodepsipeptides

Significant reasons for studying the cyclodepsipeptide valinomycin based on possible transport of potassium ion through biological membranes by this substance and synthesis (SP route) of a D-Pro analog have been summarized in the preceding chapter. A solid phase synthesis of valinomycin was described in 1971 and one of the key steps, cyclization of D-Val-L-Lac-Val-D-Hyv-$)_3$ was achieved (24% yield) as earlier described by Shemyakin and colleagues using thionyl chloride.[368]

An advance received too late to include in the table of this chapter concerned the preparation of four polydepsipeptides by the Rydon group.[512] In this work, L-2-hydroxy-4-methylpentanoic acid was prepared from Leu, depside coupling was induced by DCCI in Py, and polymer formation was initiated by treating an acid chloride hydrochloride with triethylamine. By this means, for example, was prepared poly-(0-Tri-Leu-L-2-hydroxy-4-methylpentanoic acid) which corresponded to a number average molecular weight of 46,000.

Table 24 Depsipeptides and Cyclodepsipeptides

Compound	Method	Solvent	% Yield	Chromatography Col	Tl	Ion	Paper	CCD	Recrys. Sol.	M.P. °C	[α]D	Solvent Rotation	Elect.	Other	Ref.
Z-γ-Hydroxy-Leu hydrazide															140
Z-γ-Hydroxy-Leu lactone									Alc-Aq	104–105					140
Z-DL-γ-Hydroxy-Leu amide										105–113					140
N-Z-α-Amino-Val-δ-lactone										97–101				IR	253
Val-glycolic-hydroxyisovaleric-phenyllactic acid	SP				55 125										560
Z-DL-γ-Hydroxy-Leu-Ala-OMe										97–98					140
DL-γ-Hydroxy-Leu-Ala-OMe															140
Z-Trp-DL-γ-hydroxy-Leu-Ala										108–114 (dec)					140
Z-Trp-DL-γ-hydroxy-Leu-Ala-OMe															140
Boc-Val-D-Hyv-D-Val-Lac-Val-D-Hyv-OBzl										Oil	−3.8	Alc			368
Val-D-Hyv-D-Val-Lac-Val-D-Hyv-OBzl										Oil	+13	Di			368
Z-Trp-γ-hydroxy-Leu-Ala-D-Thr-OMe	3 + 1 MCA	Thf	48.5							215					140
Boc-D-Val-Lac-Val-D-Hyv-D-Val-Lac-Val-D-Hyv											−5	Alc			368
D-Val-Lac-Val-D-Hyv-D-Val-Lac-Val-D-Hyv											−23	Alc			368
Boc-D-Val-Lac-Val-D-Hyv-D-Val-Lac-Val-D-Hyv-D-Val-Lac-Val-D-Hyv					×						−9.3	Alc			368

Boc-D-Val-Lac-Val-D-Hyv-D-Val-Lac-Val-D-Hyv-D-Val-Lac-Val-D-Hyv	DCCI	MeCl													368
D-Val-Lac-Val-D-Hyv-D-Val-Lac-Val-D-Hyv-D-Val-Lac-Val-D-Hyv											+10	Chf			368
D-Val-Lac-Val-D-Hyv-D-Val-Lac-Val-D-Hyv-D-Val-Lac-Val-D-Hyv					×						+9.3	Chf			368

Chapter 18

Steroidal Peptides

Amphibian venoms of the bufadienolide type comprise a pharmacologically interesting series of naturally occurring steroids. All have in common a 5-substituted 2-pyrone ring at the 17β-position, a 14β-oxygen substituent and a *cis* A/B ring juncture. Some, such as bufotoxin and marinobufotoxin contain a suberyl arginine side chain at position 3 (Chart 16). Many of the toad venom and plant bufadienolides exhibit a variety of potentially useful medicinal activities, for example, local anesthetic (bufalin), cardiac action (scillarenin), cytotoxicity (bersaldegenin acetate), respiratory stimulant (resibufogenin), and tumor inhibitory (hellebrigenin acetate). For the purpose of making the suberyl arginine type venom constituents more readily available for biological studies, the first partial (bufotoxin)[487] and total (marinobufotoxin)[291a] syntheses in this area were recently completed in my laboratory. The approach (Chart 16) to marinobufotoxin utilizing Arg protected as the monohydrochloride in an MCA coupling reaction was found most effective.

In other new work the structure–activity requirements for antidepressant activity of amino acids coupled to the 17β-amino position has been further evaluated.[147]

Chart 16

487

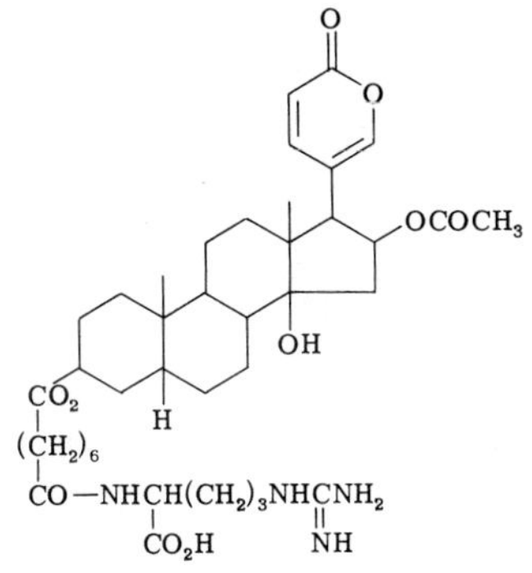

Bufotoxin

291a

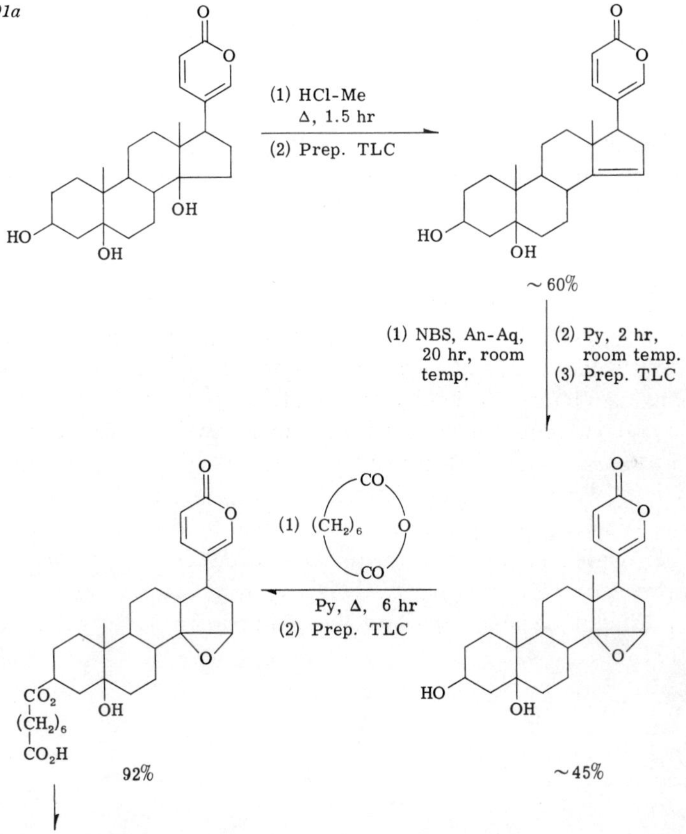

Chart 16 (cont.)

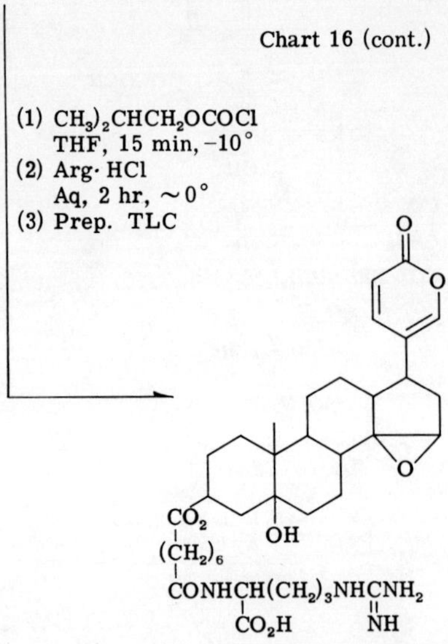

88%
Marinobufotoxin

Table 25 Steroidal Peptides

Compound	Method	Solvent	% Yield	Chromatography Col	Tl	Ion	Pape	CCD	Recrys. Sol.	M.P. °C	[α]D	Solvent Rotation	Elect.	Other	Ref.
Testosterone aspartate									An	195–200 (dec)				IR	449
Testosterone aspartate ethylene ketal									Me	203–205				IR	449
Testosterone Pht-asparate ethylene ketal									Me	225–228				IR	449
3α-(Z-Gly-amino)-5α-pregnan-20-one									Me	213–214	+58	DMF			609
3β-(Z-Gly-amino)-5α-pregnan-20-one									EtAc	150–151	+54	Chf			609
3β-(Z-Gly-amino)-5β-pregnan-20-one									Alc	188.5–189.5	+72	Chf			609
3α-(Gly-amino)-5α-pregnan-20-one										173.5–175 (dec)					609
3α-(Z-Phe-amino)-5α-pregnan-20-one									He-EtAc	184–185	+74	Chf			609
3α-(Z-Pro-amino)-5α-pregnan-20-one									He-EtAc	154–156	+30	Chf			609
3α-(Z-DL-Trp-amino)-5α-pregnan-20-one									Alc	187					609
3α-(Z-Gly-Gly-amino)-5α-pregnan-20-one	WRK		68						Me	213–214					609
3α-(Z-Phe-Gly-amino)-5α-pregnan-20-one	DCCI		36						Me	115.5–119.5					609
3β-(Z-Phe-Gly-amino)-5β-pregnan-20-one	OTCP		45						EtAc	97–99					609

(*continued*)

Table 25 Steroidal Peptides (continued)

Compound	Method	Solvent	% Yield	Chromatography				CCD	Recrys. Sol.	M.P. °C	$[\alpha]_D$	Solvent Rotation	Elect.	Other	Ref.
				Col	Tl	Ion	Paper								
Bufotoxin	MCA	THF-Me-Aq	64							209–212					487
Resibufotoxin	MCA	Aq								189–197					487
Bufalitoxin	MCA	Aq								204–211					487
Marinobufotoxin	MCA	Aq	88							176–185					291a

Chapter 19

Chromopeptides

The peptide antibiotic actinomycin D (Chart 17) is a clinically effective cancer chemotherapeutic agent useful, for example, in the treatment of Wilms' tumor. The Brockmann synthesis of actinomycin D was joined in 1970 by the Meienhofer approach.[407] From all appearances the majority of recent synthetic efforts in the chromopeptide field have been related to the actinomycins. The synthesis of actinomycin D lactam has been completed,[27] and other analogs have been obtained by Larionov and co-workers.[181] A synthesis of [4′,4′-bis-Gly-5′,5′-bis-Val]-actinomycin D has been brought to completion by Mosher and Goodman.[426] Unfortunately, the latter analog did not prove to be antibacterial in a number of screening systems or to be cytotoxic in the KB tissue culture system. However, actinomycin D lactam did show high antibacterial activity and indications of antitumor activity in a ROS mouse system. Given the earlier fundamental work with actinomycin D and current advances in peptide synthetic methods the valuable lead presented by this naturally occurring antineoplastic agent should be more easily pursued.

Chart 17

27

Actinomycin D

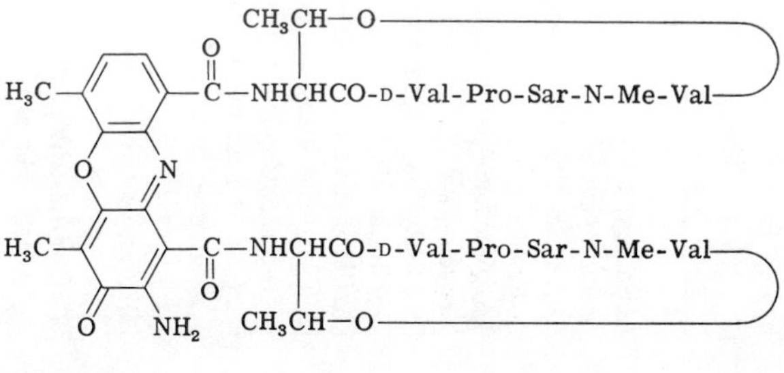

Table 26 Chromopeptides

Compound	Method	Solvent	% Yield	Chromatography: Col	Tl	Ion	Paper	CCD	Recrys. Sol.	M.P. °C	$[\alpha]_D$	Solvent Rotation	Elect.	Other	Ref.
Actinomycin C_1															343
Actinomycin C_1-carboxylic acid															343
4,6-Didemethyl-4,6-dibromoactinomycin C_1							×			255–257					552
4,6-Dimethyl-4,6-dibromoactinomycin C_1										195–198					552
[N-Methyl($Sar_{\alpha,\beta}$),8-heptadeutero]-actinomycin C_1									EtAc-Me-Cy	239–245 (dec)	−328	Me			343
[2,2(Sar_β)-dideutero]-actinomycin C_1															343
[2,2(Sar_β)-dideutero]-actinomycin C_1-carboxylic acid-(α)-lactone															343
[2,2($Sar_{\alpha,\beta}$),8-pentadeutero]-actinomycin C_1-carboxylic acid				Sephadex LH-20						182–190 (dec)	−109	Me			343
[2,2(Sar_α),8-trideutero]-actinomycin C_1									EtAc-Me-Cy	239–246 (dec)	−320	Me			343
[2,2(Sar_α),8-trideutero]-actinomycin C_1-carboxylic acid-(β)-lactone															343
[Actinocin-(α)-yl-(β)-OMe]-Thr-OMe										255–258					344
[Actinocin-(β)-yl-(α)-OMe]-Thr-OMe									EtAc-Pe	265–267 (dec)					344
[2-Desamino-2-hydroxy-actinocin-(α)-yl-(β)-OMe]-Thr-OMe					20				Chf-EtAc	236–238 (dec)					344

[2-Desamino-2-hydroxy-actinocinyl]-bis-Thr-OMe					21				Chf-Me-Bz	207–212 (dec)					344
[2-Desamino-2-hydroxy-8-deutero-actinocin-(β)-yl-(α)-OMe]-Thr-OMe									Chf-Me-EtAc	239–242 (dec)					344
[8-Deutero-actinocin-(β)-yl-(α)-OMe]-Thr-OMe										262–264					344
[2(Thr_{β})-deutero]-actinomycin C_1															343
[2(Thr_{α}),8-dideutero]-actinomycin C_1									EtAc-Me-Cy	241–246 (dec)	−314	Me			343
[2($Thr_{\alpha,\beta}$),8-trideutero]-actinomycin C_1										241–245 (dec)	−319	Me			343
[2(Val_{α},Pro_{α}),N-methyl($MeVal_{\alpha}$),8-hexadeutero]-actinomycin C_1										239–245 (dec)	−331	Me			343
[2(Val_{β},Pro_{β}),N-methyl($MeVal_{\beta}$)-pentadeutero]-actinomycin C_1															343
[2($Val_{\alpha,\beta}$,$Pro_{\alpha,\beta}$),N-methyl($MeVal_{\alpha,\beta}$),8-undecadeutero]-actinomycin C_1							×			239–244	−315	Me			343
[2(Val_{β}),N-methyl(Sar_{β},-$MeVal_{\beta}$)-heptadeutero]-actinomycin C_1															343
[2(Val_{β}),N-methyl(Sar_{β},$MeVal_{\beta}$)-heptadeutero]-actinomycin C_1 carboxylic acid-(β)-lactone										207–215 (dec)	−195	Me			343
[2(Val_{α}),N-methyl(Sar_{α},-$MeVal_{\alpha}$),8-octadeutero]-actinomycin C_1									EtAc-Me-Cy	241–245 (dec)	−328	Me			343
[2(Val_{α}),N-methyl(Sar_{α},$MeVal_{\alpha}$),8-octadeutero]-actinomycin C_1-carboxylic acid-(α)-lactone										204–213 (dec)	−131	Me			343
[2($Val_{\alpha,\beta}$),N-methyl($Sar_{\alpha,\beta}$,$MeVal_{\alpha,\beta}$),8-pentadecadeutero]-actinomycin C_1									Me-Cy	239–246 (dec)	−321	Me			343

(*continued*)

Table 26 Chromopeptides (continued)

Compound	Method	Solvent	% Yield	Chromatography Col	Tl	Ion	Paper	CCD	Recrys. Sol.	M.P. °C	$[\alpha]_D$	Solvent Rotation	Elect.	Other	Ref.
Actinocinyl-(α)-[Thr-D-allo-Ile-OMe]-(β)-[Thr-D-Val-OMe]															344
8-Deutero-actinocinyl-(α)-[Thr-D-Val-OMe]-(β)-[Thr-D-allo-Ile-OMe]										110–114					344
2-Amino-4,6-dimethyl-phenoxazinone(3)-1,9-bis[carbonyl-α,β-Dbu-D-Val-Pro-Sar-MeVal(Nβ-diamino-butyric acid) lactam]										260–267	−206	Me			27
8-Deutero-actinocinyl-bis-[O-Ac-Thr-D-Val-Pro-Sar-N-MeVal-OMe]				Seph-adex LH-20					Chf-Pe	140–146	−136 ±3	Me			344
Pteroyl-hepta-Glu															184
N-[2-Nitro-3-OBzl-4-methyl-6-deutero-Bz]-cyclo-[Thr-D-Val-Pro-N-trideuteromethyl-Gly-N-MeVal-O_{Thr}										157–163	−13	Me			343
N-[2-Nitro-3-OBzl-4-methyl-6-deutero-Bz]-cyclo-[Thr-2-deutero-D-Val-Pro-N-trideuteromethyl-Gly-N-trideuteromethyl-Val-O_{Thr}]											−13	Me			343
[Actinocin-(α)-yl-(β)-OMe]-[cyclo-(Thr-D-Val-Pro-Sar-N-MeVal-O_{Thr}]										218–223					344

[Deutero-actinocin-(β)-yl-(α)-OMe]-[cyclo-(Thr-D-Val-Pro-Sar-N-MeVal-O_{Thr})]															344
[8-Deutero-actinocin-(β)-yl-(α)-OMe]-[cyclo-(Thr-D-Val-Pro-Sar-N-MeVal-O_{Thr})]										198–203					344
N-[2-Nitro-3-OBzl-4-methyl-6-deutero-Bz]-cyclo-[2-deutero-Thr-D-Val-Pro-Sar-N-MeVal-O_{Thr}]										160–166	−14	Me			343
N-[2-Nitro-3-OBzl-4-methyl-6-deutero-Bz]-cyclo-[Thr-2-deutero-D-Val-2-deutero-Pro-Sar-N-trideutero-MeVal-O_{Thr}]										158–164	−14	Me		NMR	343
N-[2-Nitro-3-OBzl-4-methyl-6-deutero-Bz]-cyclo-[Thr-D-Val-Pro-2,2-dideutero-Sar-N-MeVal-O_{Thr}]											−14	Me			343
Actinocinyl-(α)-[cyclo-(Thr-D-Val-Pro-Sar-N-MeVal-O_{Thr})]-(β)-[O-Ac-Thr-D-Val-Pro-Sar-N-MeVal-OMe]															344

Chapter 20

Nucleopeptides

Current biochemical literature is replete with good reasons for undertaking the synthesis of various nucleopeptides. However, there still seems to be a dearth of activity in this field which is so pertinent to some very fundamental aspects of cell chemistry. Synthetic studies in 1971 ranged from an interesting approach to pyrimidinyl- and purinyl-peptides by the Pandit group[330] to analogs of the new polyoxin (Chart 18) nucleopeptide antibiotics.[253] One of the natural products, polyoxin M, (Chart 18) has been found to be a powerful inhibitor of yeast chitin synthetase, but several analogs prepared without the 5′-carboxyl and/or the 4″-hydroxyl were inactive in terms of antibiotic activity.

Recent (after July 1972) contributions to this area were directed at analogs of N-[Δ-(6-purinylthio)-Val]-Gly-Gly-Gly-OEt which had previously been observed to have some antineoplastic activity,[87] and at preparation of 2′- or 3′-O-aminoacyl nucleoside and nucleotide derivatives for use in exploring the mechanism of the transpeptidation reaction in ribosomes.[98] In the first instance, extending the tripeptide chain by one unit did not improve the antineoplastic activity.[87] In the latter program, coupling reactions employing 5-chloroquinolin-8-yl active esters OQ(Cl) were found particularly useful. The OQ(Cl) ester coupling reactions are quite selective with respect to not acylating the cytidine amino group or attacking phosphate diester linkages. Other recent advances, particularly those related to nucleopeptides derived from nucleotides have been reviewed in "Synthetic Nucleotides," Volume 1.[486]

Chart 18

253

Polyoxin M

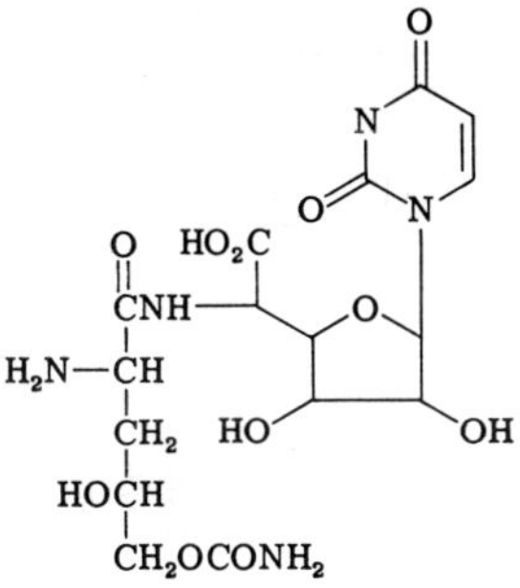

Table 27 Nucleopeptides

Compound	Method	Solvent	% Yield	Chromatography				CCD	Recrys. Sol.	M.P. °C	$[\alpha]_D$	Solvent Rotation	Elect.	Other	Ref.
				Col	Tl	Ion	Paper								
Cytidine															
Ala-(pC)															189
DL-β-(5-C)-α-Ala										360 (dec)					568
DL-β-(5-C)-α-Ala-OBu^t									DMF-DMSO	320 (dec)					568
DL-β-(5-C)-α-Ala-OBu^t perchlorate monohydrate									Bu-Aq	216–219 (dec)					568
N-Z-DL-β-(5-C)-α-Ala									DMF	316					568
N-Z-DL-β-(5-C)-α-Ala									Me	315 (dec)					568
Boc-Ala-(pC)															189
Di-Boc-Lys-(pC)															189
Lys-(pC)															189
Boc-Phe-(pC)															189
Phe-(pC)															189
N-Z-DL-β-(5-C)-α-Ala-Gly-OEt									Me	254–258 (dec)					568
N-Z-Gly-DL-β-(5-C)-α-Ala-OBu^t							×		DMSO-MEK	320 (dec)					568
Cytosine															
1-[4-Deoxy-4-(D-Ala)-amino-β-D-glucopyranosyl]cytosine									Hot Alc-Alc	233–240 (dec)	+22	Aq		NMR	658

1-[4-Deoxy-4-(N-Z-D-Ala)amino-β-D-glucopyranosyl]cytosine	ONP	DMSO				DO-1			Aq	190–194 (dec)	+24	Alc		UV NMR	658
1-[4-Deoxy-4-(N-Z-D-Phe)amino-β-D-glucopyranosyl]cytosine	ONP	DMSO							Me	>300	+20	Alc		NMR	658
1-[4-Deoxy-4-(D-Phe)amino-β-D-glucopyranosyl]cytosine										263–264	+13	Aq		NMR	658
1-[4-Deoxy-4-(N-Z-D-Ser)amino-β-D-glucopyranosyl]cytosine	DCCI	Acn				DO-1				166–170	+27	Alc-Aq		NMR	658
1-[4-Deoxy-4-(D-Ser)amino-β-D-glucopyranosyl]cytosine										271–274 (dec)	+29	Aq		NMR	658
1-[4-Deoxy-4-(N-Z-Sar-D-Ala)amino-β-D-glucopyranosyl]cytosine	ONP	DMSO				DO-1			Alc	210–212	+35	Alc		NMR	658
1-[4-Deoxy-4-(Sar-D-Ala)amino-β-D-glucopyranosyl]cytosine									Me-Alc	93–95 (dec)	+33	Aq		NMR UV	658
1-[4-Deoxy-4(N-Z-Sar-D-Phe)-amino-β-D-glucopyranosyl]cytosine										243–247 (dec)	+13	Alc		NMR	658
1-[4-Deoxy-4-(Sar-D-Phe)amino-β-D-glucopyranosyl]cytosine										221–231 (dec)	+19	Aq		NMR	658
1-[4-Deoxy-4-(N-Z-Sar-D-Ser)amino-β-D-glucopyranosyl]cytosine						DO-1 A-IRC-50				163–166 236–239 (dec)	+35	Alc-Aq		NMR	658
1-[4-Deoxy-4-(Sar-D-Ser)amino-β-D-glucopyranosyl]cytosine										136 210–250 (dec)	+44	Aq		NMR	658
1-[4-Deoxy-4-(Sar-DL-Ser)amino-β-D-glucopyranosyl]cytosine										172–189 (dec)	+11	Aq			658
Thymidine															
Ala-(dpT)															189
Boc-Ala-(dpT)															189
Uracil and uridine															
5′-Ala-amido-5′-d-U					80						+8.5	Aq		UV	253

(*continued*)

Table 27 Nucleopeptides (continued)

Compound	Method	Solvent	% Yield	Chromatography Col	Tl	Ion	Paper	CCD	Recrys. Sol.	M.P. °C	$[\alpha]_D$	Solvent Rotation	Elect.	Other	Ref.
Ala-(pU)															189
Ala-(pppU)															189
5′-N-Z-Ala-amido-5′-d-U				Si					Aq-Alc	187–190					253
Boc-Ala-(pU)															189
U(5′ → N)Ala-OEt															585
Uridylyl(5′ → N)Ala-OEt															585
Uridylyl(5′ → N)Ala-hydroxyethyl ester															585
U(5′ → N)Ala Thp-hydroxyethyl ester						DO-50								UV	585
pU-Ala-OEt													×		585
5-Arg-amido-5′-d-U					80						+14	Aq		UV IR	253
5′-N^G-Nitro-N$^\alpha$-Z-Arg-amido-5′-d-U	MCA	Di		Si						135–140					253
5-N-(Asn)-5-amino-U									Alc-Aq	135–137	+9	1 *N* HCl			260
5-N-(Asp)-5-amino-U						A-IR-50				203–206	−8.3	1 *N* HCl			260
5-N-(N-Z-β-OMe-Asp)-2′,3′,5-Tri-O-Ac-amino-U	DCCI	EtAc		Si						67–69	−62	DMSO			260
5′-N-Z-Cit-amido-5′-d-U	MCA	Di		Si						163–171				UV	253
5′-Cit-amido-5′-d-U					80						+4.5	Aq		UV IR	253

1-(3-Z-Glu-amido-γ-OMe 3-d-2′,4′,6′-Tri-O-Ac-β-D-glucopyranosyl)-uracil	MCA	Chf			25										154
1-(3-N-Z-Glu-amido-γ-OMe 3-d-β-D-glucopyranosyl)uracil					25	DO-50									154
1-(3-N-Glu-amido-γ-OMe 3-d-β-D-glucopyranosyl)uracil					132										154
5′-N-Z-Gln-amido-5′-d-U	ONP	DMF		Si					Aq-Alc	191–192	+3.9	DMF		NMR	253
5′-Gln-amido-5′-d-U					80						+26	Aq	UV NMR		253
1-(3-N,N′-Di-Z-Lys-amido-3-d-Di-O-Ac-β-D-glucopyranosyl)uracil					25					110–125					154
5′-N^{α},N^{ϵ}-Di-Z-Lys-amido-5′-d-U	ONP	DMF		Si					Alc-Aq	151–153	+3.6	DMF		NMR	253
5-N-(N^{ϵ},N^{α}-Di-Z-Lys)-2′,3′,5′-Tri-O-Ac-5-amino-U	DCCI	EtAc									−42	DMSO			260
1-(3-Lys-amido-3-d-Di-O-Ac-β-D-glucopyranosyl)uracil 2 HCl										196–203					154
5′-Lys-amido-5′-d-U					80						+18	Aq		UV IR	253
5-N-(Lys)-5-amino-U						DO-50				138–140	−59	Aq			260
5-N-(N-Z-Phe)-2′,3′,5′-Tri-O-Ac-5-amino-U	DCCI	EtAc							Eth-Hep	71–73	+35	DMSO			260
5-N-(Phe)-2′,3′,5′-Tri-O-Ac-5-aminouridine HBr										112–114					260
5-Hydroxy-U(5′ → N)-Phe-OMe															570
5-Hydroxy-U-O^{5}-phospho(P → N)-Phe-OMe															570
d-Uridylyl(5 → N)Phe															587
5′-N-Z-δ-carbamoyloxy-α-amino-Val-5′-d-U				Si	9 80									NMR IR	253

(*continued*)

Table 27 Nucleopeptides (continued)

Compound	Method	Solvent	% Yield	Chromatography Col	Tl	Ion	Paper	CCD	Recrys. Sol.	M.P. °C	$[\alpha]_D$	Solvent Rotation	Elect.	Other	Ref.
5′-δ-Carbamoyloxy-α-amino-Val-amido-5′-d-U					80						+14	Aq		UV	253
5′-N-Z-δ-hydroxy-α-amino-Val-amido-5′-d-2′,3′-O-isopropylidene-U				Si	9									UV NMR	253
5′-N-Z-δ-hydroxy-α-amino-Val-amido-5′-d-U				Si						170–174					253
5′-δ-Hydroxy-α-amino-Val-amido-5′-d-U					80						+29	Aq		IR	253
Pro-U-^{14}C															435
5-N-(Asn-Phe)-5-amino-U						DO-50				122–123	+4	HCl			260
5-N-[N-Z(β-O-methyl-Asp)-Phe]-2′,3′,5′-Tri-O-Ac-5-amino-U	DCCI	MeCl													260
5-N-[β-O-Methyl)Asp-Phe]-2′,3′,5′-Tri-O-Ac-5-amino-U HBr										118–120	+15	DMSO			260
5-N-(Orn-Leu)5-amino-U	DCCI-NHS	DMF		×						81–83	+17	HCl			260
Uridylyl(5′ → N)Phe-Gly															587
Uridylyl-(5′ → N)-Ser-(Thr)-Gly															728
5′-Ac-amido-5′-d-U						DO-50W-×8			Alc-Aq	235–236 (dec)	+23	Aq			253

Adenosine															
Ala-(pA)															189
Ala-(pppA)															189
Boc-Ala-(pA)															189
Asp-(pA)															189
β-OBu^t-Asp-(pA)															189
Boc-β-OBu^t-Asp-(pA)															189
5′-N-(N-Z-Gly)-5′-amino-2′,5′-Di-d-A										97–104				UV	521
5′-N-Gly-5′-amino-2′-5′-Di-d-A										192–196	−36	DMF		UV	521
5′-N-(N-Z-Leu)-5′-amino-2′,5′-Di-d-A				Si						205–208				UV	521
5′-N-(Leu)-5′-amino-2′,5′-d-A										98–103	−33	DMF		UV	521
N$^\epsilon$-Z-N-[1-(9-Adenyl)-2,3-O-isopropylidene-β-D-ribofuranuronosyl]-Lys									Alc-Aq	224–226 (dec)	−16	DMSO			310
N$^\epsilon$-Z-N$^\alpha$[1-(9-Adenyl)-2,3-O-isopropylidene-β-D-ribofuranuronosyl]-Lys-OMe											−9.0	Chf			310
N$^\alpha$-[1-(9-Adenyl)-β-D-ribofuranuronosyl]-Lys									Alc-Aq		−19	Aq			310
N$^\epsilon$-Z-N$^\alpha$-[1-(9-Adenyl)-β-D-ribofuranuronosyl]-Lys									DMF-Aq	244–246 (dec)	−24	DMSO			310
5′-N-(N$_\alpha$,N$_\epsilon$-Z-Lys)-5′-amino-5′-D-A									Me-*i*-Pr						521
Di-Boc-Lys-(dpA)															189
Di-Boc-Lys-(pA)															189
Lys-(dpA)															189
Lys-(pA)															189
5′-(N-Lys)-5′-amino-5′-d-A									Me-*i*-Pr						521
d-(Adenylyl(5′ → N)Phe															587
5′-N-(N-Z-Phe)-5′-amino-5′-d-A										187–189				UV	521
Boc-Phe-(pA)															189

(*continued*)

Table 27 Nucleopeptides (continued)

Compound	Method	Solvent	% Yield	Chromatography				CCD	Recrys. Sol.	M.P. °C	$[\alpha]_D$	Solvent Rotation	Elect.	Other	Ref.
				Col	Tl	Ion	Paper								
Phe-(pA)															189
5′-N-(N-Z-Phe)-5′-amino-2′,5′-Di-d-A				Si						113–117				UV	521
5′-N-(Phe)-5′-amino-2′,5′-Di-d-A										114–125	−37	DMF		UV	521
5′-N-(Phe)-5′-amino-5′-d-A									Me	208–209	−55	DMF-Aq		UV	521
Val-(pA)															189
5′-N-(N-Z-Gly-Gly)-5′-amino-2′,5′-Di-d-A										145–146				UV	521
5′-N-Gly-Gly-5′-amino-2′,5′-Di-d-A										154–157	−41	DMF		UV	521
N-[1-(9-Adenyl)-2,3-O-isopropylidene-β-D-ribofuranuronosyl]-Gly-Phe-OBzl					103						−16	Chf			310
N-[1-(9-Adenyl)-β-D-ribofuranuronosyl]-Gly-Phe									Alc-Aq	242–244 (dec)	−31	DMSO			310
N-[1-(9 Adenyl)-β-D-ribofuranuronosyl]-Gly-Phe-OBzl					102				Alc	138–140	−37	DMSO			310
5′-N-(Lys-Gly-Gly)-5′-amino-5′-d-A 2 HCl										151–153					521
Guanosine															
Ala-(pG)															189
Boc-Ala-(pG)															189
5′-N-(N-Z-Leu)-5′-amino-5′-d-G									Me-EtAc	244–245				UV	521

5′-N-(N-Z-Leu-5′-amino-2′,3′-O-isopropylidene-5′-d-G										145–155				UV	521
5′-N-(Leu)-5′-amino-5′-d-G										236–237	−7.0	Aq		UV	521
5′-N-(N-Z-Phe)-5′-amino-5′-d-G										156–157				UV	521
5′-N-(N-Z-Phe)-5′-amino-2′,3′-O-isopropylidene-5′-d-G									*i*-Pr-An	255–257 (dec)				UV	521
5′-N-(Phe)-5′-amino-5′-d-G									Alc-Aq	236–238	−11	Aq		UV	521
Guanylyl(5′ → N)Phe-Gly															587
Guanylyl(5′ → N)Phe-Val															587
Guanylyl(5′ → N)Phe-Val-Ala															587
Miscellaneous phosphates, pyramidines, and purines															
P^1,P^2-Di(N-Z-aminoacetyl-Ser)P^1,P^2-Di-Bzl-pyro-phosphate															30
P^1,P^2-Di(N-Z-Ser)P^1,P^2-Di-Bzl Pyrophosphate															30
5-Amino-4-chloro-6-[3-(homo-Cys)-*n*-propyl]-aminopyrimidine					83					Oil					105
[2-Z-Amino-6-[2,4-dioxo-1,2,3,4-tetrahydro-1-pyrimidinyl)-caproyl]Gly										202–204				UV NMR	330
Ethyl[2-Z-amino-6(2,4-dioxo-1,2,3,4-tetrahydro-1-pyrimidinyl)-caproyl]Gly														UV NMR	330
Ethyl[2-Z-amino-6-(5-methyl)-2,4-dioxo-1,2,3,4-tetrahydro-1-pyrimidinyl)caproyl]Gly														UV NMR	330
α-(2-Iminohexahydropyrimid-4-yl)Gly										166–169 not sharp					82

(*continued*)

Table 27 Nucleopeptides (continued)

Compound	Method	Solvent	% Yield	Chromatography				CCD	Recrys. Sol.	M.P. °C	$[\alpha]_D$	Solvent Rotation	Elect.	Other	Ref.
				Col	Tl	Ion	Paper								
9-{R-[3R-(Z-*p*-OMe-phenyl-Ala-amino)-2R-hydroxyl]-cyclopentyl}-6-Dimethylaminopurine									Me	170–172 200–205					116
6-Dimethylamino-9-{R-[2R-hydroxy-3R-(*p*-OMe-phenyl-Ala-amino)]cyclopentyl}purine									Alc	161–161.5					116
6-Dimethylamino-9-{S-[2S-hydroxy-3S-(*p*-OMe-phenyl-Ala-amino)]cyclopentyl}purine															116
9-{S-[3S-(Z-*p*-OMe-phenyl-Ala-amino)-2S-hydroxy]cyclopentyl}-6-dimethylaminopurine															116
N-[δ-(6-Purinylthio)valeryl]-Asp									Aq	197–199					88
N-[δ-(6-Purinylthio)valeryl]Gly-OEt									Aq	154–156					88
N-[δ-(6-Purinylthio)valeryl]-Glu									Aq	205–206					88
N-[δ-(6-Purinylthio)valeryl]-Glu-Di-OEt									An-He	100–101					88
N-[δ-(6-Purinylthio)valeryl]-DL-Val									Aq	103–105					88
N-[δ-(6-Purinylthio)valeryl]-DL-Val-OEt									An	116–118					88
N-[δ-(6-Purinylthio)valeryl]Gly-Gly-OEt									Aq-Alc	227–229					88
N-[δ-(6-Purinylthio)valeryl]Gly-Gly-Gly-OEt									Aq-Alc	236–239					88

Val-tRNA															734
Gly_n-Val-tRNA															734
Lys-Gly-Val-tRNA															734
Phe-Gly-Val-tRNA															734
Val-Gly-Val-tRNA															734

Appendix

The journals reviewed between October 1970 and the first part of January 1973 which contained papers describing peptide syntheses have been included in the following list. A frequency of appearance survey and relative importance of synthetic peptide papers in nearly 500 journals has recently been contributed by Jones.[281] Of the 73 journals containing such papers only about one-third contained about 95% of the most important advances. This reviewer's experience to January 1973 has been similar and those journals found most useful have been marked here with an asterisk.

Acta Chem. Scand.
Agr. Biol. Chem.
Angew. Chem. Int. Ed. Engl.
Agnew. Makromol. Chem.
Ann. N. Y. Acad. Sci.
Arch. Biochem. Biophys.
**Aust. J. Chem.*
Biochem. Biophys. Res. Commun.
**Biochemistry*
Biochem. J.
**Bioorg. Chem.*
Biopolymers
**Bull. Chem. Soc. Jap.*
Bull. Soc. Chim. Fr.
Can. J. Chem.
Carbohyd. Res.
**Chem. Ber.*
Chem. Commun.
Chem. Heterocycl. Compounds
Chem. Ind. (London)
Chem. Natur. Compounds
**Chem. Pharm. Bull.*
**Collect. Czech. Chem. Commun.*
C. R. Acad. Sci.
Endocrinology
Eur. J. Pharmacol.
**Experientia*
FEBS Lett.
Gazz. Chim. Ital.
**Helv. Chim. Acta*
Hoppe-Seyler's Z. Physiol. Chem.
**Int. J. Peptide Protein Res.*
Isr. J. Chem.
Ital. J. Biochem.
**J. Amer. Chem. Soc.*
J. Antibiot.
J. Biochem. (Tokyo)
J. Biol. Chem.
**J. Chem. Soc., C*
J. Chromatogr.
J. Chromatogr. Sci.
**J. Gen. Chem. USSR*
J. Heterocycl. Chem.
J. Indian Chem. Soc.
**J. Med. Chem.*
**J. Org. Chem.*
J. Org. Chem. USSR
J. Pharm. Sci.
J. Pharmacol. Exp. Ther.
J. Prakt. Chem.
**Justus Liebigs Ann. Chem.*
Macromolecules
Makromol. Chem.
Mol. Pharmacol.
Monatsh. Chem.
Nature (London)
Nature (London), New Biol.
Naturwissenschaften
Ohio J. Sci.
Org. Mass Spectrom.
Org. Prep. Proced. Int.
Proc. Nat. Acad. Sci. U. S.
Proc. Soc. Exp. Biol. Med.
Rec. Chem. Progr.
**Rec. Trav. Chim. Pays-Bas*
Rocz. Chem.
Science
Syn. Commun.
Synthesis
Tetrahedron
**Tetrahedron Lett.*
Z. Chem.

Bibliography

1. J. L. Abernethy, E. Albano, and J. Comyns, *J. Org. Chem.* **36**, 1580 (1971).
2. C. J. Abshire and G. Planet, *J. Med. Chem.* **15**, 226 (1972).
3. G. A. Agafonova, N. E. Gerasimova, M. V. Guseva, B. L. Krainova, T. V. Petrova, V. F. Pozdnev, and E. S. Chaman, *J. Gen. Chem. USSR* **40**, 2488 (1970).
4. L. V. Alekseeva, N. L. Burde, and G. V. Bilim, *J. Org. Chem. USSR* **7**, 653 (1971).
5. L. V. Alekseeva, N. L. Burde, and Z. V. Pushkareva, *J. Org. Chem. USSR* **7**, 656 (1971).
6. L. V. Alekseeva, N. L. Burde, and G. P. Tatarinova, *J. Org. Chem. USSR* **7**, 1442 (1971).
7. B. Aleksiev, P. Niranjan, S. Stoev, and V. Doseva, *Hoppe-Seyler's Z. Physiol. Chem.* **352**, 1411 (1971).
8. A. Ali, F. Fahrenholz, and B, Weinstein, *Angew. Chem., Int. Ed. Engl.* **11**, p. 289 (1972).
9. A. Ali, P. M. Hardy, and H. N. Rydon, *J. Chem. Soc. Perkin Trans.* **1**, 1070 (1972).
10. A. Ali and B. Weinstein, *J. Org. Chem.* **36**, 3022 (1971).
11. A. F. Al-Sayyab, A. T. Atto, and F. Y. Sarah, *J. Chem. Soc., C* p. 3260 (1971).
12. N. Anand, N. S. R. K. Murthy, F. Naider, and M. Goodman, *Macromolecules* **4**, 564 (1971).
13. V. S. Ananthanarayanan, R. H. Andreatta, D. Poland, and H. A. Scheraga, *Macromolecules* **4**, 417 (1971).
14. G. W. Anderson and F. M. Callahan, *J. Amer. Chem. Soc.* **80**, 2902 (1958).
15. G. W. Anderson, F. M. Callahan, and J. E. Zimmerman, *J. Amer. Chem. Soc.* **89**, 178 (1967).
16. G. W. Anderson, J. E. Zimmerman, and F. M. Callahan, *J. Amer. Chem. Soc.* **89**, 5012 (1967).
17. R. H. Andreatta and H. A. Scheraga, *J. Med. Chem.* **14**, 489 (1971).
18. K. Anzai, *Bull. Chem. Soc. Jap.* **43**, 2539 (1970).
19. Y. Ariyoshi and N. Sato, *Bull. Chem. Soc. Jap.* **45**, 942 (1972).
20. Y. Ariyoshi and N. Sato, *Bull. Chem. Soc. Jap.* **44**, 3435 (1971).
21. Y. Ariyoshi and N. Sato, *Bull. Chem. Soc. Jap.* **44**, 2787 (1971).
22. H. Arold, *Justus Liebigs Ann. Chem.* **739**, 194 (1970).
23. K. I. Arsenin and I. A. Sheka, *J. Gen. Chem. USSR* **41**, 195 (1971).
24. R. S. Asquith, M. S. Otterburn, and K. L. Gardner, *Experientia* **27**, 1388 (1971).
25. Z. A. Atare, V. S. Vilka, and G. I. Chipens, *Chem. Natur. Compounds* **5**, 241 (1969).

26. E. Atherton, H. D. Law, S. Moore, D. F. Elliott, and R Wade, *J. Chem. Soc., C* p. 3393 (1971).
27. E. Atherton and J. Meienhofer, *J. Amer. Chem. Soc.* **94,** 4759 (1972).
28. E. Atherton and J. Meienhofer, *J. Antibiot.* **25,** 539 (1972).
29. S. M. Avaeva, S. N. Kara-Murza, N. V. Ras'kova, and M. M. Botvinik, *Chem. Natur. Compounds* **3,** 277 (1967).
30. S. M. Avaeva, I. F. Syromyatnikova, V. M. Grischenko, and M. M. Botvinik *Chem. Natur. Compounds* **3,** 105 (1967).
31. L. Baláspiri, B. Penke, J. Petres, and K. Kovács, *Monatsh. Chem.* **101,** 1177 (1970).
32. G. C. Barrett, *J. Chem. Soc., C* p. 1380 (1971).
33. N. M. Bath and R. I. Gregerman, *Biochemistry* **11,** 2845 (1972).
34. K. Bauer, R. Roskoski, Jr., H. Kleinkauf, and F. Lipmann, *Biochemistry* **11,** 3266 (1972).
35. E. Bayer, E. Breitmaier, G. Jung, and W. Parr, *Hoppe-Seyler's Z. Physiol. Chem.* **352,** 759 (1971).
36. E. Bayer, P. Hunziker, M. Mutter, R. E. Sievers, and R. Uhmann, *J. Amer. Chem. Soc.* **94,** 265 (1972).
37. E. Bayer, G. Jung, I. Halász, and I. Sebestian, *Tetrahedron Lett.* p. 4505 (1970).
38. E. Bayer and M. Mutter, *Nature (London)* **237,** 512 (1972).
39. J. Beacham, G. Dupuis, F. M. Finn, H. T. Storey, C. Yanaihara, N. Yanaihara, and K. Hofmann, *J. Amer. Chem. Soc.* **93,** 5526 (1971).
40. R. L. Beamer and W. D. Brown, *J. Pharm. Sci.* **60,** 583 (1971).
41. V. M. Belikov, S. D. Sokolov, N. M. Naidenova, and M. M. Dolgaya, *Chem. Natur. Compounds* **3,** 34 (1967).
42. C. D. Bello, F. Filira, and F. D'Angeli, *J. Org. Chem.* **36,** 1818 (1971).
43. C. Di Bello, F. Filira, V. Giormani, and F. D'Angeli, *Gazz. Chim. Ital.* **102,** 180 (1972).
44. L. Bernardi, G. Bertaccini G. Bosisio, R. Bucci, R. de Castiglione, V. Erspamer, O. Goffredo, and M. Impicciatore, *Experientia* **28,** 7 (1972).
45. H. Berndt, H. Klostermeyer, and H. Zahn, *Justus Liebigs Ann. Chem.* **759,** 114 (1972).
46. C. Berse and P. Bessette, *Can. J. Chem.* **49,** 2610 (1971).
47. R. I. Bersimbaev, S. V. Argutinskaya, and R. I. Salganik, *Experientia* **27,** 1389 (1971).
48. Zh. D. Bespalova, *J. Gen. Chem. USSR* **41,** 205 (1971).
49. H. C. Beyerman, P. Kranenburg, and J. L. M. Syrier, *Rec. Trav. Chim. Pays-Bas* **90,** 791 (1971).
50. H. C. Beyerman, E. W. B. de Leer and W. van Vossen, *J. Chem. Soc. Chem. Commun.* p. 929 (1972).
51. H. C. Beyerman, L. Maat, and A. van Zon, *Rec. Trav. Chim. Pays-Bas* **91,** 246 (1972).
52. C. Birr and W. Lochinger, *Synthesis* 319 (1971).
53. K. Bláha, *Collect. Czech. Chem. Commun.* **37,** 1861 (1972).
54. K. Bláha, I. Frič, Z. Bezpalova, and O. Kaurov, *Collect. Czech. Chem. Commun.* **35,** 3557 (1970).
55. G. G. Blinova and E. G. Sochilin, *J. Gen. Chem. USSR* **40,** 2745 (1970).
56. J. F. Blount, H. C. Freeman, R. V. Holland, and G. H. W. Milburn, *J. Biol. Chem.* **245,** 5177 (1970).
57. G. Boccalon, A. S. Verdini, and G. Giacometti, *J. Amer. Chem. Soc.* **94,** 3639 (1972).
58. M. Bodanszky, N. Chaturvedi, D. Hudson, and M. Itoh, *J. Org. Chem.* **37,** 2303 (1972).
59. M. Bodanszky and L. E. Conklin, *Chem. Commun.* p. 773 (1967).
60. M. Bodanszky, Y. S. Klausner, and V. Mutt, *Bioorg. Chem.* **2,** 30 (1972).
61. M. Bodanszky and G. Lindeberg, *J. Med. Chem.* **14,** 1197 (1971).
62. M. Bodanszky and V. du Vigneaud, *J. Amer. Chem. Soc.* **81,** 5688 (1959).
63. G. Bohman-Lindgren and U. Ragnarsson, *Tetrahedron* **28,** 4631 (1972).
64. J. Bøler, J.-K. Chang, F. Enzmann, and K. Folkers, *J. Med. Chem.* **14,** 475 (1971).
65. R. F. Borch and S. R. Levitan, *J. Org. Chem.* **37,** 2347 (1972).
66. E. Bordignon, G. G. Aloisi, and A. Signor, *Gazz. Chim. Ital.* **100,** 802 (1970).

67. G. Borin, F. Marchiori, R. Rocchi, L. Moroder, and E. Scoffone, *Gazz. Chim. Ital.* **101**, 426 (1971).
68. G. Borin, C. Toniolo, L. Moroder, F. Marchiori, R. Rocchi, and E. Scoffone, *Int. J. Protein Res.* **4**, 37 (1972).
69. H. R. Bosshard, *Helv. Chim. Acta* **54**, 951 (1971).
70. C. Y. Bowers, A. V. Schally, A. Kastin, A. Arimura, D. S. Schalch, C. Gual, E. Castineda, and K. Folkers, *J. Med. Chem.* **14**, 477 (1971).
71. A. Brack and G. Spach, *Biopolymers* **11**, 563 (1972).
72. A. Brack and G. Spach, *Bull. Soc. Chim. Fr.* [6] p. 4485 (1971).
73. D. Brandenburg, *Hoppe-Seyler's Z. Physiol. Chem.* **352**, 8 (1971).
74. D. Brandenburg, H. G. Gattner, L. Herbertz, G. Krail, M. Weinert, and H. Zahn, *Biochem. J.* **125**, 51 (1971).
75. J. Brown and R. E. Williams, *Can. J. Chem.* **49**, 3765 (1971).
76. P. Brown and G. R. Pettit, *Org. Mass Spectrom.* **3**, 67 (1970).
77. M. Brugger, *Helv. Chim. Acta* **54**, 1261 (1971).
78. M. Brugger, P. Barthe, and P. A. Desaulles, *Experientia* **26**, 1050 (1970).
79. T. C. Bruice and D. M. McMahon, *Biochemistry* **11**, 1273 (1972).
80. K. Brunfeldt, T. Christensen, and P. Roepstorff, *FEBS Lett.* **25**, 184 (1972).
81. J. Burton, G. A. Fletcher, and G. T. Young *Chem. Commun.* p. 1057 (1971).
82. B. W. Bycroft, D. Cameron, and A. W. Johnson, *J. Chem. Soc., C* p. 3040 (1971).
83. R. Camble, G. Dupuis, K. Kawasaki, H. Romovacek, N. Yanaihara, and K. Hofmann, *J. Amer. Chem. Soc.* **94**, 2091 (1972).
84. C. V. Carpenter and F. C. Neuhaus, *Biochemistry* **11**, 2594 (1972).
85. J. F. Carson and L. E. Boggs, *J. Org. Chem.* **36**, 611 (1971).
86. M. E. Celis, S. Taleisnik, and R. Walter, *Biochem. Biophys. Res. Commun.* **45**, 564 (1971).
87. A. Černý, M. Semonský, and V. Jelínek, *Collect, Czech. Chem. Commun.* **37**, 2077 (1972).
88. A. Černý, M. Semonský, and V. Jelínek, *Collect. Czech. Chem. Commun.* **36**, 2248 (1971).
89. T. H. Chan and L. T. L. Wong, *J. Org. Chem.* **36**, 850 (1971).
90. J-K. Chang, H Sievertsson, C. Bogentoft, B. Currie, K. Folkers, and G. D. Daves, Jr., *J. Med. Chem.* **14**, 481 (1971).
91. J-K. Chang, H. Sievertsson, B. L. Currie, C. Bogentoft, K. Folkers, and C. Y. Bowers, *J. Med. Chem.* **15**, 623 (1972).
92. J-K. Chang, H. Sievertsson, B. Currie, K. Folkers, and C. Bowers, *J. Med. Chem.* **14**, 484 (1971).
93. V. P. Chernyshev, E. I. Filippovich, and R. P. Evstigneeva, *J. Gen. Chem. USSR* **41**, 1626 (1971).
94. V. P. Chernyshev, E. I. Filippovich, R. P. Evstigneeva, and N. A. Preobrazhenskii, *J. Gen. Chem. USSR* **40**, 1634 (1970).
95. F. Chillemi, A. Aiello, and A Pecile, *Nature (London), New Biol* **238**, 243 (1972).
96 F. Chillemi and A. Chersi, *Gazz. Chim. Ital.* **101**, 695 (1971).
97. F. Chillemi and A. Pecile, *Experientia* **27**, 385 (1971).
98. S. Chládek, *J. Org. Chem.* **37**, 2863 (1972).
99. F. C-H. Chou, R. K. Chawla, R. F. Kibler, and R. Shapira, *J. Amer. Chem. Soc.* **93**, 267 (1971).
100. N. M. Chung, R. Lohrmann, and L. E. Orgel, *Tetrahedron* **27**, 1205 (1971).
101. J. R. Coggins and N. L. Benoiton, *Can. J. Chem.* **49**, 1968 (1971).
102. G. Conio, E. Patrone, and F. Salaris, *Macromolecules* **4**, 283 (1971).
103. L. Corley, D. H. Sachs, and C. B. Anfinsen, *Biochem. Biophys. Res. Commun.* **47**, 1353 (1972).
104. A. Cosani, E. Peggion, M. Terbojevich, and M. Acampora, *Macromolecules* **4**, 390 (1971).
105. J. K. Coward and W. D. Sweet, *J. Med. Chem.* **15**, 381 (1972).
106. R. D. Cowell and J. H. Jones *Chem. Commun*, p. 1009 (1971).

107. R. D. Cowell and J. H. Jones, *J. Chem. Soc., Perkin Trans.* **1,** 2236 (1972).
108. R. D. Cowell and J. H. Jones, *J. Chem. Soc., Perkin Trans.* **1,** 1814 (1972).
109. R. D. Cowell and J. H. Jones, *J. Chem. Soc , Perkin Trans.* **1,** 1809 (1972)
110. R. D. Cowell and J. H. Jones, *J. Chem. Soc., C* p. 1082 (1971).
111. P. A Cranwell and R. D. Haworth *Tetrahedron* **27,** 1831 (1971).
112. F. Cuiban, *Tetrahedron Lett.* p. 2471 (1971).
113. M. Dabrowska, G. Kupryszewski, B. Lammek, and F. Muzalewski, *Rocz. Chem.* **46,** 1295 (1972).
114. H. Dahn and C. O'Murchu, *Helv. Chim. Acta* **53,** 1379 (1970).
115. J. Dale and K. Titlestad, *J. Chem. Soc., Chem. Commun.* p. 255 (1972).
116. S. Daluge and R. Vince, *J. Med. Chem.* **15,** 171 (1972).
117. T. P. Dang and H. B. Kagan, *Chem. Commun.* p. 481 (1971).
118. B. C. Das and R. D. Schmid, *FEBS Lett.* **25,** 253 (1972).
119. J. S. Davies, C. H. Hassall, and K. H. Hopkins, *Chem. Commun.* p. 1118 (1971).
120. G. A. Davis, *J. Amer. Chem. Soc.* **94,** 5089 (1972).
121. C. M. Deber, F. A. Bovey, J. P. Carver, and E. R. Blout, *J. Amer. Chem. Soc.* **92,** 6191 (1970).
122. C. M. Deber, D. A. Torchia, and E. R. Blout, *J. Amer. Chem. Soc.* **93,** 4893 (1971).
123. J. I. DeGraw, V. H. Brown, W A. Skinner, S. LeValley, and C. Mitoma, *J. Med. Chem.* **15,** 781 (1972).
124. V. G. Degtyar and L. A. Shchukina, *Chem. Natur. Compounds* **4,** 154 (1968).
125. R. Deslauriers, R. Walter, and I. C. P. Smith, *Biochem. Biophys. Res. Commun.* **48,** 854 (1972).
126. E. C. DeVore, S. L. Holt, R. O. Asplund, and A. W. Catalano, *Arch. Biochem. Biophys.* **146,** 658 (1971).
127. R. S. Dewey, E. F. Schoenewaldt, H. Joshua, W. J. Paleveda, Jr., H. Schwam, H. Barkemeyer, B. H. Arison, D. F. Veber, R. G. Strachan, J. Milkowski, R. G. Denkewalter, and R. Hirschmann, *J. Org. Chem.* **36,** 49 (1971).
128. Z. Deyl and J. Rosmus, *J. Chromatogr.* **69,** 129 (1972).
129. C. Djerassi and G. R. Pettit, *J. Amer. Chem. Soc.* **76,** 4470 (1954).
130. L. C. Dorman and L. D. Markley, *J. Med. Chem.* **14,** 5 (1971).
131. M. W. Draper, R. B. Merrifield, and M. A. Rizack, *J. Med. Chem.* **16,** 1326 (1973).
132. S. Dukler, M. Wilchek, and D. Lavie, *Tetrahedron* **27,** 607 (1971).
133. F. W. Dunn and J. M. Stewart, *J. Med. Chem.* **14,** 779 (1971).
134. A. S. Dutta and J. S. Morley, *J. Chem. Soc., C* p. 2896 (1971).
135. L. P. Egan, J. R. Vercellotti, and W. T. Lowry, *Carbohyd. Res.* **23,** 261 (1972).
136. D. F. Elliott, P. Moritz, and R. Wade, *J. Chem. Soc., Perkin Trans.* **1,** 1862 (1972).
137. F. Enzmann, J. Bøler, K. Folkers, C. Y. Bowers, and A. V. Schally, *J. Med. Chem.* **14,** 469 (1971).
138. F. Fahrenholz, H. Faulstich, and T. Wieland, *Justus Liebigs Ann. Chem.* **743,** 83 (1971).
139. R. Fairweather and J. H. Jones, *J. Chem. Soc., Perkin Trans.* **1,** 1908 (1972).
140. H. Faulstich and H. Trischmann, *Justus Liebigs Ann. Chem.* **741,** 55 (1970).
141. P. Fehrenbach and H. Zahn, *Chem. Ber.* **105,** 1749 (1972).
142. M. F. Ferger, W. C. Jones, Jr., D. F. Dyckes, and V. du Vigneaud, *J. Amer. Chem. Soc.* **94,** 982 (1972).
143. D. C. Fessler, F. Sipos, G. S. Denning, Jr., D. T. Pals, and F. D. Masucci, *J. Med. Chem.* **15,** 1015 (1972).
144. F. Filira, C. DiBello, A. C. Veronese, and F. D'Angeli, *J. Org. Chem.* **37,** 3265 (1972).
145. R. H. Fish and M. Friedman, *J. Chem. Soc., Chem. Commun.* p. 812 (1972).
146. G. A. Fletcher and G. T. Young, *J. Chem. Soc., Perkin Trans.* **1,** 1867 (1972).
147. G. Flouret, W. Cole, and U. Biermacher, *J. Med. Chem.* **15,** 1281 (1972).
148. G. Flouret and V. du Vigneaud, *J. Med. Chem.* **14,** 556 (1971).
149. J. M. Fréchet and C. Schuerch, *J. Amer. Chem. Soc.* **94,** 604 (1972).

150. J. M. Fréchet and C. Schuerch, *J. Amer. Chem. Soc.* **93,** 492 (1971).
151. R. J. Freer and J. M. Stewart, *J. Med. Chem.* **15,** 1 (1972).
152. M. Fridkin and H. J. Goren, *Can. J. Chem.* **49,** 1578 (1971).
153. M. Fridkin and S. Shaltiel, *Arch. Biochem. Biophys.* **147,** 767 (1971).
154. H. A. Friedman, *J. Med. Chem.* **14,** 174 (1971).
155. S.-C. J. Fu and D. S. H. Mak, *J. Chromatogr.* **54,** 205 (1971).
156. A. Fujii, K. Tanaka, Y. Tsuchiya, and E. S. Cook, *J. Med. Chem.* **14,** 354 (1971).
157. T. Fujii and S. Sakakibara, *Bull. Chem. Soc. Jap.* **43,** 3954 (1970).
158. Y. Fujimoto, F. Irreverre, J. M. Karle, I. L. Karle, and B. Witkop, *J. Amer. Chem. Soc.* **93,** 3471 (1971).
159. M. Fujino, S. Kobayashi, M. Obayashi, T. Fukuda, S. Shinagawa, I. Yamazaki, R. Nakayama, W. F. White, and R. H. Rippel, *Biochem. Biophys. Res. Commun.* **49,** 698 (1972).
160. M. Fujino, S. Kobayashi, M. Obayashi, S. Shinagawa, T. Fukuda, C. Kitada, R. Nakayama, I. Yamazaki, W. F. White, and R. H. Rippel, *Biochem. Biophys. Res. Commun.* **49,** 863 (1972).
161. M. Fujino and S. Shinagawa, *Chem. Pharm. Bull.* **20,** 1021 (1972).
162. M. Fujino, S. Shinagawa, O. Nishimura, and T. Fukuda, *Chem. Pharm. Bull.* **20,** 1017 (1972).
163. Y. Funae, N. Toshioka, I. Mita, T. Sugihara, T. Ogura, Y. Nakamura, and S. Kawaguchi, *Chem. Pharm. Bull.* **19,** 1618 (1971).
164. R. J. Gait, J. R. Langlois, and R. E. Williams, *Can. J. Chem.* **50,** 299 (1972).
165. J. Galsomias, C. Frezou, and P. Vieles, *C. R. Acad. Sci.* **274,** 1392 (1972).
166. C. H. Gaozza, B. C. Cieri, and S. Lamdan, *J. Heterocycl. Chem.* **8,** 1079 (1971).
167. H. G. Garg and R. W. Jeanloz, *Carbohyd. Res.* **23,** 437 (1972).
168. R. Garner and G. T. Young, *J. Chem. Soc., C* p. 50 (1971).
169. O. Gawron, A. Rampal, and P. Johnson, *J. Amer. Chem. Soc.* **94,** 5396 (1972).
170. R. Geiger, *Angew. Chem., Int. Ed. Engl.* **10,** 152 (1971).
171. R. Geiger, *Justus Liebigs Ann. Chem.* **750,** 165 (1971).
172. R. Geiger, W. König, H. Wissmann, K. Geisen, and F. Enzmann, *Biochem. Biophys. Res. Commun.* **45,** 767 (1971).
173. R. Geiger, H.-H. Schöne, and W. Pfaff, *Hoppe-Seyler's Z. Physiol. Chem.* **352,** 1487 (1971).
174. G. Giacometi, A. Turolla, and A. S. Verdini, snam, *Biopolymers* **11,** 215 (1972).
175. V. Giancotti, F. Quadrifoglio, and V. Crescenzi, *J. Amer. Chem. Soc.* **94,** 297 (1972).
176. D. Giles and D. W. Grant, *Chem. Ind.* (*London*) p. 1437 (1970).
177. D. Gillessen, F. Piva, H. Steiner, and R. O. Studer, *Helv. Chim. Acta* **54,** 1335 (1971).
178. B. F. Gisin and R. B. Merrifield, *J. Amer. Chem. Soc.* **94,** 6165 (1972).
179. J. D. Glass and V. du Vigneaud, *J. Med. Chem.* **15,** 486 (1972).
180. J. D. Glass, I. L. Schwartz, and R. Walter, *J. Amer. Chem. Soc.* **94,** 6209 (1972).
181. E. N. Glibin, Z. I. Korshunova, O. F. Ginzburg, and L. F. Larionov, *Chem. Heterocycl. Compounds* **5,** 445 (1972).
182. J. D. Glickson, D. W. Urry, R. T. Havran, and R. Walter, *Proc. Nat. Acad. Sci. U.S.* **69,** 2136 (1972).
183. J. D. Glickson, D. W. Urry, and R. Walter, *Proc. Nat. Acad. Sci. U.S.* **69,** 2566 (1972).
184. H. A. Godwin, I. H. Rosenberg, and C. R. Ferenz, *J. Biol. Chem.* **247,** 2266 (1972).
185. M. Goodman, L. B. Jacobsberg, and N. S. Choi, *Isr. J. Chem.* **9,** 125 (1971).
186. M. Goodman, R. Rupp and F. Naider, *Bioorg. Chem.* **1,** 294 (1971).
187. W. Gordon, R. T. Havran, I. L. Schwartz, and R. Walter, *Proc. Nat. Acad. Sci. U.S.* **60,** 1353 (1968).
188. V. A. Gorlenko, Ts. E. Agadzhanyan, A. A. Kiryushkin, and Yu. A. Ovchinnikov, *J. Gen. Chem. USSR* **40,** 1623 (1970).

189. B. P. Gottikh, A. A. Krayevsky, N. B. Tarussova, P. P. Purygin, and T. L. Tsilevich, *Tetrahedron* **26,** 4419 (1970).
190. J. S. de Graaf, A. C. A. Jansen, K. E. T. Kering, Cl. Schattenkerk, and E. Havinga, *Rec. Trav. Chim. Pays-Bas* **90,** 301 (1971).
191. O. Grahl-Nielsen, *Chem. Commun.* p. 1588 (1971).
192. O. Grahl-Nielsen and E. Solheim, *J. Chem. Soc. Chem. Commun.* p. 1092 (1972).
193. Z. Grzonka, *Rocz. Chem.* **46,** 1265 (1972).
194. Z. Grzonka and B. Liberek, *Tetrahedron* **27,** 1783 (1971).
195. Z. Grzonka, E. Rekowska, and B. Liberek, *Tetrahedron* **27,** 2317 (1971).
196. M. Guarneri, C. A. Benassi, R. Ferroni, A. Guggi, R. Tomatis, and R. Rocchi, *Gazz. Chim. Ital.* **101,** 375 (1971).
197. M. Guarneri, C. A. Benassi, R. Tomatis, A. Guggi, R. Ferroni, and R. Rocchi, *Gazz. Chim. Ital.* **101,** 363 (1971).
198. R. Guillemin, M. Amoss, R. Blackwell, J. Rivier, N. Ling, and W. Vale, *Biochem. Biophys. Res. Commun.* **48,** 1093 (1972).
199. S. L. Gurina, R. Kh. Batulina, L. V. Alekseeva, and Z. V. Pushkareva, *Chem. Heterocycl. Compounds* **4,** 321 (1968).
200. V. Gut and K. Poduška, *Collect. Czech. Chem. Commun.* **36,** 3470 (1971).
201. V. Gut, J. Rudinger, R. Walter, P. A. Herling, and I. L. Schwartz, *Tetrahedron* **24,** 6351 (1968).
202. B. Gutte, M. C. Lin, D. G. Caldi, and R. B. Merrifield, *J. Biol. Chem.* **247,** 4763 (1972).
203. B. Halpern, L. Chew, V. Close, and W. Patton, *Tetrahedron Lett.* p. 5163 (1968).
204. J. Halstrøm and K. Brunfeldt, *Hoppe-Seyler's Z. Physiol. Chem.* **353,** 1204 (1972).
205. T. Hamada and O. Yonemitsu, *Chem. Pharm. Bull.* **19,** 1444 (1971).
206. V. K. Hammerstrom, W. Lunkeheimer, and H. Zahn, *Makromol. Chem.* **133,** 41 (1970).
207. W. S. Hancock, D. J. Prescott, W. L. Nulty, J. Weintraub, P. R. Vagelos, and, G. R. Marshall, *J. Amer. Chem. Soc.* **93,** 1799 (1971).
208. B. O. Handford, T. A. Hylton, K.-T. Wang, and B. Weinstein, *J. Org. Chem.* **33,** 4251 (1968).
209. B. O. Handford, T. A. Hylton, and B. Weinstein, *Ohio J. Sci.* **68,** 104 (1968).
210. K. Harada and K. Matsumoto, *Bull. Chem. Soc. Jap.* **44,** 1068 (1971).
211. K. Harada and W. Tso, *Bull. Chem. Soc. Jap.* **45,** 2859 (1972).
212. P. M. Hardy, J. C. Haylock, D. I. Marlborough, N. H. Rydon, H. T. Storey, and R. C. Thompson, *Macromolecules* **4,** 435 (1971).
213. P. M. Hardy, J. C. Haylock, and H. N. Rydon, *J. Chem. Soc., Perkin Trans.* **1,** 605 (1972).
214. P. M. Hardy, B. Ridge, H. N. Rydon, and F. O. dos S. P. Serräo, *J. Chem. Soc., C* p. 1722 (1971).
215. P. M. Hardy H. N. Rydon, and H. T. Storey, *J. Chem. Soc., Perkin Trans.* **1,** 1523 (1972).
216. P. M. Hardy, H. N. Rydon, and R. C. Thompson, *J. Chem. Soc., C* p. 5 (1972).
217. D. N. Harpp and T. G. Back, *J. Org. Chem.* **36,** 3828 (1971).
218. D. N. Harpp and J. G. Gleason, *J. Org. Chem.* **36,** 73 (1971).
219. S. Hase S. Sakakibara, M. Wahrenburg M. Kirchberger, I. L. Schwartz, and R. Walter, *J. Amer. Chem. Soc.* **94,** 3590 (1972).
220. S. Hase, S. Sakakibara, M. Wahrenburg, and R. Walter, *J. Med. Chem.* **15,** 1017 (1972).
221. S. Hase, I. L. Schwartz, and R. Walter, *J. Med. Chem.* **15,** 126 (1972).
222. C. H. Hassall, R. B. Morton, Y. Ogihara, and D. A. S. Phillips, *J. Chem. Soc., C* p. 526 (1971).
223. C. H. Hassall, D. G. Sanger, and B. K. Handa, *J. Chem. Soc., C* p. 2814 (1971).
224. R. T. Havran, I. L. Schwartz, and R. Walter, *Mol. Pharmacol.* **5,** 83 (1969).
225. T. Hayakawa, H. Yamamoto, and N. Aoto, *Biopolymers* **11,** 185 (1972).

226. F. Heitz and G. Spach, *Macromolecules* **4,** 429 (1971).
227. L. H. Hellberg, M. J. Prodanovich, and F. Stults, *J. Heterocycl. Chem.* **9,** 401 (1972).
228. P. Hermann, I. Willhardt, K. Blaha, and I. Fric, *J. Prakt. Chem.* [4] **313,** 1092 (1971).
229. R. C. Hider and D. I. John, *J. Chem. Soc., Perkin Trans.* **1,** 1825 (1972).
230. M. Hiramoto, K. Okada, S. Nagai, and H. Kawamoto, *Chem. Pharm. Bull.* **19,** 1308 (1971).
231. M. Hirata, K. Noda, and N. Izumiya, *Bull. Chem. Soc. Jap.* **45,** 1790 (1972).
232. R. Hirschmann, H. Schwam, R. G. Strachan, E. F. Schoenewaldt, H. Barkemeyer, S. M. Miller, J. B. Conn, V. Garsky, D. F. Veber, and R. G. Denkewalter, *J. Amer. Chem. Soc.* **93,** 2746 (1971).
233. R. G. Hiskey, L. M. Beacham, III, and V. G. Matl, *J. Org. Chem.* **37,** 2472 (1972).
234. R. G. Hiskey, L. M. Beacham, III, V. G. Matl, J. N. Smith, E. B. Williams, Jr., A. M. Thomas, and E. T. Wolters, *J. Org. Chem.* **36,** 488 (1971).
235. R. G. Hiskey, G. W. Davis, M. E. Safdy, T. Inui, R. A. Upham, and W. C. Jones, Jr., *J. Org. Chem.* **35,** 4148 (1970).
236. R. G. Hiskey, E. T. Wolters, G. Ülkü, and V. Ranga Rao, *J. Org. Chem.* **37,** 2478 (1972).
237. B. T. Ho, L. W. Tansey, and J.-T. Huang, *J. Med. Chem.* **14,** 1132 (1971).
238. K. Hofmann, R. Andreatta, F. M. Finn, J. Montibeller, G. Porcelli, and A. J. Quattrone, *Bioorg. Chem.* **1,** 66 (1971).
239. K. Hofmann and C. Y. Bowers, *J. Med. Chem.* **13,** 1099 (1970).
240. M. Hooper and J. W. Robertson, *Tetrahedron Lett.* p. 2139 (1971).
241. A. Hope and B. Halpern, *Tetrahedron Lett.* p. 2261 (1972).
242. V. J. Hruby and W. Y. Chan, *J. Med. Chem.* **14,** 1050 (1971).
243. V. J. Hruby, M. F. Ferger, and V. du Vigneaud, *J. Amer. Chem. Soc.* **93,** 5539 (1971).
244. V. J. Hruby, C. W. Smith, D. K. Linn, M. F. Ferger, and V. du Vigneaud, *J. Amer. Chem. Soc.* **94,** 5478 (1972).
245. T. Ichikawa, S. Maeda, T. Okamoto, Y. Araki, and Y. Ishido, *Bull. Chem. Soc. Jap.* **44,** 2779 (1971).
246. R. Ife and E. Haslam, *J. Chem. Soc., C* p. 2818 (1971).
247. K. Ikeda, K. Kuromizu, K. Okamoto, and N. Izumiya, *Chem. Pharm. Bull.* **20,** 1849 (1972).
248. K. Inouye, K. Namba, and H Otsuka, *Bull. Chem. Soc. Jap.* **44,** 1689 (1971).
249. K. Inouye, K. Watanabe, K. Namba, and H. Otsuka, *Bull. Chem. Soc. Jap.* **43,** 3873 (1970).
250. T. Inui, *Bull. Chem. Soc. Jap.* **44,** 2515 (1971).
251. Y. Ishida, M. Onishi, T. Hiyama, and Y. Yabuuchi, *J. Pharm. Sci.* **60,** 896 (1971).
252. A. Islam and A. Darbre, *J. Chromatog.* **71,** 223 (1972).
253. K. Isono, T. Azuma, and S. Suzuki, *Chem. Pharm. Bull.* **19,** 505 (1971).
254. Y. Isowa, T. Takashima, M. Ohmori, H. Kurita, M. Sato, and K. Mori, *Bull. Chem. Soc. Jap.* **45,** 1464 (1972).
255. Y. Isowa, T. Takashima, M. Ohmori, H. Kurita, M. Sato, and K. Mori, *Bull. Chem. Soc. Jap.* **45,** 1461 (1972).
256. H. A. Itano and E. A. Robinson, *J. Biol. Chem.* **247,** 4819 (1972).
257. M. Itoh, *Chem. Pharm. Bull.* **20,** 664 (1972).
258. IUPAC-IUB Commission on Biochemical Nomenclature. (1971). *Biochemistry* **11,** 1726 (1972).
259. V. T. Ivanov, V. V. Shilin, and Yu. A. Ovchinnikov, *J. Gen. Chem. USSR* **40,** 902 (1970).
260. G. A. Ivanovics, R. J. Rousseau, and R. K. Robins, *J. Med. Chem.* **14,** 1155 (1971).
261. M. Iwai, K. Nakajima, A. Uno, S. Hase, I. Takeuchi, and K. Okawa, *Bull. Chem. Soc. Jap.* **43,** 3246 (1970).
262. K. S. N. Iyer, B. H. Iyer, and M. Sirsi, *J. Indian Chem. Soc.* **48,** 182 (1971).

263. N. Izumiya, M. Muraoka, and H. Aoyagi, *Bull. Chem. Soc. Jap.* **44,** 3391 (1971).
264. H.-D. Jakubke, *Chem. Ber.* **97,** 2816 (1964).
265. A. C. A. Jansen, K. E. T. Kerling, and E. Havinga, *Rec. Trav. Chim. Pays-Bas* **89,** 861 (1970).
266. N. G. Johansson, F. Hooper, H. Sievertsson, B. L. Currie, K. Folkers, and C. Y. Bowers, *Biochem. Biophys. Res. Commun.* **49,** 656 (1972).
267. B. J. Johnson, *J. Med. Chem.* **15,** 423 (1972).
268. B. J. Johnson, *J. Med. Chem.* **14,** 488 (1971).
269. B. J. Johnson, *J. Pharm. Sci.* **61,** 1654 (1972).
270. B. J. Johnson, *J. Pharm. Sci.* **61,** 806 (1972).
271. B. J. Johnson, *J. Pharm. Sci.* **59,** 1849 (1970).
272. B. J. Johnson and F. Chen, *J. Med. Chem.* **14,** 640 (1971).
273. B. J. Johnson and C. Cheng, *J. Med. Chem.* **14,** 1238 (1971).
274. B. J. Johnson, C. Cheng, and N. Tsang, *J. Med. Chem.* **15,** 95 (1972).
275. B. J. Johnson and T. A. Ruettinger, *J. Org. Chem.* **35,** 255 (1970).
276. B. J. Johnson and E. G. Trask, *J. Chem. Soc., C* p. 2247 (1970).
277. B. J. Johnson and E. G. Trask, *J. Med. Chem.* **14,** 251 (1971).
278. B. J. Johnson and N. Tsang, *J. Med. Chem.* **15,** 488 (1972).
279. W. H. Johnson, H. D. Law, and R. O. Studer, *J. Chem. Soc., C* p. 748 (1971).
280. D. S. Jones, *J. Chem. Soc. Perkin Trans.* **1,** 1407 (1972).
281. J. H. Jones, *Amino Acids, Peptides Proteins* **3,** 219 (1971).
282. E. C. Jorgensen, S. R. Rapaka, G. C. Windridge, and T. C. Lee, *J. Med. Chem.* **14,** 904 (1971).
283. E. C. Jorgensen, S. R. Rapaka, G. C. Windridge, and T. C. Lee, *J. Med. Chem.* **14,** 899 (1971).
284. E. C. Jorgensen, G. C. Windridge, and T. C. Lee, *J. Med. Chem.* **14,** 631 (1971).
285. K. Jošt, *Collect. Czech. Chem. Commun.* **36,** 218 (1971).
286. K. Jošt and J. Rudinger, *Collect. Czech. Chem. Commun.* **32,** 2485 (1967).
287. K. Jošt and F. Šorm, *Collect. Czech. Chem. Commun.* **36,** 2795 (1971).
288. K. Jošt and F. Šorm, *Collect. Czech. Chem. Commun.* **36,** 297 (1971).
289. K. Jošt and F. Šorm, *Collect. Czech. Chem. Commun.* **36,** 234 (1971).
290. R. Jost and A. Tun-Kyi, E. Stoll, and H. Zuber, *Helv. Chim. Acta* **55,** 1025 (1972).
291. A. Kaiser, W. Koch, M. Scheer, and U. Wölcke, *Helv. Chim. Acta* **53,** 1708 (1970).
291a. Y. Kamano and G. R. Pettit, *Experientia* **28,** 768 (1972).
292. B. Kamber, *Helv. Chim. Acta* **54,** 927 (1971).
293. B. Kamber, *Helv. Chim. Acta* **54,** 398 (1971).
294. A. Kapoor, N. Azeza, N. H. Somaiya, and M. A. Trimboli, *J. Pharm. Sci.* **60,** 956 (1971).
295. E. Kasafirek, *Tetrahdron Lett.* p. 2021 (1972).
296. R. Katakai, M. Oya, K. Uno, and Y. Iwakura, *J. Org. Chem.* **37,** 327 (1972).
297. P. G. Katsoyannis, J. Ginos, and M. Tilak, *J. Amer. Chem. Soc.* **93,** 5866 (1971).
298. P. G. Katsoyannis, J. Ginos, C. Zalut, M. Tilak, S. Johnson, and A. C. Trakatellis, *J. Amer. Chem. Soc.* **93,** 5877 (1971).
299. P. G. Katsoyannis, M. Tilak, and K. Fukuda, *J. Amer. Chem. Soc.* **93,** 5857 (1971).
300. P. G. Katsoyannis, M. Tilak, J. Ginos, and K. Suzuki, *J. Amer. Chem. Soc.* **93,** 5862 (1971).
301. P. G. Katsoyannis and C. Zalut, *Biochemistry* **11,** 3065 (1972).
302. P. G. Katsoyannis and C. Zalut, *Biochemistry* **11,** 1128 (1972).
303. P. G. Katsoyannis, C. Zalut, A. Harris, and R. J. Meyer, *Biochemistry* **10,** 3884 (1971).
304. P. G. Katsoyannis, C. Zalut, A. Tometsko, M. Tilak, S. Johnson, and A. C. Trakatellis, *J. Amer. Chem. Soc.* **93,** 5871 (1971).
305. O. A. Kaurov, V. F. Martynov, and Yu. D. Mikhailov *J. Gen. Chem. USSR* **41,** 1422 (1971).

306. O. A. Kaurov, V. F. Martynov, and V. B. Morozov, *J. Gen. Chem. USSR* **40,** 885 (1970).
307. O. A. Kaurov, V. F. Martynov, Yu. V. Natochin, O. A. Popernatskii, and E. I. Shakhmatova *J. Gen. Chem. USSR* **41,** 207 (1971).
308. O. A. Kaurov, V. F. Martynov and O. A. Popernatskii, *J. Gen. Chem. USSR* **40,** 882 (1970).
309. O. A. Kaurov, V. F. Martynov, and M. P. Smirnova, *J. Gen. Chem. USSR* **40,** 880 (1970).
310. M. Kawana, R. J. Rousseau, and R. K. Robins, *J. Org. Chem.* **37,** 288 (1972).
311. J. Kay, B. Kassell, and H. L. Friedman, *J. Med. Chem.* **14,** 1124 (1971).
312. D. S. Kemp, Z. Bernstein, and J. Rebek Jr., *J. Amer. Chem. Soc.* **92,** 4756 (1970).
313. G. W. Kenner, J. R. McDermott, and R. C. Sheppard, *Chem. Commun.* p. 636 (1971).
314. G. W. Kenner and J. H. Seely, *J. Amer. Chem. Soc.* **94,** 3259 (1972).
315. M. C. Khosla, S. Kumar, R. R. Smeby, and F. M. Bumpus, *J. Med. Chem.* **15,** 627 (1972).
316. M. C. Khosla, R. A. Leese, W. L. Maloy, A. T. Ferreira, R. R. Smeby, and F. M. Bumpus, *J. Med. Chem.* **15,** 792 (1972).
317. M. C. Khosla, R. R. Smeby, and F. M. Bumpus, *J. Amer. Chem. Soc.* **94,** 4721 (1972).
318. L. B. Kier and J. M. George, *J. Med. Chem.* **15,** 385 (1972).
319. Y. Kikuchi, D. Fujimoto, and N. Tamiya, *Biochem. J.* **124,** 695 (1971).
320. R. Kimura and T. Murata, *Chem. Pharm. Bull.* **19,** 1301 (1971).
321. A. A. Kiryushkin, H. M. Fales, T. Axenrod, E. J. Gilbert, and G. W. A. Milne, *Org. Mass Spectrom.* **5,** 19 (1971).
322. L. Kisfaludy, J. E. Roberts, R. H. Johnson, G. L. Mayers, and J. Kovacs, *J. Org. Chem.* **35,** 3563 (1970).
323. Y. Kiso and H. Yajima, *J. Chem. Soc., Chem. Commun.* p. 942 (1972).
324. Y. S. Klausner, V. Mutt, and M. Bodanszky, *Bioorg. Chem.* **2,** 87 (1972).
325. E. Klieger, *Justus Liebigs Ann. Chem.* **724,** 204 (1969).
326. H. Klostermeyer, *Hoppe-Seyler's Z. Physiol. Chem.* **352,** 2 (1971).
327. Y. Knobler, S. Bittner, and M. Frankel, *Isr. J. Chem.* **8,** 639 (1970).
328. W. König, *Hoppe-Seyler's Z. Physiol. Chem.* **352,** 2 (1971).
329. W. König and R. Geiger, *Chem. Ber.* **105,** 2872 (1972).
330. H. de Koning and U. K. Pandit, *Rec. Trav. Chim. Pays-Bas* **90,** 1069 (1971).
331. M. Konishi, N. Yoshida, and N. Izumiya, *Bull. Chem. Soc. Jap.* **44,** 2801 (1971).
332. K. D. Kopple, A. Go, R. H. Logan, Jr., and J. Šavrda, *J. Amer. Chem. Soc.* **94,** 973 (1972).
333. L. I. Korsuntseva, E. I. Filippovich, and R. P. Evstigneeva, *J. Gen. Chem. USSR* **41,** 1629 (1971).
334. A. N. Kost, R. S. Sagitullin, and M. A. Yuroskaya *Chem. Natur. Compounds* **3,** 29 (1967).
335. A. K. Koul, B. Prashad, J. M. Bachhawat, N. S. Ramegowda, and N. K. Mathur, *Syn. Commun.* **2,** 383 (1972).
336. J. Kovacs, C. Kalita, and U. R. Ghatak, *J. Org. Chem.* **37,** 30 (1972).
337. J. Kovacs, L. Kisfaludy, M. Q. Ceprini, and R. H. Johnson, *Tetrahedron* **25,** 2555 (1969).
338. G. Krail, D. Brandenburg, and H. Zahn, *Hoppe-Seyler's Z. Physiol. Chem.* **352,** 1595 (1971).
339. B. L. Krainova, G. A. Agafonova, and E. S. Chaman, *J. Gen. Chem. USSR* **41,** 1623 (1971).
340. N. Kundu and S. Roy, *J. Chromatogr.* **53,** 598 (1970).
341. S. M. Kupchan, T. J. Giacobbe, I. S. Krull, A. M. Thomas, M. A. Eakin, and D. C. Fessler, *J. Org. Chem.* **35,** 3539 (1970).
342. K. Kuromizu and N. Izumiya, *Bull. Chem. Soc. Jap.* **43,** 2944 (1970).
343. H. Lackner, *Chem. Ber.* **104,** 3653 (1971).
344. H. Lackner, *Chem. Ber.* **103,** 2476 (1970).

345. S. Lande, *J. Org. Chem.* **36**, 1267 (1971).
346. D. R. Lauren and R. E. Williams, *Tetrahedron Lett.* p. 2665 (1972).
347. R. Ledger and F. H. C. Stewart, *Aust. J. Chem.* **21**, 1101 (1968).
348. R. Ledger and F. H. C. Stewart, *Aust. J. Chem.* **20**, 2509 (1967).
349. S. Lee, R. Ohkawa, and N. Izumiya, *Bull. Chem. Soc. Jap.* **44**, 158 (1971).
350. M. S. Lehmann, T. F. Koetzle, and W. C. Hamilton, *Int. J. Peptide Protein Res.* **4**, 229 (1972).
351. M. Leplawy and J. Zabrocki, *Z. Chem.* **11**, 16 (1971).
352. C. S. Levine and C. H. Stammer, *J. Org. Chem.* **37**, 148 (1972).
353. A. Levitzki, I. Pecht, and A. Berger, *J. Amer. Chem. Soc.* **94**, 6844 (1972).
354. J. Lewalter and C. Birr, *Justus Liebigs Ann. Chem.* **740**, 48 (1970).
355. A. Lewis and H. A. Scheraga, *Macromolecules* **4**, 539 (1971).
356. C. C. Leznoff and J. Y. Wong, *Can. J. Chem.* **50**, 2892 (1972).
357. C. H. Li and B. Hemmasi, *J. Med. Chem.* **15**, 697 (1972).
358. C. H. Li and B. Hemmasi, *J. Med. Chem.* **15**, 217 (1972).
359. C. H. Li and D. Yamashiro, *J. Amer. Chem. Soc.* **92**, 7608 (1970).
360. B. Liberek and K. Steporowska, *Chem. Ind.* (*London*) p. 1263 (1970).
361. G. Lindeberg, J. Kynčl, P. Dreyfuss, and M. Bodanszky, *J. Med. Chem.* **15**, 629 (1972).
362. S. H. Lipton and R. C. Dutky, *Chem. Ind.* (*London*) p. 1099 (1971).
363. Y. Liwschitz, E. Nemes, and Z. Neiman, *J. Chem. Soc., C* p. 223 (1971).
364. L. A. Ljublinskaya and V. M. Stepanov, *Tetrahedron Lett.* p. 4511 (1971).
365. K. Lloyd and G. T. Young, *J. Chem. Soc., C* p. 2890 (1971).
366. A. Loffet and C. Dremier, *Experientia* **27**, 1003 (1971).
367. G. P. Lorenzi, B. B. Doyle, and E. R. Blout, *Biochemistry* **10**, 3046 (1971).
368. G. Loose and H. Klengel, *Tetrahedron* **27**, 1423 (1971).
369. G. Losse and U. Krychowski, *Tetrahedron Lett.* p. 4121 (1971).
370. G. Losse and U. Krychowski, *J. Prakt. Chem.* [4] **312**, 1097 (1970).
371. L. Lubiewska-Nakonieczna, B. Rzeszotarska, and E. Taschner, *Justus Liebigs Ann. Chem.* **741**, 157 (1970).
372. G. Lucente, G. Fiorentini, and D. Rossi, *Gazz. Chim. Ital.* **101**, 109 (1971).
373. G. Lucente and P. Frattesi, *Tetrahedron Lett.* p. 4283 (1972).
374. U. Ludescher and R. Schwyzer, *Helv. Chim. Acta* **55**, 2052 (1972).
375. W. Lunkenheimer and H. Zahn, *Justus Liebigs Ann. Chem.* **740**, 1 (1970).
376. V. W. Lunkenheimer and H. Zahn, *Angew. Makromol. Chem.* **10**, 69 (1970).
377. S. D. L'vova, V. M. Kozhukhovskaya, T. A. Volkova, and R. P. Evstigneeva, *J. Gen. Chem. USSR* **41**, 1398 (1971).
378. S. D. L'vova, I. N. Sologubovskaya, and R. P. Evstigneeva, *J. Gen. Chem. USSR* **40**, 210 (1971).
379. J. A. MacLaren, *Aust. J. Chem.* **25**, 1293 (1972).
380. J. A. MacLaren, *Aust. J. Chem.* **24**, 1695 (1971).
381. J. J. Maher, M. E. Furey, and L. J. Greenberg, *Tetrahedron Lett.* p. 1581 (1972).
382. J. J. Maher, M. E. Furey, and L. J. Greenberg, *Tetrahedron Lett.* p. 29 (1971).
383. J. J. Maher, M. E. Furey, and L. J. Greenberg, *Tetrahedron Lett.* p. 27 (1971).
384. S. Makisumi, S. Matsuura, M. Waki, and N. Izumiya, *Bull. Chem. Soc. Jap.* **44**, 210 (1971).
385. S. Makisumi, M. Waki, and N. Izumiya, *Bull. Chem. Soc. Jap.* **44**, 143 (1971).
386. P. Maldonado, C. Richard, J. P. Aune, and J. Metzger, *Bull. Soc. Chim.* [6] p. 2933 (1971).
387. J. M. Manning, *J. Amer. Chem. Soc.* **92**, 7449 (1970).
388. M. Manning, J. W. M. Baxter, T. C. Wuu, V. Smart-Abbey, K. Morton, E. J. Coy, and W. H. Sawyer, *J. Med. Chem.* **14**, 1143 (1971).
389. A. Marglin, *Tetrahedron Lett.* p. 3145 (1971).
390. B. Marinier and M. Bérubé, *Can. J. Chem.* **50**, 1633 (1972).

391. G. Marino, L. Valente, R. A. W. Johnstone, F. Mohammedi-Tabrizi, and G. C. Sodini, *J. Chem. Soc., Chem. Commun.* p. 357 (1972).
392. N. Marks and R. Walter, *Proc. Soc. Exp. Biol. Med.* **140**, 673 (1972).
393. D. I. Marlborough and H. N. Rydon, *J. Chem. Soc., C* p. 1 (1972).
394. G. R. Marshall and R. B. Merrifield *in* "Biochemical Aspects of Reactions on Solid Supports" (G. R. Stark, ed.), p. 111. Academic Press, New York, 1971.
395. R.-P. Martin, L. Mosoni, and B. Sarkar, *J. Biol. Chem.* **246**, 5944 (1971).
396. T. Matoba and T. Hata, *Agr. Biol. Chem.* **36**, 1423 (1972).
397. R. Matsueda, H. Maruyama M. Ueki, and T. Mukaiyama, *Bull. Chem. Soc. Jap.* **44**, 1373 (1971).
398. H. Matsuo, A. Arimura, R. M. G. Nair, and A. V. Schally, *Biochem. Biophys. Res. Commun.* **45**, 822 (1971).
399. H. Matsuo, Y. Baba, R. M. G. Nair, A. Arimura, and A. V. Schally, *Biochem. Biophys. Res. Commun.* **43**, 1334 (1971).
400. H. Matsuo, Y. Kawazoe, M. Sato, M. Ohnishi, and T. Tatsuno, *Chem. Pharm. Bull.* **18**, 1788 (1970).
401. S. Matsuura, M. Waki, and N. Izumiya, *Bull. Chem. Soc. Jap.* **45** 863 (1972).
402. W. L. Mattice and L. Mandelkern, *Macromolecules* **4**, 271 (1971).
403. A. B. Mauger, R. B. Desai, I. Rittner, and W. J. Rzeszotarski, *J. Chem. Soc., Perkin Trans.* **1**, 2146 (1972).
404. R. H. Mazur, A. H. Goldkamp, P. A. James, and J. M. Schlatter, *J. Med. Chem.* **13**, 1217 (1970).
405. F. McCapra and M. Roth, *J. Chem. Soc., Chem. Commun.* p. 894 (1972).
406. S. V. McKinley and J. W. Rakshys, Jr., *J. Chem. Soc., Chem. Commun.* p. 134 (1972).
407. J. Meienhofer, *J. Amer. Chem. Soc.* **92**, 3771 (1970).
408. J. Meienhofer, R. Cotton, and E. Atherton, *J. Org. Chem.* **36**, 3746 (1971).
409. J. Meienhofer, J. Czombos, and H. Maeda, *J. Amer, Chem. Soc.* **93**, 3080 (1971).
410. J. Meienhofer, P. M. Jacobs, H. A. Godwin, and I. H. Rosenberg, *J. Org. Chem.* **35**, 4137 (1970).
411. J. Meienhofer, A. Trzeciak, R. T. Havran, and R. Walter, *J. Amer. Chem. Soc.* **92**, 7199 (1970).
412. J. P. Meraldi, R. Schwyzer, A. Tun-Kyi, and K. Wüthrich, *Helv. Chim. Acta* **55**, 1962 (1972).
413. C. Meyers, R. T. Havran, I. L. Schwartz, and R. Walter, *Chem. Ind. (London)* p. 136 (1969).
414. Yu. V. Mitin and O. V. Glinskaya, *J. Gen. Chem. USSR* **41**, 1152 (1971).
415. Yu. V. Mitin and G. P. Vlasov, *J. Gen. Chem. USSR* **41**, 419 (1971).
416. Y. Miura, H. Sugiyama, Y. Maki, and S. Seto, *Chem. Pharm. Bull.* **20**, 215 (1972).
417. M. Miyoshi, T. Kimura, and S. Sakakibara, *Bull. Chem. Soc. Jap.* **43**, 2941 (1970).
418. M. Mokotoff and B. S. Parikh, *J. Org. Chem.* **37**, 2786 (1972).
419. M. W. Monahan and J. Rivier, *Biochem. Biophys. Res. Commun.* **48**, 1100 (1972).
420. M. W. Monahan, J. Rivier, W. Vale, R. Guillemin, and R. Burgus, *Biochem. Biophys. Res. Commun.* **47**, 551 (1972).
421. L. Moroder, G. Borin, F. Marchiori, and E. Scoffone, *Biopolymers* **11**, 2191 (1972).
422. E. A. Morozova and M. A. Zevail', *J. Gen. Chem. USSR* **40**, 1365 (1970).
423. E. A. Morozova, M. A. Zevail', and G. F. Zhukova, *J. Gen. Chem. USSR* **40**, 1362 (1970).
424. H. R. Morris, D. H. Williams, and R. P. Ambler, *Biochem. J.* **125**, 189 (1971).
425. S. E. Moschopedis and C. Mumford, *Can. J. Chem.* **49**, 2158 (1971).
426. C. W. Mosher and L. Goodman, *J. Org. Chem.* **37**, 2928 (1972).
427. C. W. Moss, M. A. Lambert, and F. J. Diaz, *J. Chromatogr.* **60**, 134 (1971).
428. A. Mostad, H. M. Nissen, and C. Rømming, *Tetrahedron Lett.* p. 2131 (1971).
429. T. Mukaiyama, *Syn. Commun.* **2**, 243 (1972).

430. A. Munoz, M. Koenig, B. Garrigues, and R. Wolf, *C. R. Acad. Sci.* **274,** 1413 (1972).
431. M. Muraki and T. Mizoguchi, *Chem. Pharm. Bull.* **19,** 1708 (1971).
432. I. Murakoshi, S. Ohmiya, and J. Haginiwa, *Chem. Pharm. Bull.* **20,** 609 (1972).
433. I. Muramatsu, S. Sōfuku, and A. Hagitani, *J. Antibiot.* **25,** 189 (1972)
434. M. Mutter, H. Hagenmaie, and E. Bayer, *Angew. Chem.* **10,** 811 (1971).
435. H. T. Nagasawa, J. G. Kohlhoff, P. S. Fraser, and A. A. Mikhail, *J. Med. Chem.* **15,** 483 (1972).
436. V. K. Naithan, H. Klostermeyer, H. R. Lange, U. Nehen, P. Fehrenbach, E. Engels, H. Berndt, D. von Wacktendon, and H. Zahn, *Hoppe-Seyler's Z. Physiol. Chem.* **352,** 2 (1971).
437. Y. Nakagawa, T. Tsuno, K. Nakajima, M. Iwai, H. Kawai, and K. Okawa, *Bull. Chem. Soc. Jap.* **45,** 1162 (1972).
438. M. Nakayama, G. Maeda, T. Kaneko, and H. Katsura, *Bull. Chem. Soc. Jap.* **44,** 1150 (1971).
439. P. K. Nandi and D. R. Robinson, *J. Amer. Chem. Soc.* **94,** 1308 (1972).
440. P. K. Nandi and D. R. Robinson, *J. Amer. Chem. Soc.* **94,** 1299 (1972).
441. L. Neelakantan, *J. Org. Chem.* **36,** 2253 (1971).
442. A. D. Neklyudov, L. A. Shchukina, N. N. Suvorov, M. D. Mashkovskii, T. K. Trubitsina, V. Z. Gorkin, and L. V. Tat'yanenko, *Chem. Natur. Compounds* **3,** 90 (1967).
443. M. Nieto, H. R. Perkins, and P. E. Reynolds, *Biochem. J.* **126,** 139 (1972).
444. L. N. Nikolenko, M. A. Ponomareva-Stepnaya, and R. Badalova, *Chem. Natur. Compounds* **4,** 27 (1968).
445. L. N. Nikolenko and L. D. Shustov, *J. Gen. Chem. USSR* **41,** 452 (1971).
446. H. Nishimura, S. Tahara, H. Okuyama, and J. Mitzutani, *Tetrahedron* **28,** 4503 (1972).
447. K. Nishioka, A. Constantopoulos, P. S. Satoh, and V. A. Najjar, *Biochem. Biophys. Res. Commun.* **47,** 172 (1972).
448. D. E. Nitecki, B. Halpern, and J. W. Westley, *J. Org. Chem.* **33,** 864 (1968).
449. A. Noda, Y. Hirakawa, N. Fujii, and S. Iguchi, *Chem. Pharm. Bull.* **19,** 196 (1971).
450. K. Noda, S. Terada, N. Mitsuyasu, M. Waki, T. Kato, and N. Izumiya, *Naturwissenschaften* **58,** 147 (1971).
451. J. Noguchi and Y. Fujiwara, *Bull. Chem. Soc. Jap.* **43,** 2515 (1970).
452. J. Noguchi, S. Tokura, T. Komai, K. Kokazi, and T. Azuma, *J. Biochem. (Tokyo)* **69,** 1033 (1971).
453. M. Ohno, K. Kuromizu, H. Ogawa, and N. Izumiya, *J. Amer. Chem. Soc.* **93,** 5251 (1971).
454. M. Ohno, S. Tsukamoto, and N. Izumiya, *J. Chem. Soc., Chem. Commun.* p. 663 (1972).
455. M. Ohno, S. Tsukamoto, S. Makisumi, and N. Izumiya, *Bull. Chem. Soc. Jap.* **45,** 2852 (1972).
456. K. Oki, K. Suzuki, S. Tuchida, T. Saito, and H. Kotake, *Bull. Chem. Soc. Jap.* **43,** 2554 (1970).
457. K. Okumura, T. Iwasaki, M. Matsuoka, and K. Matsumoto, *Chem. Ind. (London)* p. 929 (1971).
458. M. H. O'Leary and M. D. Kluetz, *J. Amer. Chem. Soc.* **94,** 665 (1972).
459. M. Olomucki, P. Hébrard, and J.-Y. Le Gall, *Bull. Soc. Chim. Fr.* [6] p. 4524 (1971).
460. V. M. Orlov, Ya. M. Varshavsky, and A. A. Kiryushkin, *Org. Mass Spectrom.* **6,** 9 (1972).
461. C. G. Overberger and K.-H. David, *Macromolecules* **5,** 373 (1972).
462. C. G. Overberger and K.-H. David, and J. A. Moore, *Macromolecules* **5,** 368 (1972).
463. N. I. A. Overweg, I. L. Schwartz, B. M. Dubois, and R. Walter, *J. Pharmacol. Exp. Ther.* **161,** 342 (1968).
464. M. Oya, M. Tomizawa, K. Uno, and Y. Iwakura, *Bull. Chem. Soc. Jap.* **43,** 3486 (1970).
465. T. B. Paiva, A. C. M. Paiva, R. J. Freer, and J. M. Stewart, *J. Med. Chem.* **15,** 6 (1972).
466. C. S. Pande, J. Rudick, L. Ornstein, I. L. Schwartz, and R. Walter, *Mol. Pharmacol.* **5,** 227 (1969).

467. C. A. Panetta and M. Pesh-Imam, *J. Org. Chem.* **37**, 302 (1972).
468. C. A. Panetta and Altaf-Ur-Rahman, *J. Org. Chem.* **36**, 2250 (1971).
469. E. B. Paniago, D. C. Weatherburn, and D. W. Margerum, *Chem. Commun.* p. 1427 (1971).
470. G. T. Panse and D. A. Laufer, *Tetrahedron Lett.* p. 4181 (1970).
471. L. Paolillo, P. A. Temussi, E. M. Bradbury, and C. Crane-Robinson *Biopolymers* **11**, 2043 (1972).
472. W. Parr and K. Grohmann, *Angew. Chem. Int. Ed. Engl.* **11**, 314 (1972).
473. W. Parr and K. Grohmann, *Tetrahedron Lett.* p. 2633 (1971).
474. W. Parr and G. Holzer, *Hoppe-Seyler's Z. Physiol. Chem.* **352**, 1043 (1971).
475. W. Parr and P. Y. Howard, *Angew. Chem. Int. Ed. Engl.* **11**, 529 (1972).
476. W. Parr, C. Yang, and G. Holzer, *Tetrahedron Lett.* p. 101 (1972).
477. J. R. Parrish Jr. and E. R. Blout *Biopolymers* **11**, 1001 (1972).
478. A. Patchornik, B. Amit, and R. B. Woodward, *J. Amer. Chem. Soc.* **92**, 6333 (1970).
479. A. P. Pavar, Z. P. Auna, and G. I. Chipens, *J. Gen. Chem. USSR* **41**, 1867 (1971).
480. A. P. Pavar and G. I. Čhipens, *J. Gen. Chem. USSR* **41**, 459 (1971).
481. J. W. Payne, *Biochem. J.* **123**, 255 (1971).
482. J. W. Payne, *Biochem. J.* **123**, 245 (1971).
483. E. Peggion, A. Cosani, M. Terbojevich, and E. Scoffone, *Macromolecules* **4**, 725 (1971).
484. W. E. Pereira, Jr. and B. Halpern, *Aust. J. Chem.* **25**, 667 (1972).
485. G. E. Perlmann and K. Grizzuti, *Biochemistry* **10**, 4168 (1971).
486. G. R. Pettit, "Synthetic Nucleotides" Vol. 1. Van Nostrand-Reinhold, Princeton, New Jersey, 1972.
486a. G. R. Pettit, S. K. Gupta, and R. H. Ode, *J. Chem. Soc., C* p. 950 (1973).
486b. G. R. Pettit and W. R. Jones, *J. Org. Chem.* **36**, 870 (1971).
487. G. R. Pettit and Y. Kamano, *Chem. Commun.* p. 45 (1972).
488. I. Photaki, *J. Chem. Soc., C* p. 2687 (1970).
489. I. Photaki, J. Taylor-Papadimitriou, G. Sakarellos, P. Mazarakis, and L. Zervas, *J. Chem. Soc., C* p. 2683 (1970).
490. P. Pietta and P. Cavallo, *Gazz. Chim. Ital.* **100**, 863 (1970).
491. P. G. Pietta, P. Cavallo, and G. R. Marshall, *J. Org. Chem.* **36**, 3966 (1971).
492. N. A. Poddubnaya, G. N. Balandina, and O. A. Tambieva, *J. Gen. Chem. USSR* **41**, 428 (1971).
493. N. A. Poddubnaya and N. Ya. Krasnobrizhii, *J. Gen. Chem. USSR* **42**, 938 (1972).
494. N. A. Poddubnaya and N. Ya. Krasnobrizhii, *J. Gen. Chem. USSR* **41**, 432 (1971).
495. N. A. Poddubnaya and N. Ya. Krasnobrizhii, *J. Gen. Chem. USSR* **40**, 1133 (1970).
496. N. A. Poddubnaya and N. Ya. Krasnobrizhii, *J. Gen. Chem. USSR* **40**, 1128 (1970).
497. M. L. Polan, W. J. McMurray, S. R. Lipsky, and S. Lande, *J. Amer. Chem. Soc.* **94**, 2847 (1972).
498. K. P. Polzhofer, *Tetrahedron* **28**, 855 (1972).
499. K. P. Polzhofer, *Tetrahedron* **28**, 1721 (1972).
500. K. P. Polzhofer and K. H. Ney, *Tetrahedron* **27**, 1997 (1971).
501. K. P. Polzhofer and K. H. Ney, *Tetrahedron* **26**, 3787 (1970).
502. Z. Procházka, K. Jošt, and F. Šorm, *Collect. Czech. Chem. Commun.* **37**, 289 (1972).
503. J. Rabinowitz, *Helv. Chim. Acta* **54**, 1483 (1971).
504. J. Rabinowitz, *Helv. Chim. Acta* **53**, 1350 (1970).
505. U. Ragnarsson, S. Karlsson, and G. Lindeberg, *Acta Chem. Scand.* **24**, 2821 (1970).
506. U. Ragnarsson, S. Karlsson, and B. Sandberg, *Acta Chem. Scand.* **25**, 1487 (1971).
507. F. J. M. Rajabalee, *Synthesis* 318 (1972).
508. E. V. Ramenskii, M. M. Botvinik, and R. U. Beisembaeva, *Chem. Natur. Compounds* **4**, 18 (1968).
509. D. W. Repke and J. E. Zull, *J. Biol. Chem.* **247**, 2189 (1972).
510 C Ressler, *J. Org. Chem.* **37**, 2933 (1972).

511. C. Ressler, G. R. Nagarajan, M. Kirisawa, and D. V. Kashelikar, *J. Org. Chem.* **36,** 3960 (1971).
512. B. Ridge, H. N. Rydon, and C. R. Snell, *J. Chem Soc., Perkin Trans.* **1,** 2041 (1972).
513. M. Rimpler, *Justus Liebigs Ann. Chem.* **745,** 8 (1971).
514. A. Rinaldi and C. de Marco, *Ital. J. Biochem.* **20,** 2 (1971).
515. B. Riniker, P. Sieber, and W. Rittel, *Nature (London) New Biol.* **235,** 114 (1972).
516. W. B. Rippon, J. L. Koenig, and A. G. Walton, *J. Amer. Chem. Soc.* **92,** 7455 (1970).
517. P. Rivaille and G. Milhaud, *Helv. Chim. Acta* **55,** 1617 (1972).
518. P. Rivaille and G. Milhaud, *Helv. Chim. Acta* **54,** 355 (1971).
519. P. Rivaille, A. Robinson, M. Kamen and G. Milhaud, *Helv. Chim. Acta* **54,** 2772 (1971).
520. J. Rivier, W. Vale, M. Monahan, N. Ling, and R. Burgus, *J. Med. Chem.* **15,** 479 (1972).
521. M. J. Robins, L. N. Simon, M. G. Stout, G A. Ivanovics, M. P. Schweizer, R. J. Rousseau, and R. K. Robins, *J. Amer. Chem. Soc.* **93,** 1474 (1971).
522. R. L. Rodin and J. Kovacs, *Bioorg. Chem.* **2,** 65 (1972).
523. H. Rodríguez, C. Chuaqui, S. Atala, and A. Márquez, *Tetrahedron* **27,** 2425 (1971).
524. I. K. Romanovskaya and G. I. Chipens, *J. Gen. Chem. USSR* **41,** 1864 (1971).
525. L. I. Rostovtseva and A. A. Kiryushkin, *Org. Mass Spectrom.* **6,** 1 (1972).
526. L. I. Rostovtseva, A. A. Kiryushkin, and A. S. Khokhlov, *J. Gen. Chem. USSR* **41,** 1618 (1971).
527. L. I. Rostovtseva, A. A. Kiryushkin, and A. S. Khokhlov, *J. Gen. Chem. USSR* **41,** 1385 (1971).
528. L. I. Rostovtseva, E. A. Markelova, A. A. Kiryushkin, and A. S. Khokhlov, *J. Gen. Chem. USSR* **41,** 1611 (1971).
529. M. Rothe and J. Mazánek, *Tetrahedron Lett.* p. 3795 (1972).
530. M. Rothe, R. Theysohn, and K.-D. Steffen, *Tetrahedron Lett.* p. 4063 (1970).
531. G. L. Rowley, A. L. Greenleaf, and G. L. Kenyon, *J. Amer. Chem. Soc.* **93,** 5542 (1971).
532. M. A. Ruttenberg, *Science* **177** 623 (1972).
533. K. Sakai and J.-P. Anselme, *J. Org. Chem.* **36** 2387 (1971).
534. M. Satō, T. Tatsuno, and H. Matsuo, *Chem. Pharm. Bull.* **18** 1794 (1970).
535. M. A. Saunders and D. L. Rohlfing, *Science* **176,** 172 (1972).
536. W. E. Savige, *Aust. J. Chem.* **24,** 1285 (1971).
537. J. Šavrda and D. H. Veyrat, *J. Chem. Soc., C* p. 2180 (1970).
538. D. J. Schafer, *J. Chem. Soc., Chem. Commun.* p. 276 (1972).
539. D. J. Schafer, *J. Chem. Soc., Perkin Trans.* **1,** 1452 (1972).
540. D. J. Schafer, G. T. Young, D. F. Elliott, and R. Wade, *J. Chem. Soc., C* p. 46 (1971).
541. A. V. Schally, A. Arimura, W. H. Carter, T. W. Redding, R. Geiger, W. König, H. Wissman, G. Jaeger, J. Sandow, N. Yanaihara, C. Yanaihara, T. Hashimoto, and M. Sakagami, *Biochem. Biophys. Res. Commun.* **48,** 366 (1972).
542. A. V. Schally, R. M. G. Nair, T. W. Redding, and A. Arimura, *J. Biol. Chem.* **246,** 7230 (1971).
543. C. Schattenkerk I, Holtkamp, J. G. M. Hessing, K. E. T. Kerling, and E. Havinga, *Rec. Trav. Chim. Pays-Bas* **90,** 1320 (1971).
544. E. Schnabel, *Justus Liebigs Ann. Chem.* **702,** 188 (1967).
545. E. Schnabel, G. Schmidt, and E. Klauke, *Justus Liebigs Ann. Chem.* **743,** 69 (1971).
546. E. Schnable, J. Stoltefuss, H. A. Offe, and E. Klauke, *Justus Liebigs Ann. Chem.* **743,** 57 (1971).
547. C. H. Schneider and W. Wirz, *Helv. Chim. Acta* **55,** 1062 (1972).
548. A. Schöberl M. Rimpler, and E. Clauss, *Justus Liebigs Ann. Chem.* **742,** 68 (1970).
549. A. Schöberl, M. Rimpler, and E. Clauss, *Chem. Ber.* **103** 3159 (1970).
550. R. Schwyzer and P. W. Schiller, *Helv. Chim. Acta* **54,** 897 (1971).
551. R. P. W. Scott, K. K. Chan, P. Kucera, and S. Zolty, *J. Chromatogr. Sci.* **9,** 577 (1971).
552. F. Seela, *J. Med. Chem.* **15,** 684 (1972).
553. H. Seki, K. Koga, and S.-I. Yamada, *Chem. Pharm. Bull.* **20,** 361 (1972).

554. E. P. Semkin, N. D. Gafurova, and L. A. Shchukina, *Chem. Natur. Compounds* **3,** 182 (1967).
555. M. A. E. Shabahn and R. W. Jeanloz, *Carbohydr. Res.* **26,** 315 (1973).
556. M. A. E. Shabahn and R. W. Jeanloz, *Carbohydr. Res.* **23,** 243 (1972).
557. L. A. Shchukina, L. A. Davankova, A. D. Neklyudov, and N. N. Suvorov, *J. Gen. Chem. USSR* **41,** 1633 (1971).
558. L. A. Shchukina and V. G. Degtyar, *Chem. Natur. Compounds* **4,** 32 (1968).
559. L. A. Shchukina, V. G. Degtyar, and E. I. Boltyanskaya, *Chem. Natur. Compounds* **3,** 31 (1967).
560. L. A. Shchukina, E. P. Semkin, and A. P. Smirnova, *Chem. Natur. Compounds* **3,** 304 (1967).
561. L. A. Shchukina, L.-Yu. Sklyarov, and V. I. Gorbunov, *Chem. Natur. Compounds* **4,** 102 (1968).
562. J. C. Sheehan and R. E. Chandler, *J. Amer. Chem. Soc.* **83,** 4795 (1961).
563. J. C. Sheehan, D. W. Chapman, and R. W. Roth, *J. Amer. Chem. Soc.* **74,** 3822 (1952).
564. J. C. Sheehan and J. J. Hlavka, *J. Org. Chem.* **28,** 635 (1958).
565. J. C. Sheehan, J. Preston, and P. A. Cruickshank, *J. Amer. Chem. Soc.* **87,** 2492 (1965).
566. J. C. Sheehan and W. L. Richardson, *J. Amer. Chem. Soc.* **76,** 6329 (1954).
567. J. C. Sheehan, R. M. Wilson, and A. W. Oxford, *J. Amer. Chem. Soc.* **93,** 7222 (1971).
568. M. Shimada, K. Takeuchi, Y. Asai, S. Kuwata, and H. Watanabe, *Bull. Chem. Soc. Jap.* **43,** 2874 (1970).
569. Y. Shimonishi, *Bull. Chem. Soc. Jap.* **43,** 3251 (1970).
570. N. G. Shinskii, N. N. Preobrazhenskaya, Z. A. Shabarova, and M. A. Prokof'ev, *J. Gen. Chem. USSR* **40,** 1114 (1970).
571. T. Shioiri, K. Ninomiya, and S. Yamada, *J. Amer. Chem. Soc.* **94,** 6203 (1972).
572. P. Sieber, M. Brugger, and W. Rittel, *in* "Peptides 1969," p. 60. North-Holland Publ. Amsterdam, 1971.
573. P. Sieber and B. Iselin, *Helv. Chim. Acta* **51,** 622 (1968).
574. P. Sieber, B. Riniker, M. Brugger, B. Kamber, and W. Rittel, *Helv. Chim. Acta* **53,** 2135 (1970).
575. P. Sieber, W. Rittel, and B. Riniker, *Helv. Chim. Acta* **55,** 1243 (1972).
576. H. Sievertsson, J.-K. Chang, C. Bogentoft, B. L. Currie, K. Folkers, and C. Y. Bowers, *Biochem. Biophys. Res. Commun.* **44,** 1566 (1971).
577. H. Sievertsson, J.-K. Chang, K. Folkers, and C. Y. Bowers, *J. Med. Chem.* **15,** 219 (1972).
578. H. Sievertsson, J.-K. Chang, K. Folkers, and C. Y. Bowers, *J. Med. Chem.* **15,** 8 (1972).
579. H. Sievertsson, J.-K. Chang, A. von Klaudy, C. Bogentoft, B. Currie K. Folkers, and C. Bowers, *J. Med. Chem.* **15,** 222 (1972).
580. A. Singerman, Y. Liwschitz, and Y. Wiesel, *Isr. J. Chem.* **9,** 105 (1971).
581. F. Sipos and D. W. Gaston, *Synthesis* 321 (1971)
582. T. D. Skalaban, I. M. Nazimov, S. S. Pankova, E. N. Zvonkova, R. P. Evstigneeva, and N. A. Preobrazhenskii, *J. Org. Chem. USSR* **7,** 47 (1971).
583. W. A. Skinner and J. G. Johansson, *J. Med. Chem.* **15,** 427 (1972).
584. J. A. Skorcz, *J. Med. Chem.* **14,** 775 (1971).
585. V. D. Smirnov, N. G. Shinskii, Z. A. Shabarova, and M. A. Prokof'ev, *Chem. Natur. Compounds* **4,** 93 (1968).
586. E. E. Smissman and V. D. Warner, *J. Med. Chem.* **15,** 681 (1972).
587. N. I. Sokolova, R. K. Ledneva, P. P. Purygin, and Z. A. Shabarova, *Chem. Natur. Compounds* **3,** 245 (1967).
588. G. L. Southard, G. S. Brooke, and J. M. Pettee, *Tetrahedron* **27,** 2701 (1971).
589. G. L. Southard, G. S. Brooke, and J. M. Pettee, *Tetrahedron* **27,** 1359 (1971).
590. G. L. Southard, B. R. Zaborowsky, and J. M. Pettee, *J. Amer. Chem. Soc.* **93,** 3302 (1971).

591. R. Spangenberg, P. Thamm, and E. Wünsch, *Hoppe-Seyler's Z. Physiol. Chem.* **352,** 655 (1971).
592. J. Sperling and D. Elad, *J. Amer. Chem. Soc.* **93,** 3839 (1971).
593. J. Sperling and D. Elad, *J. Amer. Chem. Soc.* **93,** 967 (1971).
594. F. H. C. Stewart, *Aust. J. Chem.* **24,** 2739 (1971).
595. F. H. C. Stewart, *Aust. J. Chem.* **24,** 2193 (1971).
596. F. H. C. Stewart, *Aust. J. Chem.* **24,** 1749 (1971).
597. F. H. C. Stewart, *Aust. J. Chem.* **24,** 1743 (1971).
598. F. H. C. Stewart, *Aust. J. Chem.* **24,** 1277 (1971).
599. F. H. C. Stewart, *Aust. J. Chem.* **24,** 1267 (1971).
600. F. H. C. Stewart, *Aust. J. Chem.* **23,** 2147 (1970).
601. F. H. C. Stewart, *Org. Prep. Proced.* **1,** 25 (1969).
602. H. T. Storey, J. Beacham, S. F. Cernosek, F. M. Finn, C. Yanaihara, and K. Hofmann, *J. Amer. Chem. Soc.* **94,** 6170 (1972).
603. P. T. Sullivan and S. J. Norton, *J. Med. Chem.* **14,** 557 (1971).
604. P. T. Sullivan, C. B. Sullivan, and S. J. Norton, *J. Med. Chem.* **14,** 211 (1971).
605. I. A. Surova and V. M. Stepanov, *Chem. Natur. Compounds* **3,** 59 (1967).
606. H. Suter and H. Zutter, *Helv. Chim. Acta.* **54,** 2551 (1971).
607. K. Takahashi, *J. Biochem. (Tokyo)* **70,** 617 (1971).
608. O. Takenaka, Y. Tamaura, Y. Nishimura, and Y. Inada, *J. Biochem. (Tokyo)* **69,** 1139 (1971).
609. N.-D. Tâm, *Bull. Soc. Chim. Fr.* [5] p. 3805 (1967).
610. H. Taniyama, Y. Sawada, K. Miyazeki, and F. Miyoshi, *Chem. Pharm. Bull.* **20,** 601 (1972).
611. E. Taschner, M. Smulkowski, and L. Lubiewska-Nakonieczna, *Justus Liebigs Ann. Chem.* **739,** 228 (1970).
612. S Terada, S. Yoshida, and N. Izumiya, *J. Biochem. (Tokyo)* **70,** 133 (1971).
613. G. I. Tesser and J. T. Buis, *Rec. Trav. Chim. Pays-Bas* **90,** 444 (1971).
614. G. I. Tesser and J. W. van Nispen, *Syn. Commun.* **1,** 285 (1971).
615. J. W. Thanassi, *J. Org. Chem.* **36,** 3019 (1971).
616. M. A. Tilak and C. S. Hollinden, *Org. Prep. Proced. Int.* **3,** 183 (1971).
617. M. A. Tilak, R. Russell, and M. L. Hendricks, *Org. Prep. Proced. Int.* **3,** 17 (1971).
618. M. I. Titov and Z. A. Ardemasova, *J. Gen. Chem. USSR* **41,** 1408 (1971).
619. M. I. Titov, L. I. Leont'eva, and Zh. D. Bespalova, *J. Gen. Chem. USSR* **40,** 218 (1971).
620. A. M. Tometsko, J. Tischio, and J. Garden, II, *J. Pharm. Sci.* **59,** 1655 (1970).
621. A. E. Tonelli and A. I. Brewster, *J. Amer. Chem. Soc.* **94,** 2851 (1972).
622. C. Toniolo, *Tetrahedron* **26,** 5479 (1970).
623. C. Toniolo, D. Nisato, L. Biondi, and A. Signor, *J. Chem. Soc. (London)* p. 1182 (1972).
624. C. Toniolo, D. Nisato, L. Biondi, and A. Signor, *J. Chem. Soc., Perkin Trans.* **1,** 1179 (1972).
625. D. A. Torchia, *Biochemistry* **11,** 1462 (1972).
626. D. A. Torchia, A. di Corato, S. C. K. Wong, C. M. Deber, and E. R. Blout, *J. Amer. Chem. Soc.* **94,** 609 (1972).
627. D. A. Torchia, S. C. K. Wong, C. M. Deber, and E. R. Blout, *J. Amer. Chem. Soc.* **94,** 616 (1972).
628. Y. Trudelle and G. Spach, *Bull. Soc. Chim.* [6] p. 4489 (1971).
629. Y. Trudelle and G. Spach, *Tetrahedron Lett.* p. 3475 (1972).
630. M. Ueki, H. Maruyama, and T. Mukaiyama, *Bull. Chem. Soc. Jap.* **44,** 1108 (1971).
631. I. Ugi, *Rec. Chem. Progr.* **30,** 289 (1969).
632. J. C. Unkeless and P. Goldman, *Mol. Pharmacol.* **6,** 46 (1970).
633. W. Vaalburg, J. Strating, M. G. Woldring, and H. Wynberg, *Syn. Commun.* **2,** 423 (1972).

634. W. Vale, G. Grant, J. Rivier, M. Monahan, M. Amoss, R. Blackwell, R. Burgus, and R. Guillemin, *Science* **176,** 933 (1972).
635. R. J. Vavrek, M. F. Ferger, G. A. Allen, D. H. Rich, A. T. Blomquist, and V. du Vigneaud, *J. Med. Chem.* **15,** 123 (1972).
636. R. G. Vdovina, A. K. Gracheva, M. G. Poznyak, and Yu. P. Shvachkin, *J. Gen. Chem. USSR* **41,** 237 (1971).
637. R. G. Vdovina and A. V. Karpova, *Chem. Natur. Compounds* **4,** 35 (1968).
638. D. F. Veber, J. D. Milkowski, S. L. Varga, R. G. Denkewalter, and R. Hirschmann, *J. Amer. Chem. Soc.* **94,** 5456 (1972).
639. V. du Vigneaud, J. D. Meador, M. F. Ferger, G. A. Allen, and A. T. Blomquist, *Bioorg. Chem.* **1,** 123 (1971).
640. E. I. Vinogradova, Yu. B. Alakhov, V. M. Lipkin, N. A. Aldanova, M. Yu, Feigina, Yu. B. Shvetsov, and L. A. Fonina, *J. Gen. Chem. USSR* **40,** 1371 (1970).
641 S. Visser and K. E. T. Kerling, *Rec. Trav. Chim. Pays-Bas* **89,** 880 (1970).
642. S. Visser, K. E. T. Kerling, and E. Havinga, *Rec. Trav. Chim. Pays-Bas* **89,** 876 (1970).
643. S. Visser, J. Raap, K. E. T. Kerling, and E. Havinga, *Rec. Trav. Chim. Pays-Bas* **89,** 865 (1970).
644. G. P. Vlasov and V. R. Glushenkova, *J. Gen. Chem. USSR* **41,** 425 (1971).
645. G. P. Vlasov, V. R. Glushenkova, and T. N. Prokopyuk, *J. Gen. Chem. USSR* **41,** 422 (1971).
646. H. Vorbrüggen and K. Krolikiewicz, *Chem. Ber.* **105,** 1168 (1972).
647. H. Wakamatsu, J. Uda, and N. Yamakami, *Chem. Commun.* p. 1540 (1971).
648. R. Walter, J. D. Glickson, I. L. Schwartz, R. T. Havran, J. Meienhofer, and D. W. Urry, *Proc. Nat. Acad. Sci. U.S.* **69,** 1920 (1972).
649. R. Walter and R. T. Havran, *Experientia* **27,** 645 (1971).
650. R. Walter and J. Roy, *J. Org. Chem.* **36,** 2561 (1971).
651. R. Walter, I. L. Schwartz, O. Hechter, T. Douša, and P. L. Hoffman, *Endocrinology* **91,** 39 (1972).
652. R. Walter, I. L. Schwartz, and J. Roy, *Ann. N.Y. Acad. Sci.* **192,** 175 (1972).
653. R. Walter and H. Shlank, *Endocrinology* **89,** 990 (1971).
654. M. Wälti and D. B. Hope, *J. Chem. Soc., Perkin Trans.* **1,** 1946 (1972).
655. M. Wälti and D. B. Hope, *J. Chem, Soc., C* p. 2326 (1971).
656. K.-T. Wang and C. H. Li, *J. Org. Chem.* **36,** 2419 (1971).
657. S. Wang, *Biochem. Biophys. Res. Commun.* **48,** 1511 (1972).
658. K. A. Watanabe, E. A. Falco, and J. J, Fox, *J. Org. Chem.* **37,** 1198 (1972).
659. R. F. Weaver, K. V. Rajagopalan, and P. Handler, *J. Biol. Chem.* **246,** 2015 (1971).
660. R. F. Weaver, K. V. Rajagopalan, and P. Handler, *J. Biol. Chem.* **246,** 2010 (1971).
661. M. Weinert, K. Kircher, D. Brandenburg, and H. Zahn, *Hoppe-Seyler's Z. Physiol. Chem.* **352,** 719 (1971).
662. K. Weinges, G. Graab, D. Nagel, and B. Stemmle, *Chem. Ber.* **104,** 3594 (1971).
663. R. J. Weinkam and E. C. Jorgensen, *J. Amer. Chem. Soc.* **93,** 7038 (1971).
664. R. J. Weinkam and E. C. Jorgensen, *J. Amer. Chem. Soc.* **93,** 7033 (1971).
665. R. J. Weinkam and E. C. Jorgensen, *J. Amer. Chem. Soc.* **93,** 7028 (1971).
666. B. Weinstein and A. E. Pritchard, *J. Chem. Soc., Perkin Trans.* **1,** 1015 (1972).
667. G. Weitzel, K. Eisele, H. Guglielmi, W. Stock, and R. Renner, *Hoppe-Seyler's Z. Physiol. Chem.* **352,** 1735 (1971).
668. G. Weitzel, R. Renner, and H. Guglielmi, *Hoppe-Seyler's Z. Physiol. Chem.* **352,** 1617 (1971).
669. A. Wendel, E. Schaich, U. Weber, and L. Flohé, *Hoppe-Seyler's Z. Physiol. Chem.* **353,** 514 (1972).
670. F. C. Westall, J. Scotchler, and A. B Robinson, *J. Org. Chem.* **37,** 3362 (1972).
671. F. Weygand and E. Frauendorfer, *Chem. Ber.* **103,** 2437 (1970).
672. F. Weygand, A. Prox, L. Schmidhammer, and W. König, *Angew. Chem.* **75,** 282 (1963).

673. D. N. J. White and J. D. Dunitz, *Isr. J. Chem.* **10**, 249 (1972).
674. T. Wieland, C. Birr, and A. von Dungen, *Justus Liebigs Ann. Chem.* **747**, 207 (1971).
675. T. Wieland, H. Faulstich, and F. Fahrenholz, *Justus Liebigs Ann. Chem* **743**, 77 (1971).
676. T. Wieland, L. Lapatsanis, J. Faesel, and W. Konz, *Justus Liebigs Ann. Chem.* **747**, 194 (1971).
677. T. Wieland, J. Lewalter, and C. Birr, *Justus Liebigs Ann. Chem.* **740**, 31 (1970).
678. T. Wieland and C. Rietzel, *Justus Liebigs Ann. Chem.* **754**, 107 (1971).
679. T. Wieland and A. Seeliger, *Chem. Ber.* **104**, 3992 (1971).
680. M. Wilchek and T. Miron, *Biochem. Biophys. Res. Commun.* **47**, 1015 (1972).
681. M. A. Wille, V. du Vigneaud, and W. Y. Chan, *J. Med. Chem.* **15**, 11 (1972).
682. A. W. Williams and G. T. Young, *J. Chem. Soc., Perkin Trans.* **1**, 1194 (1972).
683. M. W. Williams and G. T. Young, *J. Chem. Soc.*, (*London*) p. 881 (1963).
684. G. C. Windridge and E. C. Jorgensen, *J. Amer. Chem. Soc.* **93**, 6318 (1971).
685. H. Wissmann and R. Geiger, *Angew. Chem. Int. Ed., Engl.* **9**, 908 (1970).
686. D. T. Witiak, S. K. Seth, E. R. Baizman, S. L. Weibel, and H. H. Wolf, *J. Med. Chem.* **15**, 1117 (1972).
687. H. Witte and W. Seeliger, *Justus Liebigs Ann. Chem.* **755**, 163 (1972).
688. Y. Wolman and Y. S. Klausner, *Isr. J. Chem.* **9**, 229 (1971).
689 Y. Wolman and Y. S. Klausner, *Isr. J. Chem.* **9**, 217 (1971).
690. Y. Wolman and Y. S. Klausner, *Isr. J. Chem.* **9**, 211 (1971).
691. Y. Wolman and Y. S. Klausner, *Isr. J. Chem.* **9**, 205 (1971).
692. Y. Wolman, D. Ladkany, and M. Frankel, *Isr. J. Chem.* **8**, 59 (1970).
693. Y. Wolman, A. Schejter, and M. Sokolovsky, *J. Amer. Chem. Soc.* **94**, 1720 (1972).
694. Y. Wolman, M. Schwarzberg, and M. Frankel, *Isr. J. Chem.* **8**, 53 (1970).
695. E. Wünsch, *Naturwissenschaften* **59**, 239 (1972).
696. E. Wünsch and K.-H. Deimer, *Hoppe-Seyler's Z. Physiol. Chem.* **353**, 1255 (1972).
697. E. Wünsch and K.-H. Deimer, *Hoppe-Seyler's Z. Physiol. Chem.* **353**, 1246 (1972).
698. E. Wünsch and F. Drees, *Chem. Ber.* **99**, 110 (1966).
699. E. Wünsch, A. Högel-Betz, and E. Jaeger, *Hoppe-Seyler's Z. Physiol. Chem.* **352**, 1553 (1971).
700. E. Wünsch and E. Jaeger, *Hoppe-Seyler's Z. Physiol. Chem.* **352**, 1584 (1971).
701. E. Wünsch, E. Jaeger, M. Deffner, R. Scharf, and P. Lehnert, *Chem. Ber.* **105**, 2515 (1972).
702. E. Wünsch, E. Jaeger, and G. Schönsteiner-Altmann, *Hoppe-Seyler's Z. Physiol. Chem.* **352**, 1580 (1971).
703. E. Wünsch, E. Jaeger, and G. Schönsteiner-Altmann, *Hoppe-Seyler's Z. Physiol. Chem.* **352**, 1568 (1971).
704. E. Wünsch, E. Jaeger, and G. Schönsteiner-Altman, *Hoppe-Seyler's Z. Physiol. Chem.* **352**, 1560 (1971).
705. E. Wünsch, G. Schönsteiner-Altman, and E. Jaeger, *Hoppe-Seyler's Z. Physiol. Chem.* **352**, 1424 (1971).
706. E. Wünsch and R. Spangenberg, *Chem. Ber.* **104**, 2427 (1971).
707. E. Wünsch and R. Spangenberg, *in* "Peptides 1969," p. 30. North-Holland Publ., Amsterdam 1971.
708. E. Wünsch and P. Thamm, *Chem. Ber.* **104**, 2454 (1971).
709. E. Wünsch and G. Wendlberger, *Chem. Ber.* **105**, 2508 (1972).
710. E. Wünsch, G. Wendlberger, and A. Högel, *Chem. Ber.* **104**, 2430 (1971).
711. E. Wünsch, G. Wendlberger, and P. Thamm, *Chem. Ber.* **104**, 2445 (1971).
712. H. Yajima and H. Kawatani, *Chem. Pharm. Bull.* **19**, 1905 (1971).
713. H. Yajima and Y. Kiso, *Chem. Pharm. Bull.* **19**, 420 (1971).
714. H. Yajima, N. Shirai, and Y. Kiso, *Chem. Pharm. Bull.* **19**, 1900 (1971).
715. H. Yajima, F. Tamura, and Y. Kiso, *Chem. Pharm. Bull.* **18**, 2574 (1970).
716. H. Yajima, H. Watanabe, and M. Okamoto, *Chem. Pharm. Bull.* **19**, 2185 (1971).

717. S. Yamada and Y. Takeuchi, *Tetrahedron Lett.* p. 3595 (1971).
718. S.-I. Yamada, M. Wagatsuma, Y. Takeuchi, and S. Terashima, *Chem. Pharm. Bull.* **19**, 2380 (1971).
719. H. Yamamoto and T. Hayakawa, *Biopolymers* **11**, 1259 (1972).
720. M. Yamamoto, R. K. Turker, P. A. Khairallah, and F. M. Bumpus, *Eur. J. Pharmacol.* **18**, 316 (1972).
721. T. Yamanaka, S. Hase, S. Sakakibara, I. L. Schwart, B. M. Dubois, and R. Walter, *Mol. Pharmacol.* **6**, 474 (1970).
722. D. Yamashiro, J. Blake, and C. H. Li, *J. Amer. Chem. Soc.* **94**, 2855 (1972).
723. K. Yamauchi, M. Kinoshita, and M. Imoto, *Bull. Chem. Soc. Jap.* **45**, 2528 (1972).
724. N. Yanaihara, N. Sakura, C. Yanaihara, and T. Hashimoto, *J. Amer. Chem. Soc.* **94**, 8243 (1972).
725. J. A. Yankeelov Jr. and C. J. Jolley, *Biochemistry* **11**, 159 (1972).
726. A. Yaron, M. C. Otey, H. A. Sober, E. Katchalski, S. Ehrlich-Rogozinski, and A. Berger, *Biopolymers* **11**, 607 (1972).
727. N. Yoshida, T. Kato, and N. Izumiya, *Bull. Chem. Soc. Jap.* **43**, 2912 (1970).
728. B. A. Yuodka, I. V. Obruchnikov, Z. A. Shabarova, and M. A. Prokof'ev, *Chem. Natur. Compounds* **4**, 274 (1968).
729. M. Zaoral and M. Flegel, *Collect. Czech. Chem. Commun.* **37**, 1539 (1972).
730. M. Zaoral, J. Kolc, and F. Šorm, *Collect. Czech. Chem. Commun.* **35**, 1716 (1970).
731. K. Y. Zee-Cheng and C. C. Cheng, *J. Med. Chem.* **15**, 13 (1972).
732. M. A. Zevail' and E. A. Morozova, *J. Gen. Chem. USSR* **40**, 2758 (1970).
733. Ziauddin, K. D. Kopple, and C. A. Bush, *Tetrahedron Lett.* p. 483 (1972).
734. E. Ziv, N. de Groot, S. Rappoport, and Y. Lapidot, *Isr. J. Chem.* **9**, 243 (1971).

A 4
B 5
C 6
D 7
E 8
F 9
G 0
H 1
I 2
J 3